Second Edition

Fundamentals
of
Soil
Ecology

ks to be returned on or

LIVERPOOL JMU LIBRARY

3 1111 01100 8206

Second Edition

Fundamentals of Soil Ecology

David C. Coleman
D. A. Crossley, Jr.
Paul F. Hendrix

Institute of Ecology
University of Georgia
Athens, Georgia

ELSEVIER
ACADEMIC
PRESS

Amsterdam • Boston • Heidelberg • London
New York • Oxford • Paris • San Diego
San Francisco • Singapore • Sydney • Tokyo

Acquisition Editor: *David Cella*
Editorial Coordinator: *Kelly Sonnack*
Project Manager: *Brandy Palacios*
Full Service Provider: *Graphic World Publishing Services*
Composition: *SNP Best-set Typesetter Ltd., Hong Kong*
Printer: *The Maple-Vail Book Manufacturing Group*

Elsevier Academic Press
200 Wheeler Road, Burlington, MA 01803, USA
525 B Street, Suite 1900, San Diego, California 92101-4495, USA
84 Theobald's Road, London WC1X 8RR, UK

This book is printed on acid-free paper. ∞

Copyright © 2004, Elsevier Inc. All rights reserved.

No part of this publication may be reproduced or transmitted in any form or by any means, electronic or mechanical, including photocopy, recording, or any information storage and retrieval system, without permission in writing from the publisher.

Permissions may be sought directly from Elsevier's Science & Technology Rights Department in Oxford, UK: phone: (+44) 1865 843830, fax: (+44) 1865 853333, e-mail: permissions@elsevier.com.uk. You may also complete your request on-line via the Elsevier homepage (http://elsevier.com), by selecting "Customer Support" and then "Obtaining Permissions."

Library of Congress Cataloging-in-Publication Data
Coleman, David C., 1938–
     Fundamentals of soil ecology / David C. Coleman, D.A. Crossley, Jr., Paul F. Hendrix.—2nd ed.
       p. cm.
     Includes bibliographical references and index.
     ISBN 0-12-179726-0 (alk. paper)
     1. Soil ecology.   2. Soil biology.   I. Crossley, D. A.   II. Hendrix, Paul F.   III. Title.
     QH541.5.S6C65 2004
     577.5′7—dc22

                                                                    2004046994

British Library Cataloguing in Publication Data
A catalogue record for this book is available from the British Library

ISBN: 0-12-179726-0

For all information on all Academic Press publications
visit our website at www.academicpressbooks.com

PRINTED IN THE UNITED STATES OF AMERICA
04 05 06 07 08 09 9 8 7 6 5 4 3 2 1

Contents

5   Decomposition and Nutrient Cycling   187

6   Soil Food Webs: Detritivory and Microbivory in Soils   227

Preface to the Second Edition

We endorse all of the comments and observations made in the Preface to the First Edition of this book. Over the last 8 years, considerable progress has been made in opening soil processes up for scientific inquiry, indeed, viewing soils "through a ped darkly" (Coleman, 1985) and getting away from the simplistic approaches of the "black box" that prevailed in much of the 20th century.

In the midst of the wonder and awe surrounding the pictures that have been transmitted across 100 million miles to Earth during 2004 from the two Mars rovers, it is important to point out a basic fallacy in the discussions over the findings on the surface of Mars. The engineers and physical scientists in charge of the study persist in calling the Mars surface material "soil." As we note many times in our book, biology is the leading characteristic of soil. Organisms are one of the five major soil-forming factors, and life itself characterizes a true soil. Anything found on the surface of Mars—barring totally unexpected news to the contrary—is no doubt complex and interesting, but it is essentially weathered parent material, not soil. Arthur C. Clarke came closer with the title of his science fiction novel *Sands of Mars*.

On the biological side of soil studies, much progress has been made recently in elucidating not only biotic function, especially in the case of bacteria and fungi, but also the identity of which species is performing what process. We focus primarily on the biological aspects, and devote a smaller proportion of our total coverage to soil physics and chemistry, largely because they are discussed extensively in recent treatises by Hillel (1997) and Brady and Weil (2000).

As a reflection of these new developments, we have singled out Soil Biodiversity and Linkages to Soil Processes for coverage in its own chapter (Chapter 7) to identify and emphasize one of the areas of burgeoning research and conservation interest. Also included is a final chapter (Chapter 9) on laboratory and field exercises that have proven useful in our course in Soil Ecology at the University of Georgia. We hope they will be helpful to faculty and students who use this book. We invite our readers to become "Earth rovers," and participate in the wonder and excitement of studying the ecology of soils, a marvelously complex milieu. We hope that this textbook, along with other recent ones, such as the extensive compendium of Lavelle and Spain (2001), will provide the interested scientist with some of the background necessary to work in this often difficult but always fascinating field of research. Two colleagues who were instrumental in critiquing our first edition, Eugene P. Odum and Edward T. Elliott, are now deceased, but their influence is still felt by the soil ecology community and by us. A new generation of students and postdoctoral fellows from the University of Georgia and other universities have contributed ideas and inspiration to this effort, including: Sina Adl, Mike Beare, Heleen Bossuyt, George Brown, Weixin Cheng, Charles Chiu, Greg Eckert, Christien Ettema, Shenglei Fu, Jan Garrett, Randi Hansen, Liam Heneghan, Nat Holland, Coeli Hoover, Shuijin Hu, John Johnston, Keith Kisselle, Sharon Lachnicht, Karen Lamoncha, Stephanie Madson, Rob Parmelee, Mitchell Pavao-Zuckerman, Kitti Reynolds, Chuck Rhoades, Breana Simmons, Guanglong Tian, Petra van Vliet, Thaïs Winsome, Christina Wright, David Wright, Qiangli Zhang, and our soil ecology colleagues at the University of Georgia, Colorado State University, Oregon State University, University College Dublin, and at many LTER sites around the world. Any errors are of course ours, and we would appreciate comments from readers pointing them out.

We thank our helpful secretary and colleague, Linda Lee Enos, for her tireless efforts in compiling the tables and figures. Our spouses, Fran, Dot, and Cathy, deserve credit for their tolerance of this further foray into the arcane but now ever-more-relevant world of soil biology and ecology.

David C. Coleman
D. A. Crossley, Jr.
Paul F. Hendrix
Athens, Georgia, February 2004

Preface to the First Edition

One of the last great frontiers in biological and ecological research is in the soil. Civilizations, so dependent on soils as a source of nutrients for food—consumed directly as vegetables, fruit, and grain, or animals that feed on plants—owe a considerable debt to soils.

Over the span of several millennia, there has been concern about the use and misuse of soils. There is ample evidence that numerous civilizations, from ancient Sumeria and Babylonia to those that use modern high-intensity, high-input agriculture, have suffered with problems of long-term sustainability (Whitney, 1925; Pesek, 1989).

Indeed, one prominent soil physicist has been moved to comment that "the plow has caused more destruction to civilizations than the sword" (Hillel, 1991). Perhaps the adage of "beating swords into plowshares" needs rethinking. As we will discover in the course of this book, it is truly a time to be working with nature and to cease treating soils as a "black box."

Soil is a unique entity. It has its origins in physical, chemical, and biological interactions between the parent materials and the atmosphere. The simplest definitions of soil follow the most common understanding, such as "the upper layer of earth which can be dug or plowed and in which plants grow." (*Webster's New Universal Unabridged Dictionary*, 1983). The soil scientist defines it as "a natural body, synthesized in profile form from a variable mixture of broken and weathered minerals and decaying organic matter, which covers the earth in a thin layer and which supplies, when containing the proper amounts of air and water, mechanical support and, in part, sustenance for plants" (Buckman and Brady, 1970). This definition recognizes that soil has vertical structure, is composed of a variety of materials, and has a biological nature as well; it is derived in part from decaying organic matter. Nevertheless, uncertainties emerge when more restrictive definitions are attempted. How deep is soil or when is nonsoil encountered? Working definitions of soil depth range from 1 meter to many meters, depending on the ecosystem and the nature of the investigations. Are barren, rocky areas excluded if they do not allow growth of higher plants? Lindeman (1942) considered the substratum of a lake as a benthic soil (see also Jenny, 1980). When do simple sediments become soil? When can they support plant growth? Only after biological, physical, and chemical interactions convert sediments into an organized profile?

Soils are composed of a variable combination of four key constituents: minerals, organic matter, water, and air. Of the Greco-Roman concepts of fundamental constituents—earth, air, fire, and water—three of the four are contained with the broad concept of soil. Indeed, if the energetic process of life ("the fire of life") (Kleiber, 1961) is included within the soil, then all four of the ancient "elements" are present therein.

Are living organisms part of soil? We would include the phrase "with its living organisms" in the general definition of soil. Thus, from our viewpoint soil is alive and is composed of living and nonliving components having many interactions.

It is as a part of that larger unit, the terrestrial ecosystem, that soil must be studied and conserved. The interdependence of terrestrial vegetation and animals, soils, atmosphere, and hydrosphere is complex, with many feedback mechanisms. When we view the soil system as an environment for organisms, we must remember that the biota have been involved in its creation, as well as adapting to life within it. The principles by which organisms in soils are distributed, interact, and carry on their lives are far from completely known, and the importance of the biota for soil processes is not often appreciated.

This book on soil ecology emphasizes the interdisciplinary nature of studies in ecology as well as soils. Considerable "niche overlap" (similarity in what they do, i.e., their "profession") (Elton, 1927) exists between the two disciplines of ecology and soil science. Ecology, which is heavily organism-oriented, is concerned with all forms of life in relation to their environment. Soil science, in contrast, contains several other aspects in addition to soil biology, such as soil genesis and classification, soil physics, and soil chemistry. A broader view of ecology asks: How do systems work? From that perspective, ecology and soil science share similar objectives.

The overlap between ecology and soil science is both extensive and interesting. Aspects of soil physics, chemistry, and mineralogy have a great impact on how many different kinds of organisms coexist in the opaque, complex, semiaquatic milieu that we call soil. We first describe what soils are and how they are formed, and then discuss some of the current research being done in soil ecology.

With a rising tide of interest worldwide in soils, and in belowground processes in general, numerous types of studies using tools in all ranges of the size and electromagnetic energy spectra, and encompassing from microsites to the biosphere, are now possible. Significant achievements during the past 5 to 10 years make a book of this sort both timely and useful. This book is intended primarily as a source of ideas and concepts and thus is intended as a supplemental reference for courses in ecology, soil science, and soil microbiology.

We hope that we will interest a new generation of ecologists and soil scientists in the world of soil ecology: the interface between biology, chemistry, and physics of soil systems.

David C. Coleman
D. A. Crossley, Jr.
Paul F. Hendrix
Athens, Georgia, February 2004

Historical Overview of Soils and the Fitness of the Soil Environment

1

THE HISTORICAL BACKGROUND OF SOIL ECOLOGY

The "roots" of human understanding of soil biology and ecology can be traced into antiquity and probably even beyond the written word. We can only imagine hunter–gatherer societies attuned to life cycles of plant roots, fungi, and soil animals important to their diets, their welfare or their cultures, and particularly to environmental conditions favorable to such organisms. Indeed, early agriculture must certainly have developed, at least in part, from a practical knowledge of soils and their physical and biological characteristics.

Soil is so fundamental to human life that it has been reflected for millennia in our languages. The Hebrew word for soil is *adama*, from which comes the name Adam—the first man of the Semitic religions. The word "human" itself has its roots in the Latin *humus*, the organic matter in soil (Hillel, 1991). Early civilizations had obvious relationships with soils. The Mesopotamian region encompasses present-day Iraq and Kuwait, occupying the valley of the Euphrates and Tigris rivers from their origin as they come out from the high tablelands and mountains of present-day Armenia to their mouth at the Persian Gulf. It had one of the earliest recorded civilizations, the Sumerian, dating from about 3300 years BCE (Hillel, 1991). An inventory taken in the time of the early Caliphates showed 12,500,000 acres (nearly 5,100,000 hectares) under cultivation in the southern half of Mesopotamia (Whitney, 1925). With many centuries of irrigation, this so-called "hydraulic civilization" was plagued with problems of siltation and salinization, which was written about at the time of King Hammurabi (1760 BCE) (Hillel, 1991).

An impressive sequence of civilizations waxed and waned over the millennia: Sumerian, Akkadian, Babylonian, and Assyrian, as cultivation shifted from the lower to central and upper regions of Mesopotamia. Siltation and salinization continue to beset present-day civilizations that practice extensive irrigation.

To the east of Mesopotamia, past the deserts of southern Iran and of Baluchistan, lies the Indus River Valley. Another irrigation-based civilization developed here, probably under the influence of the Mesopotamian civilization. The Indus River civilization probably encompassed a total land area far exceeding that of either Sumeria or Egypt; little is known about it. No written records have been discovered, but its fate, like that of the Sumerian, succumbed to environmental degradation, exacerbated by the extensive deforestation which occurred to provide fuel to bake the bricks used in construction (Hillel, 1991). The bricks in Mesopotamian cities were sun-baked, similar to the adobe style of construction used in the desert of the southwestern United States.

In contrast, the Egyptian civilization persisted more or less in place, as a result of the annual floods of the Nile River, which renewed soil fertility in vast areas along the river's length as it flowed northward. Over the millennia, from one to three million people lived along the Nile, and produced enough grain to export wheat and barley to many countries around the Mediterranean rim. Now that the population is some 30 times greater, it must import some foodstuffs and is economically in questionable condition, in spite of the vast areas being irrigated with water from the Aswan high dam.

The ancient Chinese concept of fundamental elements included earth, air, fire, water, and moon. In the Yao dynasty from 2357 to 2261 BCE, the first attempt was made at soil classification surveying. The Emperor established nine classes of soils in as many provinces of China, with a taxation system based upon this system. These classes included the yellow and mellow soils of Young Chow (Shensi and Kansu); the red, clayey, and rich soils of Su Chow (Shantung, Kiangsu, and Anhwei); the whitish, rich salty soils of Tsing Chow (Shantung); the mellow, rich, dark and thin soils of Yu Chow (Honan); the whitish and mellow soils of Ki Chow (Chili and Shansi); the black and rich soils of Yen Chow (Chili and Shantung); the greenish and light soils of Liang Chow (Szechuan and Shensi); and the miry soils of King Chow (Hunan and Hupeh) and Yang Chow (Kiangsu) (Whitney, 1925). This system reflects a sophisticated knowledge within early Chinese civilization of soils and their relationship with plant growth. Interestingly, in recognition of the importance of biological activity in soils, the ancient Chinese termed earthworms as "angels of the soil" (Blakemore, 2002).

The Greeks believed there were four basic elements: earth, air, fire, and water; and Aristotle, understanding the role of earthworms in organic matter decomposition, considered earthworms to be the "intestines of the earth" (Edwards and Lofty, 1977). The Greeks and Romans also had a clear differentiation of the productive capacities of different types of soils. They referred to good soils as "fat," and soils of lower quality as "lean" (Whitney, 1925). For the Roman writers, "humus" referred to soil or earth. Virgil (79–19 BC), in his *Georgics*, named the loamy soil *pinguis humus* and used the words *humus*, *solum*, or *terra* more or less interchangeably for the notions of soil and earth. Columella in the first century AD noted, "wheat needs two feet of good *humus*" (Feller, 1997; italics added).

The word *humus* seems to have entered the European scientific vocabulary in the 18th century. Thus in Diderot and d'Alembert's Encyclopaedia (vol. 8) in 1765: "Humus, natural history, this Latin word is often borrowed by naturalists (even into French) and denotes the mould, the earth of the garden, the earth formed by plant decomposition. It refers to the brown or darkish earth on the surface of the ground. Refer to the mould or vegetable mould" (translation in Feller, 1997).

By the beginning of the 19th century, the leading authorities with a biological view of soils were Leeuwenhoek, Linnaeus, and other pre-Darwinians, and then Darwin himself (1837, 1881), who "fathered" the modern era. Müller (1879, 1887), cited in Feller (1997), laid the groundwork for the present-day scientific bases of the different forms of humus, and even included a general survey of soil genetic processes in cold and temperate climates. Müller developed terms for the three humus types—Mull, Mor, and Mullartiger Torf—the latter equivalent to Moder. *Mull* is mould and *Torf* is peat in Danish. Thus Mullartiger Torf is *mould peat* in Danish, and it is viewed as an intermediate form between the two extremes (see Feller, 1997, for more details on the history of these fascinating substances).

The first scientific view of soils as natural bodies that develop under the influence of climate and biological activity acting on geological substrates arose in Russia with the work of Dokuchaev and his followers (Zonn and Eroshkina, 1996; Feller, 1997) and in Europe with Müller's (1887) descriptions of soil horizon development (Tandarich *et al.*, 2002). The ecological basis of the Russian tradition is clear in the words of Glinka (1927; cited in Jenny, 1941), a disciple of Dokuchaev, whose view of soil included ". . . not only a natural body with definite properties, but also its geographical position and surroundings, i.e., climate, vegetation, and animal life." This Russian perspective predates the formal statement of the ecosystem concept by several decades (Tansley, 1935).

During this early period of theoretical development across the Atlantic, soil science in the United States was more concerned with practical matters of agriculture, such as soil productivity and crop growth (Tandarich *et al.*, 2002) and, later, on restoration of soils badly degraded from poor management (e.g., the "dust bowl" in the Great Plains and the severely eroded croplands of the southeastern United

States). It was not until the 1920s that ideas of pedogenesis gained wide recognition in the United States. Within the next decade, Hans Jenny (1941) published his classic work on soil formation, drawing heavily from Dokuchaev's ideas to synthesize pedological and ecological perspectives into the concept of a "... soil system [that] is only a part of a much larger system . . . composed of the upper part of the lithosphere, the lower part of the atmosphere, and a considerable part of the biosphere." He formulated this concept into the now famous "fundamental equation of soil-forming factors":

$$s = f(cl, o, r, p, t, \ldots)$$

where s refers to the state of a body of soil at a point in time; f refers to function; cl to climate; o to organisms; r to relief or topography; p to parent material; and t to time. Jenny, probably more than any North American soil scientist of his era, emphasized the importance of the biota in and upon soils. His last major work, *The Soil Resource* (1980), is now a classic in the literature on ecosystem ecology.

Since Jenny's work, research in soil ecology has experienced a "renaissance" as the significance of biological activity in soil formation, organic matter dynamics, and nutrient cycling have become widely recognized. The post–World War II scientific boom was an important impetus for science generally, including soil science. In the United States, the Atomic Energy Commission (later the Department of Energy), through the national laboratories, funded soil biology in relation to nutrient and radioisotope recycling in soil systems (Auerbach, 1958); more recently, the National Science Foundation's Division of Environmental Biology and the United States Department of Agriculture (USDA) National Research Initiative in Soils and Soil Biology have supported a wide array of research in soil ecology. The International Biological Program (IBP) on the international scene greatly expanded methodologies in soil ecology and increased our knowledge of ecological energetics and soil biological processes (Golley, 1993).

In sum, all of these developments and advances in knowledge, from the ancient to the modern, have led to a vast literature upon which is based our current understanding of the soil beneath our feet and the vital role that this living milieu plays in sustaining life on a thin, dynamic, fragile planetary crust.

WATER AS A CONSTITUENT OF SOIL

"The occurrence of water is, moreover, not less important and hardly less general upon the land. In addition to lakes and streams, water is almost everywhere present in large quantities in the soil, retained there mainly by capillary action, and often at greater depths." (Henderson, 1913).

Lawrence J. Henderson, a noted physical chemist and physiologist, published a book (*The Fitness of the Environment*, 1913), which was a landmark among books on biological topics. Henderson's thesis is that one substance, water, is responsible for the characteristics of life, and the biosphere as we know it. The highly bipolar nature of water, with its twin hydrogen bonds, leads to a number of intriguing characteristics (e.g., high specific heat), which have enabled life in the thin diaphanous veil of the biosphere (Lovelock, 1979, 1988) to extend and proliferate almost endlessly through the air, water, soil, and several kilometers into the earth's mantle (Whitman *et al.*, 1998).

A central fact of soil science is that certain physicochemical relationships of matter in all areas of the biosphere are mediated by water. Thus soil, which we normally think of as opaque and solid, from the wettest organic muck soil to the parched environs of the Atacama, Kalahari, Gobi, or Mojave deserts, is dominated by the amount and availability of water.

Consider water in each of its phases—solid, liquid, and gaseous:

1. *Solid:* In aquatic ecosystems, water freezes from the top down, because it has its greatest density at 4°C. This allows for organismal activity to continue at lower depths and in sediments as well. In soil, the well-insulated nature of the soil materials and water with its high specific heat means that there is less likelihood of rapid freezing. Water expands when it freezes. In more polar climates (and in some temperate ones), soil can be subjected to "frost heaving," which can be quite disruptive, depending on the nature of the subsurface materials.

2. *Liquid:* Water's high specific heat of 1 calorie per gram per degree Celsius increase in temperature has a significant stabilizing influence in bodies of water and soil (Table 1.1; Hadas, 1979). The effect of the high specific heat is to reduce fluctuations in temperature. The location of the liquid, in various films, or in empty spaces, has a marked influence on the soil biota.

3. *Vapor:* It is somewhat counterintuitive but true that the atmosphere within air-dry soil (gravimetric water content of 2% by weight) has a relative humidity of 98%. The consequences of this humidity for life in the soil are profound. Most soil organisms spend their lives in an atmosphere saturated with water. Many soil animals absorb and lose water through their integuments, and are entirely dependent upon saturated atmospheres for their existence.

From the pragmatic viewpoint of the soil physicist, we can consider aqueous and vapor phases of water conjointly. Following a moisture release curve, one can trace the pattern of water, in volume and location in the soil pore spaces, in the following manner (Vannier, 1987). Starting

TABLE 1.1.   Specific Heats of Various Substances

Substance	Specific Heat[A]
Lead	0.03
Iron	0.10
Quartz	0.19
Sugar	0.30
Chloroform	0.24
Hexane	0.50
Water	
Liquid	1.0
Solid	0.5
Gas	0.3–0.5
Ammonia, liquid	1.23

[A]Calories (= 4.18 Joule) to raise 1 gram by 1°C. Modified from Hadas, 1979.

with freestanding, or gravitational, water at saturation, the system is essentially subaquatic (Fig. 1.1). With subsequent evaporation and plant transpirational water losses from the soil, the freestanding water disappears, leaving some capillary-bound water (Fig. 1.2), which has been termed the edaphic system. Further evaporation then occurs, resulting in the virtual absence of any capillary water, leaving only the adsorbed water at a very high negative water tension (Fig. 1.3).

The implications of this complex three-dimensional milieu are of fundamental importance for a very diverse biota. Vannier (1973) proposed the term "porosphere" for this intricate arrangement of sand, silt, clay, and organic matter. Primitive invertebrates first successfully undertook the exploitation of aerial conditions at the beginning of the Paleozoic era (Vannier, 1987). This transition probably took place via the soil medium, which provided the necessary gradient between the fully aquatic and aerial milieus. This water-saturated environment, so necessary for such primitive, wingless (Apterygote) forms as the Collembola, or springtails (Fig. 1.1), is equally important for the transient life-forms such as the larval forms of many flying insects, including Diptera and Coleoptera. In addition, many of the micro- and mesofauna, (described in Chapter 4,) could be considered part of the "terrestrial nannoplankton" (Stout, 1963). Stout included all of the water-film inhabitants, namely: bacteria and yeasts, protozoa, rotifers, nematodes, copepods, and enchytraeids (the small oligochaetes also called potworms). Raoul Francé, a German sociologist, made analogies between aquatic plankton and the small and medium-sized organisms that inhabit the water films and water-filled pores in soils, terming them: "Das Edaphon" (Francé, 1921).

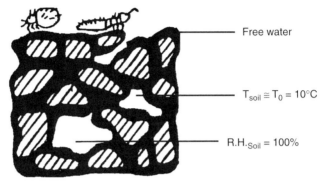

First phase of soil water evaporation;
surrounding conditions
Air temperature = 15°C
Wet temperature = 10°C
Relative humidity = 51%

Natural displacement of mites and springtails

Free water

$T_{soil} \cong T_0 = 10°C$

R.H.$_{Soil}$ = 100%

2.5 > pF > 0

FIGURE 1.1. Diagram of gravitational moisture (the subaquatic system) in the soil framework (from Vannier, 1987). pF = −log cm H_2O suction; R.H. = relative humidity; 2.5 pF = field capacity.

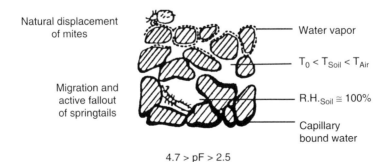

Second phase of soil water evaporation;
surrounding conditions
Air temperature = 15°C
Wet temperature = 10°C
Relative humidity = 51%

Natural displacement of mites

Water vapor

$T_0 < T_{Soil} < T_{Air}$

Migration and active fallout of springtails

R.H.$_{Soil}$ ≅ 100%

Capillary bound water

4.7 > pF > 2.5

FIGURE 1.2. Diagram of capillary moisture (the edaphic system) in the soil framework (from Vannier, 1987). 4.7 pF = −5 mPa; 2.5 pF = −0.03 mPa = field capacity.

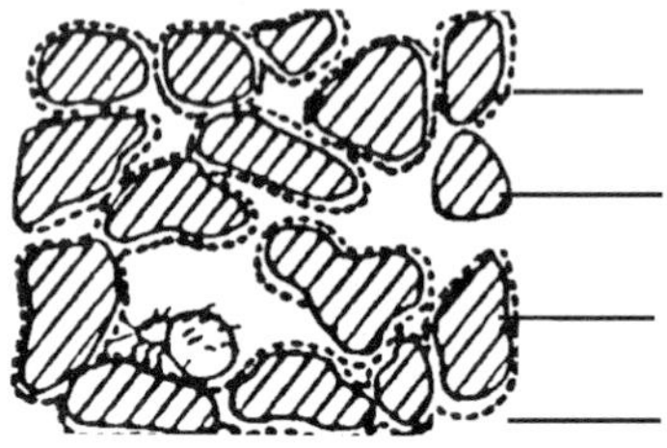

Third phase of soil water evaporation;
surrounding conditions
Air temperature = 15°C
Wet temperature = 10°C
Relative humidity = 51%

$R.H._{Soil} = R.H._{Air} = 51\%$

$T_{Soil} \cong T_{Air} = 15°C$

$R.H._{Soil} < 100\%$

Water vapor

Migration and
active fallout
of mites

$7 > pF > 4.7$

FIGURE 1.3.   Diagram of adsorptional moisture (the aerial system) in the soil framework (from Vannier, 1987). $7\,pF = -1000\,mPa$; $4.7\,pF = -5\,mPa$; permanent wilting point $= -1.5\,mPa = 4.18\,pF$.

As noted in Figures 1.1 to 1.3, there is a marked difference in moisture requirements of some of the soil microarthropods. Thus another major group, the Acari, or mites, are often able to tolerate considerably more desiccation than the more sensitive Collembola. In both cases, the microarthropods make a gradual exit from the soil matrix as the desiccation sequence described above continues.

Other organisms, more dependent on the existence of free water or water films, include the protozoa and nematoda, the life histories and feeding characteristics of which are covered in Chapter 4. In a sense, the very small fauna, and the bacteria they feed upon, exist in a qualitatively different world from the other fauna, or from fungi. Both larger fauna and fungi move into and out of various water films and through various pores, which are less than 100% saturated with water vapor, with comparative ease (Hattori, 1994).

In conclusion, this overview of soil physical characteristics and their biological consequences notes the following: "For a physicist, porous bodies are solids with an internal surface that endows them with a remarkable set of hygroscopic properties. For example, a clay such as bentonite has an internal surface in excess of $800\,m^2g^{-1}$, and a clay soil containing 72% montmorillonite possesses an internal surface equal to $579\,m^2g^{-1}$. The capacity to condense gases on free walls of capillary spaces (the phenomenon of adsorption) permits porous bodies to reconstitute water reserves from atmospheric water vapor" (Vannier, 1987). Later, we will address the phenomenon of adsorption in other contexts, ones that are equally important for soil function as we know it.

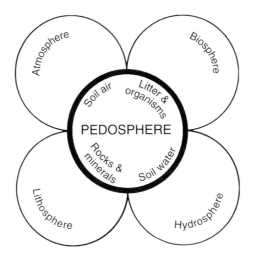

FIGURE 1.4.   The pedosphere, showing interactions of abiotic and biotic entities in the soil matrix (from FitzPatrick, 1984).

ELEMENTAL CONSTITUTION OF SOIL

Many elements are found within the earth's crust, and most of them are in soil as well. However, a few elements predominate. These are hydrogen, carbon, oxygen, nitrogen, phosphorus, sulfur, aluminum, silicon, and alkali and alkaline earth metals. Various trace elements or micronutrients are also biologically important as enzyme co-factors, and include iron, cobalt, nickel, copper, magnesium, manganese, molybdenum, and zinc.

A more functional and esthetically pleasing approach is to define soil as predominantly a sand–silt–clay matrix, containing living (biomass) and dead (necromass) organic matter, with varying amounts of gases and liquids within the matrix. In fact, the interactions of geological, hydrological, and atmospheric (Fig. 1.4) facets overlap with those of the biosphere, leading to the union of all, overlapping in part in the pedosphere. Soils, in addition to the three geometric dimensions, are also greatly influenced by the fourth dimension of time, over which the physicochemical and biological processes occur.

HOW SOILS ARE FORMED

Soils are the resultant of the interactions of several factors—climate, organisms, parent material, and topography (relief)—all acting through time (Jenny, 1941, 1980) (Fig. 1.5). These factors affect major

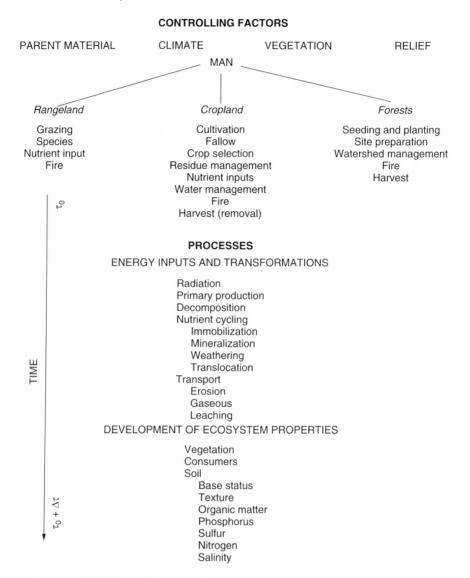

FIGURE 1.5. Soil-forming factors and processes, interaction over time (from Coleman *et al.*, 1983, modified from Jenny, 1980).

ecosystem processes (e.g., primary production, decomposition, and nutrient cycling), which lead to the development of ecosystem properties unique to that soil type, given its previous history. Thus such characteristics as cation-exchange capacity, texture, structure, organic matter status, etc., are the outcomes of the aforementioned processes operating as constrained by the controlling factors. Different arrays of processes may predominate in various ecosystems (see Fig. 1.5).

PROFILE DEVELOPMENT

The abiotic and biotic factors noted above lead to certain chemical changes down through the top few decimeters of soil [Fig. 1.6(a), 1.6(b)]. In many soils, particularly in more mesic or moist regions of the world, there is leaching and redeposition of minerals and nutrients, often accompanied by a distinct color change (profile development). Thus as one descends through the profile from the air-litter surface, one passes through the litter (L), fermentation (F), and humification (H) zones (O_i, O_e, and O_a, respectively), then reaching the mineral soil surface, which contains the preponderant amount of organic matter (A horizon). The upper portion of the A horizon is termed the topsoil, and under conditions of cultivation, the upper 12–25 centimeters (cm) is called the plow layer or furrow slice. This is followed by the horizon of maximum leaching, or eluviation, of silicate clays, Fe, and Al oxides, known as the E horizon. The B horizon is next, with deeper-dwelling organisms and somewhat weathered material. This is followed by the C horizon, the unconsolidated mineral material above bedrock. The solum includes the A, E, and B horizons plus some of the cemented layers of the C horizon. All these horizons are part of the regolith, the material that overlies bedrock. More details on soil classification and profile formation are given in soil textbooks, such as Russell (1973) and Brady and Weil (2000).

The work of the soil ecologist is made somewhat easier by the fact that the top 10–15 cm of the A horizon, and the L, F, and H horizons (O_i, O_e, and O_a) of forested soils contain the majority of plant roots, microbes, and fauna (Coleman *et al.*, 1983; Paul and Clark, 1996). Hence a majority of the biological and chemical activities occur in this layer. Indeed, a majority of microbial and algal-feeding fauna, such as protozoa (Elliott and Coleman, 1977; Kuikman *et al.*, 1990) and rotifers and tardigrades (Leetham *et al.*, 1982), are within 1 or 2 cm of the surface. Microarthropods are most abundant usually in the top 5 cm of forest soils (Schenker, 1984) or grassland soils (Seastedt, 1984a), but are occasionally more abundant at 20–25 cm and even 40–45 cm at certain times of the year in tallgrass prairie (O'Lear and Blair, 1999). This region may be "primed," in a sense, by the continual input of leaf, twig, and root materials, as well as algal and cyanobacterial production and turnover in some ecosystems, while soil mesofauna such as nematodes and microarthropods may be concentrated in the top 5 cm. Significant numbers of nematodes may be found at several meters' depth in xeric sites such as deserts in the American Southwest (Freckman and Virginia, 1989).

(a)

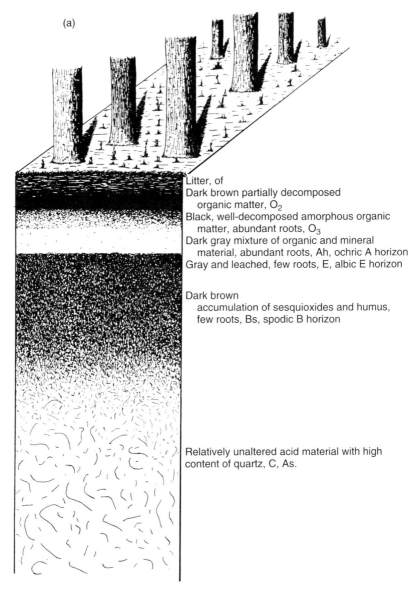

Litter, of
Dark brown partially decomposed
    organic matter, O_2
Black, well-decomposed amorphous organic
    matter, abundant roots, O_3
Dark gray mixture of organic and mineral
    material, abundant roots, Ah, ochric A horizon
Gray and leached, few roots, E, albic E horizon

Dark brown
    accumulation of sesquioxides and humus,
    few roots, Bs, spodic B horizon

Relatively unaltered acid material with high
content of quartz, C, As.

FIGURE 1.6. (a) Diagram of a Podzol (spodosol in North American soil taxonomy) profile with minerals accumulating in subsurface horizons. This is the characteristic soil of coniferous forests (from FitzPatrick, 1984). (b) Diagram of a Cambisol profile, with the organic matter well mixed in the A horizon; due to faunal mixing there is no mineral accumulation in subsurface horizons. This is the characteristic soil of the temperate deciduous forests (from FitzPatrick, 1984).

(b)

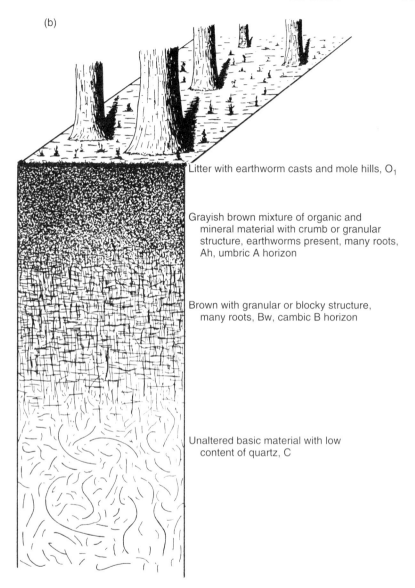

Litter with earthworm casts and mole hills, O_1

Grayish brown mixture of organic and mineral material with crumb or granular structure, earthworms present, many roots, Ah, umbric A horizon

Brown with granular or blocky structure, many roots, Bw, cambic B horizon

Unaltered basic material with low content of quartz, C

FIGURE 1.6. *Continued.*

SOIL TEXTURE

Historically, texture was a term used to describe the workability of an agricultural soil. A heavy, clay soil required more effort (horsepower) to till than a lighter, sandy loam (Russell, 1973). A more quantifiable approach is to characterize soils in terms of the sand, silt, and clay present, which are ranged on a spectrum of light–intermediate–heavy or sandy–silt–clay. The array of textural classes (Fig. 1.7) shows percentages of sand, silt, and clay, and the resulting soil types such as sandy, loamy, or clayey soils.

The origin and mineralogical composition of mineral particles in soil is a most interesting and complex one. The particles are in two major categories: (1) crystalline minerals derived from primary rock, and (2) those derived from weathering animal and plant residues. The microcrystalline forms are comprised of calcium carbonate, iron or aluminum oxides, or silica.

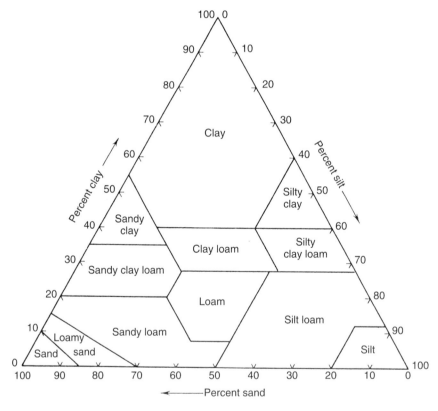

FIGURE 1.7.   Diagram by means of which the textural name of a soil may be determined from a mechanical analysis. In using the diagram, the points corresponding to the percentages of silt and clay present in the soil under consideration are located on the silt and clay lines respectively. Lines are then projected inward, parallel in the first case to the clay side of the triangle and in the second case parallel to the sand side. The name of the compartment in which the two lines intersect is the class name of the soil in question (modified from Buckman and Brady, 1980).

The clay fraction, so important in imparting specific physical properties to soils, to microbial life, and to plant activity via nutrient availability, is comprised of particles less than 2 micrometers (μm) in diameter. Unlike the sand–silt minerals, clays are weathered forms of primary minerals, and hence they are referred to as secondary minerals. Coarse clay particles (0.5 μm) often are derived from quartz and mica; finer clays (0.1 μm) are clay minerals or weathered products of these (such as hydrated ferric, aluminum, titanium, and manganese oxides).

No matter what size the particle is, microorganisms in unsaturated soil exist in a world dominated by the presence of extensive surfaces. There seems to be a general advantage to microbes living at these interfaces in terms of enhanced nutrient concentrations and the potential to use many of the physical substrates themselves as energy or nutrient sources. The thickness of water films in unsaturated conditions allows the microbes little option except to adhere to the surfaces (Mills, 2003). We discuss some of the microbial dynamics and interactions with soil organic matter in Chapter 3.

The roles of coarse and fine clays in organic matter dynamics are under intensive scrutiny in several laboratories around the world (Oades and Waters, 1991; Six *et al.*, 1999). It is possible that labile (i.e., easily metabolized) constituents of organic matter are preferentially adsorbed onto fine clay particles and may be a significant source of energy for the soil microbes (Anderson and Coleman, 1985). For more information on the environmental attributes of clays, see Hillel (1998).

CLAY MINERAL STRUCTURE

The clay minerals in soil are in the form of layer-lattice minerals, and are made up of sheets of hydroxyl ions or oxygen. The clay minerals fall into two groups: (1) those with three groups of ions lying in a plane (the 1:1 group of minerals), and (2) those with four groups of ions lying in a plane (the 2:1 group of minerals). The type mineral of the 1:1 group is kaolinite, which typically has a very low charge on it. In contrast, the 2:1 type mineral, for example illite, carries an appreciably higher negative charge per unit weight than the kaolin group. More detailed information on the clay particles, their composition, and charges upon them is given in Theng (1979) and Oades *et al.* (1989).

A key concern to the soil ecologist is the extremely high surface area found per gram of clay mineral. Surface areas can range from 50 to 100 square meters per gram for kaolinitic clays, from 300 to 500 square meters per gram for vermiculites, and from 700 to 800 square meters per gram for well-dispersed smectites (Russell, 1973). These impressively large surface areas can play a pivotal role in adsorbing and desorbing inorganic and organic constituents in soils, and have only recently been treated in an appropriately analytical fashion as an integral part of the soil nutrient system (Tisdall and Oades, 1982; Oades *et al.*, 1989).

SOIL STRUCTURE

Structure refers to the ways in which soil particles are arranged or grouped spatially. The groupings may occur at any size level on a continuum from either extreme of what are nonstructural states: single grained (such as loose sand grains) or massive aggregates of aggregates (large, irregular solid).

An additional aspect of aggregates, their stabilization once they are formed, is significant for soil ecology. Stabilization is the result of various binding agents. Plant and microbial polysaccharides and gums serve as binding agents (Harris *et al.*, 1964; Cheshire, 1979; Cheshire *et al.*, 1984). A variety of other organic compounds act as binding agents (Cheshire, 1979), and some biological agents such as roots and fungal hyphae (Tisdall and Oades, 1979, 1982; Tisdall, 1991) play a similar role.

The implications of soil structure refer not only to the particles but also extend to the pore spaces within the structure, as noted earlier. Indeed, it is the nature of the porosity that exists in a well-structured soil that leads to the most viable communities within it. This in turn has strong implications for ecosystem management, particularly for agroecosystems (Elliott and Coleman, 1988). There is a very active area of research in soil ecology related to dynamics of micro- and macroaggregates, in relationship to drying–wetting cycles and tillage management. Denef *et al.* (2001) measured marked differences in aggregate formation and breakdown as a function of amount of bacterial and fungal activity in soils with ^{13}C-labeled crop residues. They traced differences in fine intra-aggregate Particulate Organic Matter (POM) to variations in wetting and drying regimes versus those soils not experiencing such environmental fluctuations. We discuss the aggregate formation process further in Chapter 3 on microbes and their effects on ecosystems.

Several types of structural forms are found in soils. The four major types are platelike, prismlike, blocklike, and spheroidal (Fig. 1.8). All of these are "variations on a theme," as it were, of a fundamental unit of soil aggregation: the ped. A ped is a unit of soil structure, such as an aggregate, crumb, prism, block, or granule, formed by natural processes. This is distinguished from a clod, which is artificial or man-made (Brady and Weil, 2000). Soils may have peds of differing shapes, in surface and subsurface horizons. These are the result of differing tempera-

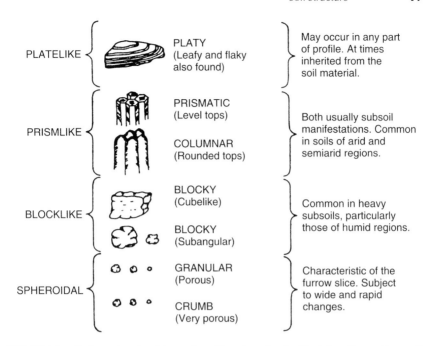

FIGURE 1.8. Various structural types found in mineral soils. Their location in the profile is suggested. In arable topsoils, a stable granular structure is prized (from Brady, 1974).

ture, moisture, and chemical and biological conditions at various levels in the soil profile.

Another concept is helpful in soil structure: the pedon. This is an area, from 1 to 10 square meters, under which a soil may be fully characterized. Later in the book, we will consider the arrangement of soil units in a landscape, and in an entire region. Next, we will examine some of the causes for the formation, or genesis, of soil structure.

Input of organic matter to soil is one of the major agents of soil structure. The organic matter comes from both living and dead sources (roots, leaves, microbes, and fauna). Various physical processes, such as deformation and compression by roots and soil fauna, and freezing–thawing or wetting–drying, also have significant influences on soil structure. It is generally recognized that plant roots and humus (resistant organic breakdown products) play a major role in the formation of aggregates (Elliott and Coleman, 1988; Paul and Clark, 1996). However, bacteria and fungi and their metabolic products play an equally prominent role in promoting granulation (Griffiths, 1965; Cheshire, 1979; Foster, 1985). We will explore organic matter dynamics in the sections on soil biology.

The interaction of organic matter and mineral components of soils has a profound influence on cation-adsorption capabilities. The interchange of cations in solution with cations on these surface-active materials is an important phenomenon for soil fertility. The capacity of soils to adsorb ions (the cation-exchange capacity) is due to the sum of exchange sites on both organic matter and minerals. However, in most soils, organic matter has the higher exchange capacity (number of exchange sites). For a more extensive account, see Paul and Clark (1996).

There is a hierarchical nature to the ways in which soil structure is achieved, and it reflects the biological interactions within the soil matrix (Elliott and Coleman, 1988; Six *et al.*, 2002). Several Australian researchers (Tisdall and Oades, 1982, 1984; Waters and Oades, 1991) have noted how the processes of structuring soils extend over many orders of magnitude, from the level of the individual clay platelet to the ped in a given soil. For most of the biologically significant interactions, one can consider changes across a range of at least six orders of magnitude from $<0.01\,\mu m$ to $<1\,cm$ (Tisdall and Oades, 1982) (Fig. 1.9). Not all soils are aggregated by biological agents; for heavily weathered Oxisols with kaolinite-oxide clays, there seems to be no hierarchy of organization below $20\,\mu m$, because only physicochemical forces predominate there (Oades and Waters, 1991). Studies in our Horseshoe Bend agroecosystem project at the University of Georgia have uncovered significant differences between tillage regimes (conventional, moldboard plowing versus no-tillage, direct drilling of the seeds into the soil). The aggregates in the $53-106\,\mu m$ and $106-250\,\mu m$ categories are most affected by fungal growth and proliferation, reflecting physical binding and the increased amounts of acid-hydrolysable carbohydrates, which are more prevalent in the no-tillage treatments as compared with the bacteria-dominated conventional tillage systems (Beare *et al.*, 1994a, 1994b, 1997).

It is the interactions between physical, chemical, and biological agents in soils that are so fascinating, complex, and important to consider as we increase the intensity of management of terrestrial ecosystems, or alter their usage in response to increased human concerns about their use, and also strive for effective sustainability of them worldwide (Coleman *et al.*, 1992, 1998). Indeed, Lavelle (2000) observed that soil ecology can be considered to have arisen from the convergence of three major approaches: (1) the development of enormous databases on communities of microorganisms and invertebrates and their energy budgets via the International Biological Program (e.g., Petersen and Luxton, 1982); (2) the placement of decomposition processes on center stage, bridging soil chemistry with soil biology (Swift *et al.*, 1979); and (3) an appreciation of the effects of soil organisms on soil structure,

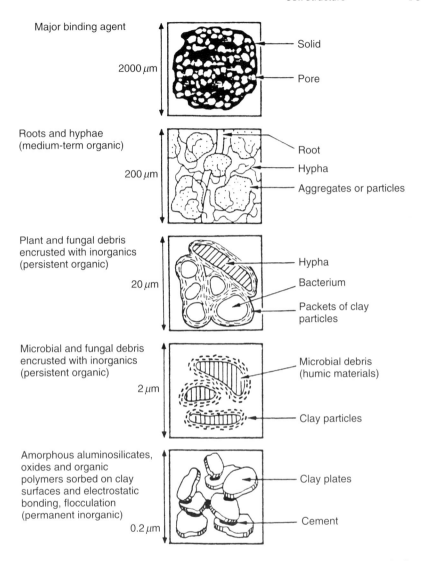

Major binding agent

2000 μm

Solid

Pore

Roots and hyphae
(medium-term organic)

200 μm

Root

Hypha

Aggregates or particles

Plant and fungal debris
encrusted with inorganics
(persistent organic)

20 μm

Hypha

Bacterium

Packets of clay
particles

Microbial and fungal debris
encrusted with inorganics
(persistent organic)

2 μm

Microbial debris
(humic materials)

Clay particles

Amorphous aluminosilicates,
oxides and organic
polymers sorbed on clay
surfaces and electrostatic
bonding, flocculation
(permanent inorganic)

0.2 μm

Clay plates

Cement

FIGURE 1.9. Soil microaggregates, across five orders of magnitude, beginning at the level of clay particles, through plant and fungal debris, up to a 2-mm diameter soil crumb (from Tisdall and Oades, 1982).

including the influence of macrofauna as ecosystem engineers (Bal, 1982, Jones *et al.*, 1994). We would add a fourth dimension, that of soils and sediments as repositories or libraries of DNA. In Siberia, several permafrost cores dating from 10,000 to 400,000 years old have yielded at least 19 different plant taxa, as well as megafauna sequences of mammoth, bison, and horse (Willerslev *et al.*, 2003). Temperate cave sediments from New Zealand yielded 29 taxa characteristic of the pre-human environment, including two species of ratite moas. These genetic records of paleoenvironments will add to our understanding of past ecosystem structure and, possibly, function.

SOILS AS SUPPLIERS OF ECOSYSTEM SERVICES

Soils are large repositories of mineral and organic wealth, available for both the use and misuse by civilizations on this planet (Hillel, 1991). Levels of soil carbon have dropped by as much as 50% after 50 to 100 years of intensive farming in the North American Great Plains (Haas *et al.*, 1957). Similar concerns were expressed about loss of organic matter and erosion of soils in the Mediterranean region at the time of Plato in the third century BCE, as noted above (Whitney, 1925).

An example of the monetary value of what soils provide is given by the costs of raising crops in intense nonsoil conditions using hydroponic culture. Construction of a modern hydroponics system in the United States, including pumps and sophisticated computer control systems, costs upward of $850,000 per hectare (FAO 1990, cited by Daily *et al.*, 1997). Soils also play significant roles in the regulation of global greenhouse gases such as carbon dioxide, methane, and nitrous oxides (Schimel and Gulledge, 1998). As we present in detail in later chapters, the cleansing and recycling role that soils play in processing organic wastes and recycling nutrients constitutes one of the major benefits provided "free" to humanity and all the biota (outside the market economy) but worth literally trillions of dollars per year as one of the major ecosystem services (Costanza *et al.*, 1997) on Earth.

SUMMARY

The physical properties of the soil are the production of continued interactions between soil biota and their abiotic milieu. Water, the "universal solvent," exerts a strong influence on the biota because many of the biota are adapted to life in a saturated atmosphere. The interplay between liquid and gaseous phases of water, in turn, is largely determined by pore size. The arrangement of particles in soils (the porosphere) is an important determinant for the ecology of the soil microbes (Archaea, bacteria, fungi) and fauna.

Soil formation—the product of climate, organisms, parent material, and topography, over time—leads to various soil types. Profile development and soil texture are the product of interactions of these factors.

The capabilities for nutrient retention, important for primary producers in all soils, are affected by both mineral content and soil organic matter, with organic matter usually having the higher number of exchange sites. The aggregate structure of soils is biologically mediated in many soil types. Soils play major roles in both recycling matter and nutrients, as well as being important sources and sinks of global greenhouse gases. It is apparent that soil ecology is being considered much more centrally in ecological studies and in ecosystem management as well.

2

Primary Production Processes in Soils: Roots and Rhizosphere Associates

INTRODUCTION

A. J. Lotka (1925), in his classic overview of ecological function, considered the system-level features of carbon gain, or anabolism, and the system-level losses of carbon reduction, or catabolism. This chapter is concerned with the primary sources of organic carbon inputs to soils, or system anabolism. These inputs have a major impact on nutrient (nitrogen, phosphorus, and sulfur) dynamics and soil food web function, as will be shown in Chapters 5 and 6.

How can we best address the problems of measurement of primary production? Some ecological studies have declared that taking accurate measurements of belowground inputs to ecosystems is virtually impossible, and assumed that belowground production equals that of production aboveground for total net primary production (NPP) (Fogel, 1985). This rule-of-thumb is clearly inadequate and often very wrong (Vogt *et al.*, 1986). Our objectives in this chapter include addressing the processes and principles underlying primary production, and indicating where the "state of the art" is now, and is likely to be, over the next several years. A wide range of new techniques is now available. We anticipate that information on and our understanding of belowground NPP will continue to increase.

THE PRIMARY PRODUCTION PROCESS

In the process of carbon reduction, there is a net accumulation of sugars, or their equivalents, in the organism's tissues. The costs of photosynthesis are extensively treated by plant physiologists and are out of the purview of this book. Other costs, related to movement of the photosynthates within the plant and allocation to symbiotic associates, are significant to the plant and to the ecosystem, and will be considered further on.

Gross primary production minus plant respiration yields net primary production. NPP is the resultant of two principal processes: (1) increases in biomass and (2) losses due to organic detritus production, which follows from or is dependent on the biomass production (Fogel, 1985). The detritus production includes leaves, branches, bark, inflorescences, seeds, and roots. Additional losses are traceable to exudation, volatilization, leaching, and herbivory (Cheng *et al.*, 1993, 1996).

Measurement of aboveground components is at times tedious, but fairly complete in many studies (see reviews by Persson, 1980, and Swank and Crossley, 1988). In contrast, measurement of belowground production processes has been fraught with errors and many difficulties. However, the total allocation of NPP belowground is often 50% or greater (Coleman, 1976; Harris *et al.*, 1977; Fogel, 1985, 1991; Kuzyakov and Domanski, 2000) (Table 2.1). A sizable portion of the total production is contributed by fine roots, which often have a high turnover rate of weeks to months (Table 2.2), which may be closely linked to nitrogen availability on a seasonal basis (Nadelhoffer *et al.*, 1985, 1992; Publicover and Vogt, 1993). In addition to production of fibrous root tissues, there are accompanying inputs of soluble compounds, namely organic acids, sugars, and other compounds. All of these have a considerable impact on rhizosphere (the zone of soil immediately surrounding the root and comprised of root secretions, exfoliations, and the microbial

TABLE 2.1.   Annual Production ($Mg \cdot ha^{-1}$) of Fine Roots (<2 mm) and Root Production as Percentage of Total NPP in Different Ecosystems

Ecosystem	Age (years)	% Contribution	Production
Coniferous forest			
Douglas fir	55	73	4.1–11.0
Loblolly pine	?	?	8.6
Scots pine	14	60	3.5
Deciduous forest			
Liriodendron	80?	40	9.0
Oak—maple	80	?	5.4
Herbaceous			
Corn	<1	25	1.2–4.2
Soybean	<1	25	0.6
Tallgrass prairie	?	50	5.1

From Fogel, 1985; Coleman *et al.*, 1976.

TABLE 2.2.   Annual Losses (Due to Consumption
and Decomposition) of Fine Root Biomass in
Different Forests

Ecosystem	Loss (% total)
Deciduous forest	
European beech	80–92
Oak	52
Liriodendron	42
Walnut	90
Coniferous forest	
Douglas fir	40–47
Scots pine	66

From Fogel, 1985.

communities contained therein) (Hiltner, 1904; Curl and Truelove, 1986) processes. We cover these later in the chapter.

When comparing across ecosystems, one needs to be aware of marked differences in root morphology and distribution, i.e., root architecture (Fitter, 1985, 1991). Thus wheat roots in a Kansas field are not markedly different in size, with primary and secondary laterals arising from root initials. In contrast, coniferous tree roots are often comprised of long, supporting lateral roots and short roots, which do the primary job of water and nutrient absorption. Ecologists often use a rather simple, pragmatic classification approach: roots with a diameter of less than 2 millimeters (mm) are classified as fine roots, and roots with a diameter greater than 2 mm are classified as structural roots (Fogel, 1991).

METHODS OF SAMPLING

There are several methods for sampling roots, many of which have been reviewed by Böhm (1979). They may be generally classified into two principal approaches (Upchurch and Taylor, 1990): (1) destructive (sampling soil cores or monoliths), and (2) nondestructive, or observational, using rhizotrons or borescopes; termed minirhizotrons (Upchurch and Taylor, 1990; Cheng *et al.*, 1990; Pregitzer *et al.*, 2002).

Destructive Techniques

The Harvest Method

This method involves taking samples, usually as soil cores, dry-sorting the organic material or rinsing it free by use of water or other flotation media, then sieving, sorting, and obtaining dry mass values.

For sorting and categorizing roots, three factors need to be considered: root diameter, spatial distribution, and also temporal distribution (Fogel, 1985). Much of the existing data have been derived from thousands of cores that have been washed, sorted, and analyzed by legions of weary researchers. Some of these data have been truly informative and worth the effort. Other efforts, perhaps a majority of the published papers, have limited value. In the course of measuring root production by the harvest method, scientists often use what is known as the "peak-trough" calculation, in which the peaks and valleys of root-standing crops through the course of a growing season as represented on a graph are successively added or subtracted about some general mean level. Unfortunately, there can be a fairly frequent occurrence of no net changes in root biomass, perhaps as often as 30% of the time in grasslands studies (Singh *et al.*, 1984); these are known as zero-sum years, which have no net production because the increases in production are canceled out by those periods which show decreases. These problems were reviewed by Singh *et al.* (1984) (Fig. 2.1). They extensively analyzed a grassland root production data set, looking for effects of sample (replicate number) size and sampling frequency, and coming to the conclusion that fairly frequently (perhaps in 3 years out of 10) one could expect to measure no significant increments to growth when using the peak-trough harvest method. In addition, they compared the amount of NPP that one would expect from the peak-trough harvest method with a multiple-year–based computer simulation model of root production and turnover. They found that the peak-trough method at times overestimated either the "true" or the simulated root production by as much as 150% because of widely varying means; this led to spuriously high "production" values. The simulated production was not more "real" than the data, of course, but the researchers raised the question that perhaps the peak-trough method, as applied usually, may often lead to some significant overestimates of root production rates.

Considerable information is available on fine root production (FRP) in forested ecosystems. Nadelhoffer and Raich (1992) compiled 59 published estimates of annual net FRP from 43 forest sites worldwide. They compared four techniques used by investigators: (1) sequential core method (calculated as differences in means of fine root biomass between sampling periods and measured across growing seasons); (2) maximum–minimum method (simpler than the first method in that it uses only the difference between annual minimum and maximum fine root biomass to estimate FRP; (3) ingrowth core method (similar to the method of Steen [1984, 1991], cited later in this chapter); and (4) the nitrogen budget method (based on annual measures of net nitrogen mineralization in soil and net nitrogen flux into aboveground tissues. Annual nitrogen allocation to fine roots is calculated from the difference

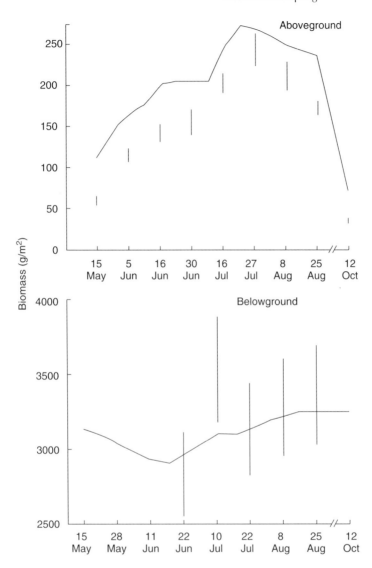

FIGURE 2.1. Comparison of values for aboveground and belowground biomass (g/m²) predicted by a simulation model with data collected in the field. The curves represent output of the model; vertical bars are means of field data plus and minus one standard error (from Singh *et al.*, 1984).

between net nitrogen mineralization and net nitrogen fluxes into aboveground tissues. FRP is then calculated as the product of annual nitrogen allocation to fine roots and the carbon–nitrogen ratio in fine roots (Nadelhoffer *et al.*, 1985). Interestingly, with this last method, total root allocation (TRA), which equals root respiration plus root production, was predicted from aboveground litterfall carbon with an $r^2 = 0.36$, where $p < 0.02$. Whereas this method leaves almost 2/3 of the variability in root production unaccounted for, the other methods are even less reliable in predicting FRP.

Isotope-Dilution Method

Another approach, using the ^{14}C-dilution technique, has been used to determine belowground biomass turnover in grasslands. Milchunas *et al.* (1985) performed a pulse labeling of plants with ^{14}C-CO_2 for a few hours, then followed the time-course of new ^{12}C label incorporated into both soluble and structural tissues of the root systems a few weeks to months later. They then calculated the subsequent production, on the assumption that any tissues lost would have a constant ratio of ^{14}C to ^{12}C in the structural tissues. An additional step was to include ^{14}C incorporated into plant cell walls between the first and second sampling times. This greatly reduced the errors of the estimates (Table 2.3). Milchunas *et al.* (1985) found that the grass roots were continuing to mobilize additional amounts of ^{14}C from storage tissues in the grasses, but then made further measurements of the labeled plants to adequately account for the translocated [14]C. Other researchers, notably Caldwell and Camp (1974), have used the isotope-dilution technique with considerable success.

Root-Ingrowth Technique

The root-ingrowth technique (Steen, 1984, 1991) involves removing long cores of soil, sieving the soil free of roots, and then replacing the root-free soil into nylon tubular mesh bags with a mesh size of 5–7 mm. The mesh bags are inserted by drawing them over a plastic tube, and the pipe plus mesh bag on the outside is inserted into the hole in the field soil. The soil is tamped down in 5-centimeter (cm) increments as the pipe is gradually withdrawn, leaving the mesh bag in position in the hole. Care must be taken to have a bulk density similar to that in the surrounding matrix. After the soil mesh bags are placed in their respective holes, roots are allowed to regrow into the bags. The bags are then recovered at various intervals, and the living and recent dead root biomass measured (Hansson *et al.*, 1991; Steen, 1991). The principal assumptions are that growth into the root-free soil is the same as the root production would have been in the normal, undisturbed soil. Some concerns one might have about this technique are: Was the bulk density of the soil in the mesh tubes identical to that in the surrounding soil? Were any significant soil aggregates broken in the soil-sieving process, which might alter rates of root growth in the bags? (Larger soil

TABLE 2.3. Comparison of Increments in the Belowground Biomass of Blue Grama and Wheat as Determined by Complete Harvest and $^{14}C/^{12}C$ Dilution Techniques[a]

Time (days)	Cell-wall carbon				Belowground production[b]				
					Uncorrected			Corrected for ^{14}C incorporation	
	$^{12}C(g)$[c]	$^{14}C(g \times 10^{-4})$[c]	$R(\times 10^{-4})$[d]	$R_C(\times 10^{-4})$[e]	Harvest	$(R_1C_1/R_2C_2-1)B$[f]	Error(%)[g]	$(R_1C_1/R_2C_2-1)B$[f]	Error(%)[g]
Blue grama									
0.15	0.70	4.37	6.23						
5	0.90	22.56	25.07	10.58	3.52	-0.65	-118	3.48	-1
8	0.95	33.48	35.26	15.70	3.64	1.22	-66	3.56	-2
11	1.19	34.57	28.99	16.21	2.70	0.53	-80	2.73	1
25	1.40	39.67	28.25	18.60	2.15	0.78	-64	2.14	0
40	2.13	48.04	22.52						
Wheat									
0.15	0.53	1.82	3.46						
5	0.52	5.46	10.46	2.52	3.45	1.02	-70	3.53	2
8	0.72	6.91	9.60	3.19	2.96	1.25	-58	2.97	0
11	0.77	7.64	9.99	3.52	2.73	1.35	-50	2.68	-2
25	1.26	8.73	6.95	4.02	1.31	0.51	-56	1.33	2
40	2.02	9.83	4.87	4.53	-0.20	-0.35	75	-0.15	-25
62	2.17	10.19	4.70						

[a]Values are expressed on an ash-free, dry-weight basis.

[b]Increment is grams of total root (structural plus labile) between that sampling time and the last sampling time.

[c]$^{14}C = $ (DPM sample) (60 s/min) [Ci(3.7 × 10^{10} DPS)] (0.22442 g/Ci).

[d]$R = ^{14}C/^{12}C$; $R_1 = ^{14}C/^{12}C$ for that sampling time; and $R_2 = ^{14}C/^{12}C$ for the last sampling time.

[e]$R_C = R_2$ corrected for the amount of ^{14}C incorporated into structural material between that sampling time and the last sampling time.

[f]$C = \%$ cell wall at time 1 or 2; $B_1 = $ biomass (structural plus labile) at time 1.

[g](Dilution−harvest)/harvest × 100.

Modified from Milchunas et al., 1985.

aggregates might be left intact, if they do not contain any roots [Steen, 1991].) What effect is caused by higher water contents in soil volumes without living roots? Advantages of the technique include being able to get a clear, more accurate measure of production of roots over discrete time intervals. Also, it is possible to obtain information on decomposition of dead roots by placing fine-mesh (<0.1 mm) cloth bags into the soil cylinders and determining the loss rates of dead roots simultaneously with measurement of new root ingrowth over time (Titlyanova, 1987). However, see papers by Reid and Goss (1982) and Cheng and Coleman (1990) for comments about live root and organic matter interactions. Some of these concerns will be addressed further in a discussion of decomposition processes in soils later in this chapter.

Nondestructive Techniques

There has been a great resurgence recently of interest in observational, nondestructive techniques for studying root-related processes. Several review volumes present detailed discussion of minirhizotron and rhizotron usage, including Taylor (1987), Box and Hammond (1990), and Fahey *et al.* (1999). In essence, the rhizotron approach involves installing a large glass plate in an observation gallery and then measuring the growth of roots against the glass over time (Fogel and Lussenhop, 1991) (Fig. 2.2). Using this technique, one can follow a large part of a given root population visible through the glass over various time periods. The disadvantages are that the soil profile must be recreated and re-tamped to an equivalent bulk density, or mass per unit volume of soil that closely approximates the density of the surrounding soil. It is also only a small fraction of an entire field or forest.

Minirhizotrons, on the other hand, have a smaller amount of surface area in one place, being tubular (5–7 cm in diameter), and are placed, as are the rhizotrons, at a 20–25° angle from the vertical (Fig. 2.3). However, being light and readily handled, they enable extensive replication in any given plot, experimental treatment, or entire field site. Tubes may be of either glass or a durable plastic such as polycarbonate. For example, Cheng *et al.*, (1990) used 12 minirhizotron tubes in each replicate and two replicates per treatment (conventional tillage and no-tillage) in a study of sorghum root growth and turnover in a southeastern United States agroecosystem. Other studies have followed the dynamics of soil mesofauna, namely collembola, in fields under various crops in Michigan agroecosystems (Snider *et al.*, 1990). A number of precautions should be employed in the usage of minirhizotrons, so as to avoid artifacts of placement. For example, total root biomasses can be underestimated in the top 7–10 cm if inadequate care is taken to shield

FIGURE 2.2.   University of Michigan Soil Biotron. (a) View of tunnel and aboveground laboratory from the south. (b) View from the west. Note white pine stump left after logging and burning in about 1917. (c) Interior of tunnel showing window bays covered with insulated shutters. (d) Close-up of glass, wire-reinforced, 6-mm-thick windowpane. Note fungal rhizomorphs. The wire grid is about 2 cm by 2 cm (from Fogel and Lussenhop, 1991).

the top of the minirhizotron tubes from transmitted light. Also, adequate soil–tube contact needs to be ensured by careful drilling and smoothing of the bored hole (preferably using a hydraulic coring apparatus), as noted by Box and Johnson (1987). If there is some open space between the outer tube surface and the soil, roots may respond as if this is a major soil crack and preferentially grow along it (van Noordwijk et al., 1993). To handle the large amounts of data and images obtained using minirhizotrons, it is necessary to use image analysis programs such as those described by Smucker et al. (1987), Hendrick and Pregitzer (1992), and Pregitzer et al. (2002). With the advent of digital analysis techniques and image storage on CD-ROM, the literally millions of bits of information per soil–root image can be manipulated and analyzed reasonably promptly and efficiently. Caution must be taken, how-

FIGURE 2.3. An auger jig system used to install angled minirhizotron tubes (from Mackie-Dawson and Atkinson, 1991).

ever, to ensure that the material used for the tube has a minimal effect on the roots being observed. Withington *et al.* (2003) compared minirhizotron data for glass, acrylic, and butyrate tubes in an apple orchard, and acrylic and butyrate tubes in a study with six forest tree species. Root phenology and morphology were generally similar among tubes. Root survivorship varied markedly between hardwood and conifer species, however, probably because of hydrolysis by fungi interacting with the plastic tubes. Comparison of data from cores of root-standing crops with data from cores of minirhizotron-standing crops showed a closer match with the acrylic than the butyrate data. Glass was considered to be the most inert, but one-third of the glass tubes were lost as a result of breakage during the winter in the Pennsylvania site.

Frank *et al.* (2002), using minirhizotrons, measured significant increases in root growth in nine higher-elevation (1635–2370 meters) mixed-grass grazing lands in Yellowstone National Park. They found that large migratory herds of elk, bison, and pronghorn, by their grazing, stimulated aboveground, belowground, and whole-grassland productivity by 21%, 35%, and 32%, respectively. This feedback effect, which was demonstrated earlier by Dyer and Bokhari (1976) and McNaughton (1976), will be addressed further in system-level effects considered in Chapter 5.

In a study of seven minirhizotron data sets, Crocker *et al.* (2003) substituted root numbers for root lengths using a regression technique. Linear regression models were fitted between root length and root number for production and mortality of a wide range of tree species from subtropical to boreal conditions. Treatments yielded r^2 values ranging from 0.79 to 0.99, indicating that changes in root numbers can be used to predict root-length dynamics reliably. Slope values for mean root segment length (MRSL) ranged from 2.34 to 8.38 mm per root segment for both production and mortality. Crocker *et al.* (2003) caution that the quantitative relationship between root lengths and numbers must be established for a particular species–treatment combination, but it will save on time needed to quantify root dynamics.

Additional approaches to calculation of fine root production and turnover continue to appear in the scientific literature. Gaudinski *et al.* (2001) used one-time measurements of radiocarbon (^{14}C) in fine roots (<2 mm in diameter) from three temperate forests in the eastern United States—a coniferous forest in Maine, a mixed hardwood forest in Massachusetts, and a loblolly pine plantation in South Carolina. Roots were sampled as either mixed (live and dead) or live. Using accelerator mass spectrometry (AMS) to analyze very small samples, Gaudinski *et al.* (2001) found that root tissues are derived from recently fixed carbon, and the storage time prior to allocation to root growth is less than 2 years and more likely less than 1 year. Live roots in the organic horizons contain carbon fixed 3–8 years ago, versus roots in mineral B horizons with carbon of 11–18 years mean age. This spatial component to root age has not been measured before, and has important implications in calculating more realistic carbon budgets for terrestrial ecosystems. This assessment of mean root ages in forest tree roots is in marked contrast to the more rapid turnover times as noted above in the studies using minirhizotrons and other direct means of observation. It does not negate the findings of the observational studies, but emphasizes the need to be aware of the wide range of age of fine roots in the entire soil profile.

ADDITIONAL SOURCES OF PRIMARY PRODUCTION

An additional contribution to net primary production comes from algal populations in the surface few millimeters or on the soil surface itself. By measuring CO_2 fixation by cyanobacteria and algae on the surface of intact cores taken from an agroecosystem, Shimmel and Darley (1985) calculated that approximately 39 grams (g) of carbon were fixed per square meter per year. This is a small proportion (5%) of total NPP for the study site (the conventional-tillage agricultural system in Georgia), which averaged 800 g per square meter per year aboveground NPP. The type of organic matter and the amount that may feed directly into detritivorous fauna could be of importance beyond the total production figures on an annual basis. For example, cryptogamic crusts can be significant agents of nitrogen fixation, providing inputs of nitrogen and carbon in the ecosystems of nutrient-poor arid lands (Evans and Belnap, 1999; Belnap, 2002).

SYMBIOTIC ASSOCIATES OF ROOTS

From the earliest origins of a land flora, more than 400 million years ago, there has been a structural–functional interaction between plant roots and arbuscular mycorrhizae (AM) [Fig. 2.4(a)] as shown in the fossil record (Pirozynski and Malloch, 1975; Malloch et al., 1980). Probably the earliest land plants arose during the Ordovician period and were similar to present-day hornworts and liverworts (Redeker, 2002). Today's hornworts and liverworts do form associations with AM fungi, but because they do not have true roots, they do not meet all the criteria of AM (Read et al., 2000). Most families of terrestrial plants have mycorrhizal symbionts; however, two families—the Cruciferae and the Chenopodiaceae are conspicuous exceptions (Allen, 1991). Indeed, recent analyses have shown that zygomycetous fungi colonize a wide range of lower land plants (hornworts, many hepatics, lycopods, Ophioglossales, Psilotales [horsetails], and Gleicheniaceae) (Read et al., 2000). These associations are structurally analogous to mycorrhizas, but their functions remain to be determined (Read et al., 2000).

The ectomycorrhizae, or ECM [Fig. 2.4(b)], are prevalent in several tree families such as the Fagaceae (including the beeches and oaks), and also in the Pinaceae within the conifers. ECM arose relatively recently, only 160 million years ago, in the Cretaceous period (St. John and Coleman, 1983).

After examining the structures of the principal types of mycorrhizae, we will consider information on carbon costs to the plant as well.

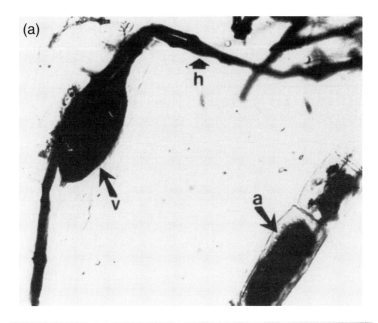

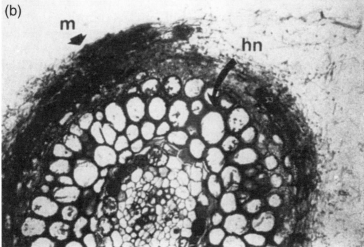

FIGURE 2.4. The internal structures of the different mycorrhizal types from ecto-mycorrhizae of *Quercus dumoa* (b) and arbuscular mycorrhizae of *Adenostoma sparsifolia* (a). Shown are arbuscules *(a)*, vesicles *(v)*, internal hyphae *(h)*, mantle *(m)*, and Hartig net *(hn)* (from Allen, 1991).

MYCORRHIZAL STRUCTURE AND FUNCTION

Arbuscular mycorrhiza (AM), the so-called endomycorrhiza, are characterized by structures within root cells; these structures are called arbuscules because they grow and ramify, treelike, within the cell (see Fig. 2.4[a]). They are members of the Phycomycete fungi family. Most, but not all AM (two exceptions are known in the Endogonaceae), also have storage structures known as vesicles, which store oil-rich products. AM send out hyphae for several centimeters (a maximum of 6–10) into the surrounding soil and are instrumental in facilitating nutrient uptake, particularly phosphate ions (Allen, 1991). AM are known only as obligate mutualists (i.e., the root provides carbon, and the mycorrhiza tap an enhanced pool of mineral nutrients) and have not been cultured yet apart from their host roots. Because AM hyphae will grow out from the germinating chlamydospore toward root surfaces, responding to soluble compounds, possibly including flavonoids, they are considered to have slight saprophytic competence (Azcón-Aguilar and Barea, 1995). Some rhizosphere microorganisms seem to stimulate AM germination and mycelial growth, functioning either by detoxifying or removing inhibitors from the growth medium or by utilizing self-inhibiting compounds from the AM fungus, enabling more growth than would be possible under axenic conditions (Azcón-Aguilar and Barea, 1995).

Ectomycorrhiza (ECM) are significantly different in physiology and ecology. These are principally Basidiomycetes and proliferate between cells, not inside them as is the case for AM. An obvious morphological alteration occurs with formation of the mantle and Hartig net (a combination of epidermal cells and ECM fungal tissues) on the exterior of the root (see Fig. 2.4[b]). ECM send hyphae out several meters into the surrounding soil. The hyphae aid in nutrient uptake, including inorganic and some organic nitrogen–phosphorus compounds (Read, 1991). The hyphae constitute a significant proportion of carbon allocated to belowground NPP in coniferous forests (Vogt et $al.$, 1982). The reproductive structures of ECM are the often-observed mushrooms in oak or pine forests. ECM will form resting stages, or sclerotia—cordlike bundles of hyphae that can persist for years. ECM, unlike AM, often can be cultured apart from their host plants. Some ECM may have considerable decomposing capabilities and can obtain a portion of their reduced carbon from decomposing substrates (i.e., leaf litter). An innovative study by Hobbie et $al.$ (2002) examined the flows of ^{13}C and ^{14}C to basidiomycete sporocarps, needles, and litter in a western Oregon forest soil, using accelerator mass spectrometry of 1- to 2-milligram samples of soils and tissues. Mycorrhizal associations were indicated by very young (0–2 years) age of ^{14}C, whereas the saprotrophic genera averaged 10 years in radiocarbon age (Hobbie et $al.$, 2002). With analytical tools now at hand, Hobbie et $al.$

(2002) suggest that needle and fungal carbohydrates should be analyzed for ^{14}C content separately from needle and fungal protein. They predict that protein will be significantly older than carbohydrates when amino acids are taken up directly from the soil, versus the faster flow of carbon from photosynthates to the current crop of needles in litter.

ECOSYSTEM-LEVEL CONSEQUENCES OF ECM FUNCTION

ECM form fungal mats in the surface soils of Douglas fir (*Pseudotsuga menziesii*) forests and probably other coniferous forests. The mats contain higher microbial biomass, with two to three times more soil mesofauna (collembola, mites, and nematodes) feeding on them (Cromack *et al.*, 1988). The fungal mats play an important role in buffering against iron and aluminum activity in the acidic soils on the west slope of the Cascade Mountains in the Pacific Northwest of the United States (Entry *et al.*, 1992). As iron and aluminum activity in soil increases, calcium oxalate crystals formed by the fungal mats dissociate, releasing calcium and chelating the iron and aluminum. Entry *et al.* (1991) also measured greater litter decomposition rates and greater mineralization of nitrogen and phosphorus in ECM mat soils.

Alternative methods for determining ECM fungal community make-up have been developed using molecular techniques. Landeweert *et al.* (2003) used a basidiomycete-specific primer pair (ITS1F–ITS4B) to amplify fungal internal transcribed spacer (ITS) from total DNA extracts of the soil horizons, followed by an amplified basidiomycete DNA cloning and sequencing procedure, to identify the ECM fungi present. The soil samples were from four distinct horizons of a spodosol profile, under coniferous (Norway spruce *[Picea abies]* and Scots pine *[Pinus sylvestris]*) vegetation. By identifying basidiomycete mycelium in the soil, the ECM fungal community was analyzed in a novel fashion (ECM root tips were excluded from the analysis). Landeweert *et al.* (2003) sampled from the O layer (0–2 cm deep); the E horizon (3–18 cm); the enriched eluvial, or B horizon (18–35 cm); and the parent material, or C horizon (deeper than 40 cm). They found 16 of the 25 total operational taxonomic units (OTUs) exclusively in the deeper mineral soil, or B horizon. The authors suggested that these distributions might be the result of somewhat higher amounts of carbon and a higher pH existing deeper in the profile. This analysis demonstrates the need to consider the full suite of ECM fungi present, and is a cautionary note that the entire profile should be considered when determining species richness in a given site.

There are other kinds of mycorrhiza—most notably Ericaceous mycorrhiza—which have some traits in common with ECM and AM.

Ericaceous mycorrhiza are symbiotic with many heathland plants; *Rhododendron* and *Kalmia* spp. are often infected with Ericaceous mycorrhiza (Dighton and Coleman, 1992). The fungus normally involved in forming the infections is the ascomycete *Hymenoscyphus ericae* or its anamorphs, and significant amounts of chitin-N can be transferred to the host plant (Kerley and Read, 1995). Ericaceous mycorrhiza are noted for the ability to facilitate direct uptake of organic nitrogen in low pH environments, and some of them produce proteases to enhance the nitrogen uptake without going through any external mineralization in the soil solution (Bending and Read, 1996). As may often be the case with residues derived from ericaceous sclerophylls, the essential elements (nitrogen and phosphorus) will be masked by skeletal materials, namely lignin or its breakdown products. In these circumstances, the ability of the mycorrhizal fungus to produce lignase or phenol oxidase activities and thus expose the nutrient-containing substrates would be just as important as production of the enzymes (e.g., phosphatases and proteases) that are directly involved in nutrient release. Also see comments in the section on future directions in mycorrhizal research at the end of this chapter. For information on these, and other less-common mycorrhiza, refer to Allen (1991, 1992), Read (1991), and Smith and Read (1997).

ACTINORHIZA

Another symbiotic associate with roots plays an important role in many forested ecosystems worldwide. It is the actinorhiza, an actinomycete (filamentous, branching, gram-positive bacteria) that forms nodules and fixes dinitrogen in a fashion analogous to that used by rhizobia. A majority of actinorhizal species are pioneers on nitrogen-poor, open sites (Baker and Schwintzer, 1990). The dominant actinorhizal genus is *Frankia*, occurring on roots of eight plant families, encompassing 24 genera and some 230 species of dicotyledons. Prominent actinorhizal plant families and genera include Betulaceae, on 47 *Alnus* species; Casuarinaceae, on 16 *Casuarina* species and 54 *Allocasuarina* species; Myricaceae, on 28 *Myrica* species; and Rhamnaceae, on 31 *Ceanothus* species. These genera are widespread in ecosystems on all continents except Australia (Baker and Schwintzer, 1990).

CARBON ALLOCATION IN THE ROOT/RHIZOSPHERE

Looking at the root–soil system as a whole, what is the totality of the resources involved, and how are these resources allocated under various conditions of stress and soil type?

Several reviewers (Coleman, 1976; Coleman *et al.*, 1983; Fogel, 1985, 1991; Martin and Kemp, 1986; Cheng *et al.*, 1993; Cheng, 1996; Kuzyakov, 2002) have noted that from 20 to 50% more carbon enters the rhizosphere from root exudates and exfoliates (sloughed cells and root hairs) than actually is present as fibrous roots at the end of a growing season. This was determined in a series of experiments using ^{14}C as a radiotracer of the particulate and soluble carbon (Shamoot *et al.*, 1968; Barber and Martin, 1976). In fact, the mere change from a hydroponic medium to a sand medium was enough to double the amount of labile carbon as an input to the medium. This difference was attributed to the abrasion of roots against sand particles. In addition, the root–rhizosphere microflora has the potential to act as a sizable carbon sink (Wang *et al.*, 1989; Helal and Sauerbeck, 1991), which can double the losses to soil as well. This is convincing proof that the combined below-ground system—roots, microbes, soil, and fauna—is governed by source-sink relationships, just as are intact plants (i.e., roots and shoots).

Extensive amounts and complexities of carbon compounds are elaborated in the rhizosphere (Rovira *et al.*, 1979; Kilbertus, 1980; Foster *et al.*, 1983; Foster, 1988; Lee and Foster, 1991; Cheng *et al.*, 1993). The extent to which this exuded carbon is integral to root and rhizosphere function is of great interest to ecologists. Nitrogen-fixing bacteria residing in the rhizosphere and the release of their nitrogen to the plant can be stimulated by root exudates (Rao *et al.*, 1998, cited in Jones *et al.*, 2003). There are numerous direct and indirect positive and negative effects of carbon flows in the rhizosphere that encompass a wide array of symbiotic associations and trophic and biochemical interactions (Jones *et al.*, 2003) (Figs. 2.5 and 2.6). Although the potential for rhizodeposition-driven N_2 fixation in the soil is small in comparison to inorganic and symbiotic fixation inputs, it may be of importance in nitrogen-limited ecosystems (Jones *et al.*, 2003). The boundary layer between roots and soil—the so-called "mucigel" (Jenny and Grossenbacher, 1963)—is jointly contributed by microbes and root surfaces. Studies of the root tip and capsule components have been most informative about the roles of signal molecules that are exuded at subnutritional rates in soil. One of the key components involved is the border cells. These cells are lost from the root tip at a rate regulated by the root and secrete compounds that alter the environment of, and gene expression in, soil microorganisms and fauna (Farrar *et al.*, 2003) (Fig. 2.7). These root-tip capsule components include high molecular weight (MW) mucilage secreted by the root cap, as well as cell-wall breakdown products resulting from the separation of thousands of border cells from each other and the root cap (see Fig. 2.7).

Much research has been conducted on ways to separate total CO_2 efflux into that from microbial respiration from soil organic matter

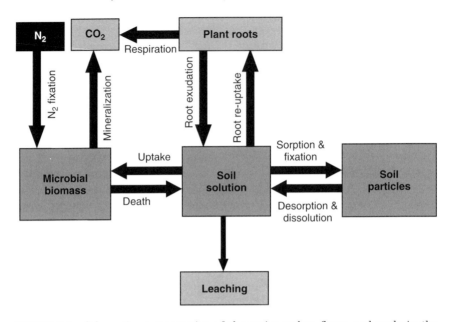

FIGURE 2.5. Schematic representation of the major carbon fluxes and pools in the rhizosphere (from Jones *et al.*, 2003).

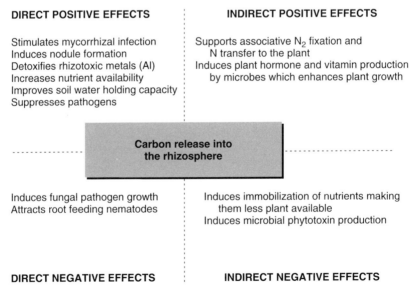

FIGURE 2.6. Schematic representation of the positive and negative direct and indirect effects of root exudates on plant growth (from Jones *et al.*, 2003).

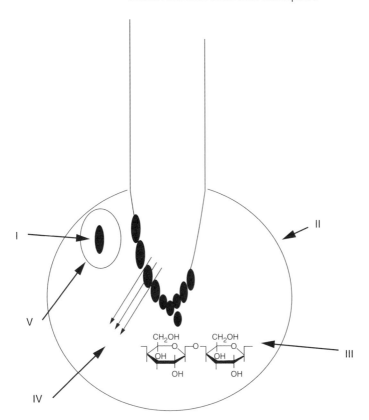

FIGURE 2.7. Root-tip capsule components: (I) border cell populations, (II) high-molecular-weight mucilage secreted by the root cap, (III) cell-wall breakdown products resulting from separation of thousands of border cells from each other and the root cap, (IV) other extracellular products secreted by the cap, (V) other extracellular products secreted by the border cells (from Farrar *et al.*, 2003).

(SOM) and that from roots and rhizosphere microbial populations. Methods employed have been summarized into three broad categories (Hanson *et al.*, 2000): component integration, root exclusion, and isotopic approaches. Component integration entails separation of the constituent soil components involved in respiring CO_2 (i.e., roots, sieved soil, and litter) followed by measurement of the specific rates of CO_2 outputs from each component part (Coleman, 1973; Trumbore *et al.*, 1995). The root exclusion method estimates root respiration indirectly by subtracting the soil respiration without roots from the soil respiration with roots (Anderson, 1973; Edwards, 1991). Isotope methods refer to the use of either radioisotopic [14]C (Cheng *et al.*, 1993; Horwath *et al.*, 1994) or stable ^{13}C isotopes to trace the origin of soil respiration (Andrew *et al.*,

1999; Robinson and Scrimgeour, 1995). Isotope methods have a significant advantage over the component integration and root exclusion methods because they permit researchers to partition carbon between rhizosphere respiration and SOM decomposition *in situ*, thus avoiding the effects of soil disturbance.

CARBON ALLOCATION COSTS OF DEVELOPMENT AND MAINTENANCE OF SYMBIOTIC ASSOCIATIONS WITH ROOTS

It is difficult to measure apportioning of energy to roots because accurate measurement of belowground NPP entails a number of precautions, as previously noted. Over the last 2 decades, there have been only a few estimates of the carbon costs that have been exacted by the fungal or rhizobial symbiont upon its root partner. The following are two examples of the sorts of measurements that have proven informative.

Pate *et al.* (1979) compared partitioning and utilization of assimilated carbon and nitrogen, using nonnodulated, nitrate-fed, and nodulated, dinitrogen-fixing plants of white lupine, *Lupinus albus* L. Pate *et al.* (1979) calculated production and losses, and noted that not only were the nodulated root microflora more active, but there was also more new root growth under the stimulation of the nodule bacteria, which were acting as a greater root sink for translocated carbon.

Kucey and Paul (1982) measured two symbionts, an AM mycorrhiza and rhizobia in seedlings of fava beans, *Vicia faba* L. The bean seedlings were arranged experimentally as either mycorrhizal- or nonmycorrhizal-infected, and also as nodulated or nonnodulated bean plants—four treatments in all. After inoculating or infecting the plants of choice, they then measured CO_2 fixation rates, translocation of ^{14}C–labeled photosynthate to roots, and nitrogen-15 fixed by the various plants. Kucey and Paul found gradually increasing amounts of labeled ^{14}C translocated and/or evolved belowground, as a function of infection complexity (Table 2.4). In addition, they obtained useful information on root and shoot weight, and rates of respiration.

The nodules of fava beans utilized 6% of the carbon fixed by singly infected (rhizobial) plants, but twice that amount, or 12%, of the carbon used by the doubly infected plants, i.e., both mycorrhiza and rhizobia symbionts (Table 2.5). Interestingly, rates of CO_2 fixation increased significantly with biotic complexity, but changes in root and shoot biomass, while opposite to that of CO_2, were statistically insignificant.

Other studies, using real-time monitoring of ^{11}C under laboratory conditions (Wang *et al.*, 1989), found that mycorrhizal infection nearly doubled the "sink strength" of the roots, hence there was greater carbon flow of translocated photosynthate in studies of African *Panicum*

TABLE 2.4. [14]C Distribution in 5- to 6-Week-Old[a] Symbiotic and Nonsymbiotic Beans

	Control	Mycorr. (MI)	Rhiz. (RI)	Mycorr.-Rhiz. (RMI)
		Plant data		
CO_2 fixation rate[b]	6.79c	6.96b	7.32b	9.34a
Shoot weight (g)	4.31a	4.40a	3.64a	3.59a
Root weight (g)	2.03a	1.65a	1.75a	1.64a
Nodule weight (g)	—	—	0.11	0.15
Mycorrhizal infection (%)	—	58.6	—	54.8
		[14]C distribution (%)		
Shoot biomass	54.6	52.20	46.8	42.0
Shoot respiration	1.7	1.0	2.0	1.1
Root biomass	20.7	20.2	25.0	16.8
Root respiration	23.0[c]	26.8[c]	24.6[c]	37.9[c]
Mycorrhizal biomass	—	ND	—	ND
Mycorrhizal respiration	—	ND	—	ND
Nodule biomass	—	—	1.61	2.24
Nodule respiration	—	—	ND	ND

[a]8-hour labeling duration.
[b]$(mg\,C\,g^{-1}$ shoot $C\,h^{-1})$ calculated using shoot weights as measured at the end of the experiment.
[c]Root plus symbiont respiration.
a–c Means followed by the same letter do not differ (P < 0.5).
From Kucey and Paul, 1982.

TABLE 2.5. N_2 Fixation by Mycorrhizal (RMI) and Nonmycorrhizal (RI) Nodulated Beans (4- to 5-Week-Old)

	Nodule wt/root wt $(mg\,g^{-1})$	% N in shoot	N fixed (mg)	N fixed per unit nodule wt $(mg\,g^{-1})$
Rhizobial	87.7	3.81	0.78	16.2
Mycorrhizal-rhizobial	104.0[b]	4.34[a]	1.06[a]	15.8

[a]Significantly different (P ≤ 0.1%) from rhizobial treatment.
[b]Significantly different (P < 0.01%) from rhizobial treatment.
From Kucey and Paul, 1982.

grasses (the same species studied by McNaughton, 1976, and McNaughton *et al.*, 1998), when compared with noninoculated control plants. Some elegant field studies in Scottish grassland soils near Edinburgh have demonstrated significant flows of carbon from roots to mycorrhiza. Within 21 hours of pulse-labeling a grassland sward in the field with $^{13}CO_2$, between 3.9 and 6.2% of the $^{13}CO_2$ passed through the external mycelium of the AM fungal symbionts to the atmosphere (Johnson *et al.*, 2002). This is the first in-the-field verification of similar results measured using pot experiments. Additional recent pot experiments have exposed mycorrhizal plants to fossil (^{14}C–depleted) carbon dioxide and collected samples of extraradical mycelium (ERM) hyphae over the following 29 days. Analyses of their ^{14}C content by accelerator mass spectrometry (AMS) revealed that most ERM hyphae of AM fungi live, on average, 5 to 6 days (Staddon *et al.*, 2003). This high turnover rate indicates the existence of a large and rapid mycorrhizal pathway of carbon in the soil carbon cycle.

Recent field research has demonstrated the significant effects of an additional flow of carbon from plant roots to mycorrhiza and into the soil. The glycoprotein glomalin, which is produced by arbuscular mycorrhizal fungi (AMF), has a marked effect on soil aggregate water stability (Wright, *et al.*, 1999). We discuss this further in the section on microbial interactions in soil (in Chapter 3).

FUTURE DIRECTIONS FOR RESEARCH ON ROOTS AND MYCORRHIZAL FUNCTION AND BIODIVERSITY

As researchers and government agencies become ever more interested in and concerned about "sustainability" and long-term management of ecosystems, they will require much more information on system-level carbon allocation and energetics of these ecosystems. A recent example is related to concerns about carbon sequestration and ecosystem carbon cycling. Root–mycorrhizal interactions have been found to be diagnostic of significant differences in potential seedling relative growth rate (RGR) (Cornelissen *et al.*, 2001). Plant species with ericoid mycorrhiza showed consistently low RGR, low foliar nitrogen and phosphorus concentrations, and poor litter decomposability. Species with ectomycorrhiza had an intermediate RGR, higher foliar nitrogen and phosphorus, and intermediate to poor litter decomposability. Plant species with AM showed comparatively high RGR, high foliar nitrogen and phosphorus, and fast litter decomposition. The incorporation of mycorrhizal associations into functional type classifications should prove useful in assessing plant-mediated controls on carbon and nutrient cycling.

Several recent studies of mycorrhizal function have noted the ways in which mycorrhiza facilitate nutrient uptake from a wide range of sources. For example, Perez-Moreno and Read (2001) measured an enhanced recycling of nitrogen and phosphorus from the necromass of nematodes in cultures with *Betula pendula*. In nonmycorrhizal treatments, the uptake of nitrogen and phosphorus was slightly greater, but with mycorrhiza of *Paxillus involutus* (an ECM) present, the nutrient uptake was significantly higher still. Ectomycorrhizal plants (e.g., lodgepole pine *[Pinus contorta]*) and mycorrhizal fungi such as *Amanita muscaria* have been shown to utilize organic forms of nitrogen directly (Abuzinadah and Read, 1989; Finlay *et al.*, 1992). In addition, ericoid mycorrhiza can produce extracellular enzymes that mineralize nitrogen from protein–tannin complexes (Leake and Read, 1989; Bending and Read, 1996). In a number of ecosystems worldwide that have ericaceous or heath vegetation, this direct pathway, by "short circuiting" the microbial mineralization pathway, would enable the ericaceous plants to survive, indeed thrive, in the absence of adequate nitrogen from usual mineralization rates (Northup *et al.*, 1995). Note that newly available analytical techniques such as accelerator mass spectrometry (see Staddon *et al.*, 2003, on previous page) have allowed detection of small quantities of carbon other than those from atmospheric sources, such as those from the direct uptake of amino acids, as noted above, and also those from anaplerotic (dark fixation) pathways, the latter providing as much as 3% of total carbon in ECM or saprophytic tissues (Hobbie *et al.*, 2002).

As noted above, the vast majority of plants have mycorrhizal associates. Little is known yet of the species richness of mycorrhiza, particularly of the arbuscular, or AM, type. Until recently, AM mycorrhiza were assumed to have little host-specificity and to generally colonize a wide range of possible host species. In an intensive study of the plant-AM fungal interactions within a 1-hectare old field in North Carolina, Bever *et al.* (2001) found that, rather than the initial estimate (in 1992) of 11 AM species in this field, they now have isolated at least 37 species, with one-third of them previously unrecorded. The ecological preference ranges of each species are quite different, reflecting significantly different optima for temperature, moisture content, host, and phenological phases of the plants in the field. The implications of these varied interactions for plant diversity are very large, and the subject is one of increasing interest among plant community ecologists; the extent of mycorrhizal growth and uptake of labile carbon may affect the overall plant community makeup in interesting, hitherto unthought-of ways (Bever, 1999).

The molecular identification of AM has increased greatly in the last decade. By taking samples from within growing roots, amplifying, and producing 18S rDNA sequences, Redeker *et al.* (2000) and Redeker (2002) found a phylogenetically deep divergence of lineages within the Glomales, one of the principal groups within the AM fungi. In addition, two or more species were found to be co-occurring within the same root, indicating the probable existence of complex interactions of the fungi involved. The possibility for various species of AM fungi to become active at various times of the year is an intriguing one for future research on root–mycorrhizal fungal ecology (Redeker, 2002).

In addition to plant–fungi interactions, there are additional interactions with underground fauna to be considered. Klironomos and Kendrick (1995) found that the hyphae of AM are generally less palatable to fungivorous fauna than the hyphae of soil-borne conidial fungi. Such species-specific differences in palatability have been observed in ectomycorrhizal fungi as well (Schultz, 1991). We consider feeding behaviors in Chapter 4 and system-level impacts in Chapters 5 and 6.

The tools and analytical skills are at hand; it is now necessary to proceed with as much care in the assessment and measurement of belowground processes as has heretofore been given to aboveground processes.

SUMMARY

Primary production processes constitute the principal biochemical motive force for all subsequent activities of heterotrophs in soils. The inputs come from two directions: (1) from aboveground onto the soil surface, as litter, and (2) from belowground, as roots, which contribute exudates and exfoliated cells while the roots are alive, and then as root litter when the roots die.

A wide range of direct measurements of root production and turnover are now in use. These include various nondestructive techniques including rhizotrons and minirhizotrons, and destructive techniques including soil coring and isotopic-labeling of roots followed by destructive sampling at specified time intervals to determine dynamics (e.g., over an entire growing season).

Of equal importance to roots themselves are their generally more efficient physiological extensions—the root–fungus mutualistic association, mycorrhiza. At a cost of 5–30% of the total photosynthate translocated belowground, mycorrhiza assist in obtaining inorganic nutrients, water, and in some cases, organic nutrients over a much wider range of the soil volume than roots alone. This symbiotic association has a significant effect on other biota, namely microbes and fauna, which inhabit all soil systems.

3

Secondary Production: Activities of Heterotrophic Organisms—Microbes

INTRODUCTION

We will now consider system-level catabolism, or dissipation and transformation of energy (the roots of catabolism, *cata bolos*, mean "breaking-down activity"). The transfer of energy from the primary producers into organisms farther along the food chain supports a wide range of heterotrophs. The production of new body tissues by heterotrophs from primary production is called secondary production. If the plant food sources are living, the linkages are called a *grazing* food chain. Conversely, if the contributions from net primary production (NPP) are dead, the sequence is termed a *detrital* food chain.

This difference in food chains has some impact on system function, in that grazing food chains have a direct feedback, whereas detrital food webs have only indirect effects. Soil food chains and webs are discussed further in Chapter 6.

The array of energy dissipators, or heterotrophs, in soil is incredibly diverse. The size range goes from less than 1 micrometer (μm) in length (bacteria) to the largest fossorial mammals, such as aardvarks or badgers, and giant earthworms that reach 2 meters (m) in length (Lee, 1985). Larger entities include ant and termite colonies, considered by some to be a "superorganism" (Emerson, 1956). Larger yet are super-colonies of one organism of uniform genetic material such as the extended mycelium of a fungus in Michigan that extended over more than 7 hectares (Smith *et al.*, 1992).

All heterotrophs, of whatever size or volume, are involved in ingesting organic carbon and associated nutrients and assimilating them into carbohydrates, lipids, and proteins. Using a portion for production of new body tissue, an extensive amount (40% or more) of the chemical bond energy is lost as metabolic heat and evolved carbon dioxide (CO_2). Assimilated NPP (e.g., plant carbohydrate) is catabolized according to the general formula:

$$C_6H_{12}O_6 + 6O_2 \rightarrow 6CO_2 + 6H_2O$$

The more general formulation thus becomes:

$$C_nH_{2n}O_n + nO_2 \rightarrow nCO_2 + nH_2O$$

Stoichiometrically, this is the reverse of photosynthesis, which was discussed in Chapter 2.

COMPOUNDS BEING DECOMPOSED

There are literally thousands of chemical and biochemical compounds involved in catabolism. Viewed in an ecological context, however, they can be classified into two functional categories: (1) Primary compounds, those which are directly derived from plant, microbial, or animal tissues, and (2) secondary compounds, those which are produced as a result of organic matter–mineral interactions, usually resulting in small or large chemical changes in chemical bonds or degree of aromaticity.

Both categories comprise a few major types (or groups) of compounds: soluble, or labile, versus relatively insoluble (in water) nonlabile, or resistant, compounds. Compounds in the former category include organic acids, amino acids, and simple sugars. Compounds in the latter category include lignin, cellulose, cutins, and waxes. One should also consider biochemical versus biological bond types as defined by McGill and Cole (1981). These reflect the differences between ester linkages, designated R–C–O–O–R, which yield energy when broken, and the carbonyl C–N, C–P, or C–S bonds, which require energy to be cleaved, yielding nutrients to the microbes (Newman and Tate, 1980).

MICROBIAL ACTIVITIES IN RELATION TO CATABOLISM IN SOIL SYSTEMS

The principal "players" in the decomposition process are the microbial populations, (i.e., the bacteria, fungi, and viruses). The bacteria and fungi are as biochemically diverse as they are diverse in phylogeny. Bacteria, currently considered to encompass more than 35 phyla, are probably the most speciose array of organisms on earth (Tiedje *et al.*, 2001; Torsvik and Øvreås, 2002). In addition, bacteria are undoubtedly the most numerous organisms, and have been estimated to total from 4 to $6 \cdot 10^{30}$ cells on Earth. A sizable proportion (more than 90%) of bacteria are in the subsurface, which includes the earth's mantle to 4 kilometers (km) in depth (Whitman *et al.*, 1998). Numbers of bacteria in soils of all biomes were estimated to be $2.5 \cdot 10^{29}$ cells, with some of the larger quantities in desert scrub and savanna lands. The foregoing counts translate into $2 \cdot 10^9$ cells g^{-1} in the top meter, and $1 \cdot 10^8$ cells g^{-1} in the 1- to 8-m soil depth, with numbers in forest soils being markedly lower (Whitman *et al.*, 1998).

In soils away from the rhizosphere, the environment for bacteria is usually stressful. A majority of bacteria exist in this low-nutrient condition and may be starving (Morita, 1997). One should note that although some bacteria can double every 20 minutes or less in growth media in the laboratory, they may undergo only two to three divisions per year, on average, in soil under field conditions because of the extreme limitations of available reduced carbon substrates. This energetic limitation is considered in detail in Chapter 6. A pictorial representation of bacterial carbon flow is given in Figure 3.1 (Scow, 1997). Note the aforementioned flows to both biomass growth and maintenance respiration. The latter requirement becomes limiting to many bacteria under conditions of nutrient limitation. In anoxic or low-redox microsites within soil aggregates and faunal guts, decomposition via microbial fermentation or anaerobic respiration with nitrate or other electron acceptors can occur. Decomposition linked to aerobic respiration would occur in regions with higher levels of oxygen, such as on the exteriors of aggregates.

Many genera of prokaryotes, including both bacteria and archaea, have evolved the highly important biochemical trait of "fixing" (rupturing the triple covalent bonds of) dinitrogen, making it available as ammonium for plant or microbial uptake (Postgate, 1987). This has important ramifications for nitrogen and phosphorus cycling and interactions with soil organic matter (Stewart and Cole, 1983; Stewart *et al.*, 1990; Giller, 2001).

Knowledge of the prokaryotes has increased greatly in the past decade, with numerous accounts of their phylogeny published (Bergey's manual at http://www.cme.msu.edu/bergeys; Torsvik and Øvreås, 2002). The principal concern of bacterial phylogeny is to trace both the extent of species of bacteria, as well as the archaea. Until recently, archaea were considered to be inhabitants of extreme environments, including deep sea trenches and vents and hot springs, but they have been found also in numerous other habitats, including fresh water lakes and forest and agricultural soils (Bintrim *et al.*, 1997; Jurgens *et al.*, 1997; Pace, 1997). For more information on soil prokaryote interactions in soils and rhizospheres, see the review by Kent and Triplett (2002).

A method often used to analyze bacterial populations is to amplify DNA extracted from environmental samples by polymerase chain

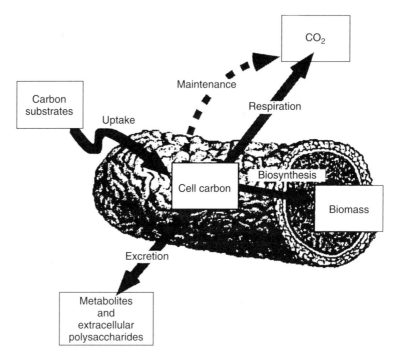

FIGURE 3.1.   Flow of carbon in a bacterial cell (from Scow, 1997).

reaction (PCR), using primers universal to the 16S ribosomal RNA (rRNA) genes of bacteria and archaea (Lane, 1991; Prosser, 2002). In both tropical (Borneman and Triplett, 1997) and arid southwestern U.S. soils (Kuske *et al.*, 1997), more than 50% of the prokaryotic DNA sequences of soil prokaryotes belonged to groups with no representatives in laboratory culture. This has marked implications for identifying prokaryotes involved in biogeochemical cycling and other environmental processes (see Chapter 7). Either DNA or RNA can be extracted from soils, but a majority of the studies have been based on DNA extraction, which is easier to accomplish efficiently because of the higher lability and turnover of RNA. The rRNA content in active cells is higher than in inactive ones, thus rRNA-based analyses are a better approach for characterizing active microbial populations in soils (Ogram and Sharma, 2002). Techniques are now available to analyze microbial community structure and function, by analyzing microbial rRNA and mRNA, respectively. Both types of RNA can be extracted from soils and converted to complementary DNA (cDNA) by the enzyme reverse transcriptase for subsequent PCR amplification. Standard PCR analyses using "universal primers for rRNA genes" are not quantitative but do provide very useful qualitative information on dominant microbial populations. As long as suitable primers are available, microbial rRNA copies or mRNA copies can be quantified using quantitative, or "real-time," PCR. These latter approaches provide an important means for linking soil microbial community structure and function.

Perhaps the greatest difference between bacteria and fungi is to be found in their mode of growth. Fungi have long strands (hyphae) that can grow into and explore many small microhabitats, secreting any of a considerable array of enzymes, decomposing material there, imbibing the decomposed monomers, and translocating the carbon and other nutrients back into the hyphal network (Fig. 3.2). In contrast to bacteria, fungi can remain active in soils at very low water potential (−7200 kPa) and are better suited than bacteria to exist in interpore spaces (Shipton, 1986). Many genera of fungi are closely associated with plants (see Chapter 2) and animals. Studies of coevolution of fungi with other eukaryotes have been summarized by Pirozynski and Hawksworth (1988). For an extensive overview of the roles that fungi play in terrestrial ecosystems, see Dighton (2003).

Although not covered in detail in this book, there is a rapidly growing area of interest in effects of plant pathogens in ecosystems. Because plant pathogens play important roles in mediating plant competition and succession, and in the maintenance of plant species diversity, they can have important feedback effects on soil communities and ecosystem processes as well. For a good review of these processes, see Gilbert (2002). We discuss effects of microarthropod grazing on fungal plant pathogens in Chapter 4.

In contrast to fungi, bacteria are usually unicellular, or in clustered colonies, occupying discrete patches of soil measuring only a few cubic micrometers (μm^3) in volume. Bacteria depend on many episodic events, such as rainfall and root growth or ingestion by various soil fauna, for passive movement to enable them to move about. When flagella are present, directed motility in the water-film is also possible.

Viruses may play significant roles in the microbial ecologies of soil environments because they can be a source of mortality, particularly for bacteria. Farrah and Bitton (1990) noted that lytic phages (viruses attacking bacteria) could act so as to restrict the growth of susceptible bacteria, and other phages could transmit genetic information between bacteria. The information on viral numbers and activities in soil in general is quite limited. Temperate phages (as distinct from virulent ones) in desert systems were inactivated on soil particles at acid pH (4.5–6). These phages had virtually no effect on populations of soil bacteria in Arizona soils, but persisted at low densities in their hosts (Pantastico-Caldas *et al.*, 1992). This contrasts markedly with the often-cited deleterious impacts of virulent phages on *Escherichia coli* in liquid cultures

FIGURE 3.2. Extensive growth of fungal mycelium (arrow) was observed when crushed microaggregate (0.50-mm diameter) from native soil was stained with water-soluble aniline blue (a); smaller-sized (0.10-mm diameter) aggregates from a crushed macro-aggregate (1.0-mm diameter) were held together by fungal hyphae (b) (from Gupta, 1989).

in the laboratory. For more information on bacteriophages and interactions with bacteria under starvation conditions, see Schrader *et al.* (1997).

Microbial Abundance and Distribution in Soil

Unfortunately for the soil ecologist, the distribution and abundance of microorganisms is so patchy that it is very difficult to make an accurate determination of their mean abundances without dealing with a very high variance about that mean, when viewed on a macro scale. Part of this variation is due to the close "tracking" of organic matter "patches" by the microbes. There are aggregations of microbes around roots (the often-cited "rhizosphere") (Lynch, 1990), around fecal pellets and other patches of organic matter (Foster, 1994), and in pore necks (Fig. 3.3) (Foster and Dormaar, 1991). In addition, microorganisms concentrate in the mucus secretions that line the burrows of earthworms (the "drilosphere," as defined by Bouché [1975] and reviewed by Lee [1985]). The phenomenon of "patches" is discussed more in Chapter 6.

A large proportion of soil ecology studies has focused on processes occurring in the O and upper A horizons because so much of the short-term dynamics occurs there. With tools of microbial community analysis, Fierer *et al.* (2003) used phospholipid fatty acid (PLFA) analysis to examine the vertical distribution of specific microbial groups and their diversity in two soil profiles down to a depth of 2 m. The number of individual PLFAs decreased by about one-third from the soil surface down to 2 m. Changes in certain ratios of fatty acid precursors and ratios of total saturated to total monounsaturated fatty acids increased with soil depth, indicating that microbes in the lower horizons were more carbon limited at greater depths. Interestingly, approximately 35% of the total amount of microbial biomass was found in soil below a depth of 25 centimeters (cm). Gram-positive bacteria and actinomycetes tended to increase in proportional abundance with depth, whereas Gram-negative bacteria, fungi, and protozoa were highest at the soil surface.

Soil is an impressively heterogeneous matrix of minerals and organic matter. Ways in which this heterogeneity in organic matter and texture can influence microbial populations have been widely studied for more than a century. A number of studies using transmission electron microscopy (TEM) and scanning electron microscopy (SEM) have revealed the intimate associations of bacteria and fungi with soil aggregates (Figs. 3.4 and 3.5) (V. V. S. R. Gupta, personal communication).

With the development of more sophisticated imaging tools and statistical analyses of data, there have been several studies of microbial spatial patterns at the field or plot scale. Unfortunately, these studies have not spanned the range of spatial variability, which may exist at levels well below the millimeter scale (Nunan *et al.*, 2002). Taking a large vol-

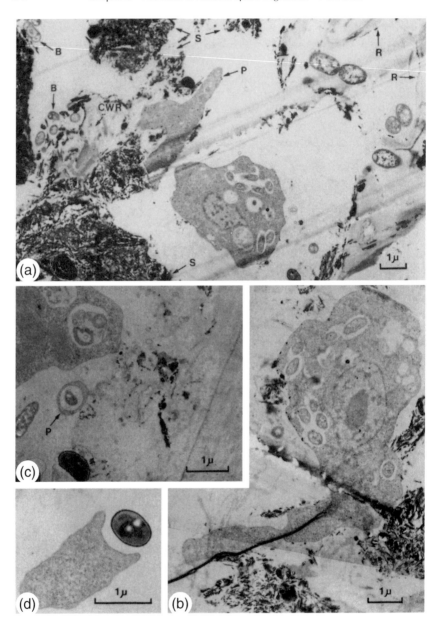

FIGURE 3.3.   (a) An amoeba probing a soil microaggregate containing cell wall remnants (CWR) and a microcolony of bacteria (B); P, pseudopodium; R, root; S, soil minerals; bar, 1 μm (from Foster and Dormaar, 1991). (b) An amoeba with an elongated pseudopodium reaching into a soil pore. The amoeba contains intact ingested bacteria in its food vacuoles (from Foster and Dormaar, 1991). (c) An amoeba with partly digested bacteria in food vacuoles; note bacterium enclosed by a pseudopodium (P) (from Foster and Dormaar, 1991). (d) A pseudopodium associated with a Gram-positive microorganism (from Foster and Dormaar, 1991).

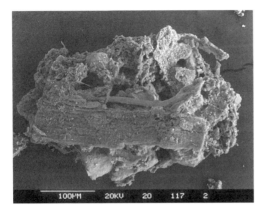

FIGURE 3.4. Scanning electron microscopy (SEM) picture of a macroaggregate (250- to 500-μm diameter) with particulate organic matter and hyphae (V. V. S. R. Gupta, with permission).

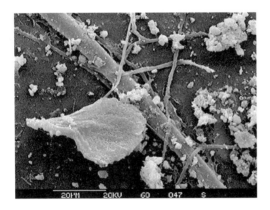

FIGURE 3.5. Amoebae feeding on fungi (V. V. S. R. Gupta, with permission).

ume of soil, both topsoil and subsoil ($3 \times 3 \times 0.9$ m) from an arable field, Nunan *et al.* (2002) prepared subsampled cores and biological thin sections in which the *in situ* distribution of bacteria could be quantified. They acquired spatially referenced RGB digital images using epifluorescence microscopy at 630× magnification. Average bacterial numbers per thin section were calculated using nine replicate images captured from each thin section (Fig. 3.6). Analysis of spatial dependence or continuity of soil bacterial density was performed using geostatistical tools at three scales: (1) centimeter to meter, (2) millimeter to centimeter, and (3) micrometer to millimeter scale, using appropriate semivariogram formulas (for more information on use of semivariograms, see Robertson and Gross [1994]). Spatial structure was found only at the micrometer

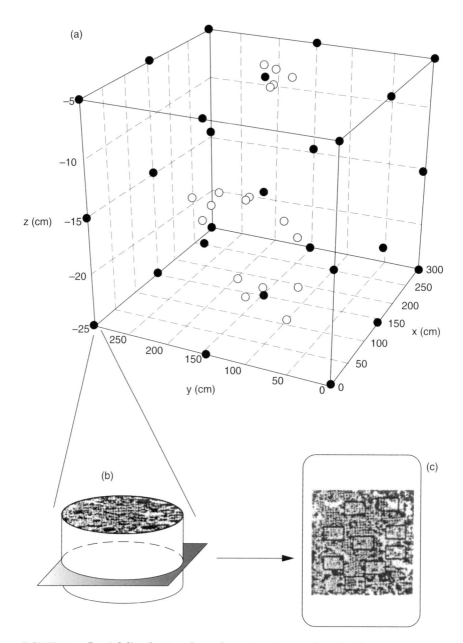

FIGURE 3.6. Spatial distribution of sampling points in topsoil (a). Solid circles form systematic random lattice and open circles form a biased random cluster. An undisturbed core (b) was sampled at each point and a thin section (c) cut from the horizontal plane. Nine spatially referenced images, in which bacteria were mapped, were acquired from each thin section. Average bacterial density per thin section was calculated and the values used to study large-scale variability. Bacterial maps were divided into 100 quadrats and bacterial density in each quadrat calculated. There were 900 quadrats per thin section and these bacterial density values were used to study microscale spatial variability (from Nunan *et al.*, 2002).

scale in the topsoil, whereas evidence for nested scales of spatial structure was found in the subsoil at both the micrometer scale and at the centimeter to meter scales. Evidence for spatial aggregation in bacteria was stronger in the topsoil and decreased with depth in the subsoil. Nunan *et al.* (2002) suggest that factors that regulate the distribution of bacteria in the subsoil operate at two scales, in contrast to one scale in the topsoil, and that bacterial patches are larger and more prevalent in the topsoil.

Textbooks such as those by Swift *et al.* (1979) and Paul and Clark (1996) cover a number of methodological approaches for estimating microbial numbers and turnover in considerable detail. In this book, we present a few principal techniques for measuring numbers and identifying members of the microbial communities. We then relate them to studies of nutrient immobilization and mineralization, covered later in Chapter 5 on decomposition processes.

TECHNIQUES FOR MEASURING MICROBIAL COMMUNITIES

Techniques for measuring populations and biomass of microorganisms are either direct (by counting) or by inference (from chemical and physical measurements). The following are a few of the more commonly used techniques for studies of microbial standing crops (biomass at the time of sampling) and activity in a community and ecosystem context.

Direct Measures of Numbers and Biomass

Total counts of microbes are made by preparation of soil (about 10 milligrams [mg]) suspensions spread in thin agar films on microscope slides (Jones and Mollison, 1948). The films are then stained, often with fluorescent dyes, and scanned. More recently, there have been improvements to the direct count technique such as the membrane filtration technique, which enables one to quickly count fungal hyphae against the filter or a stained background. This approach is generally much faster and easier than the more laborious agar film technique. Other more classical techniques such as viable counts on nutrient-containing agar media are discussed by Parkinson *et al.* (1971) and Parkinson and Coleman (1991). The viable culture techniques usually recover only 1% or less of the total viable cells, so they are useful only for comparative purposes when one is focusing on a few readily culturable species of bacteria.

Other direct measures include sampling for extractable DNA (Torsvik *et al.*, 1990a, 1990b; Torsvik *et al.*, 1994; Torsvik and Øvreås, 2002), and using the PCR to multiply specific genes (such as the 16S rRNA) to determine the identities of the organisms of interest. Another approach to microbial community analysis uses signature lipid biomarkers (SLB). This technique, pioneered by Dr. David White and colleagues at the University of Tennessee (Tunlid and White, 1992), measures ester-linked polar lipid fatty acids and steroids to determine microbial biomass and community structure. Further comments on these techniques are given in Paul and Clark (1989, 1996). For microbial community characterizations, various biomarkers are used. Prominent among these are the PLFAs. Phospholipids are found in the membranes of all living cells but not in the storage products of microorganisms. They may be extracted and characterized using GC/MS (gas chromatography/mass spectrometry) (Zelles and Alef, 1995).

A variety of techniques has been developed for the isolation and identification of DNA from soil. Techniques include the cell extraction method and the direct lysis method (Saano and Lindström, 1995, Zhou *et al.*, 1996). In recent studies of microbial community makeup in agroecosystem field soils, Furlong *et al.* (2002) compared the microbial community composition of earthworm- and non-earthworm–influenced soils at the Horseshoe Bend field site in Athens, GA. The objective was to compare microbial communities from worm casts and open soil; this was done by creating clone "libraries" of the 16S rRNA genes, which were prepared from DNA isolated directly from the soil and earthworm casts. In the cast soils, representatives of the genus *Pseudomonas*, as well as the Actinobacteria and Firmicutes, increased in number (Furlong *et al.*, 2002). The results were consistent with a model where a large portion of the microbial population in soils passed through the gastrointestinal tract of the earthworm unchanged while representatives of some bacterial phyla increased in abundance. In Chapter 4, we will consider the various faunal groups in soil, and their life-history attributes that have impact on microbial community makeup and turnover.

As noted in the introduction to prokaryotes above, we are only now becoming aware of the phylogenetic richness of archaea in soil communities. PCR amplification using primers specific for archaeal 16S rRNA genes allows detection of archaea in diverse habitats (Bomberg *et al.*, 2003). The abundance of crenarchaeal (one of the two kingdoms comprising the archaeal domain) 16S rRNA in both cultivated and native field soils has been estimated to be from 1 to 2% of the total 16S rRNA in these soils (Buckley *et al.*, 1998).

The above procedures are primarily used in determining bacterial community structure. For fungi, several recent studies have made use of the fact that ergosterols are specific to fungi, and the amounts of ergosterols can be quantified to determine the amount of fungal tissues (biomass) present in soils (Newell and Fallon, 1991; Eash *et al.*, 1994, Zelles and Alef, 1995). Molecular phylogeny has been an equally powerful tool for describing fungal communities (see, for example, Husband *et al.*, 2002).

Indirect Measures of Biomass

Chemical Methods

Jenkinson (1966) cited an earlier suggestion of Störmer (1908) that a flush of CO_2, evolved after fumigation, was due to the decomposition of organisms killed during fumigation by the surviving microorganisms remaining after fumigation. This relates to the extensive work done on "partial sterilization" of soils, in Great Britain and elsewhere (Russell and Hutchinson, 1909; Powlson, 1975), under the misguided assumption that most soil microorganisms were somehow deleterious to subsequent plant growth, particularly in agricultural fields. Many of the soil heterotrophs are now considered generally beneficial, particularly when viewed in a whole-system nutrient cycling context.

The Chloroform Fumigation and Incubation (CFI) Technique

We now consider ways to use fumigants to measure microbial biomass. Using a fumigant such as chloroform, and incubating the soil for 10 or 20 days, the size of the flush of CO_2 output can be related to the size of the microbial biomass by the expression $B = F/k_c$; where B = soil biomass C (in $\mu g\ C \cdot g^{-1}$ soil); F = carbon dioxide carbon (CO_2—C) evolved by fumigated soil minus CO_2 evolved by unfumigated soil over the same time period; and k_c = fraction of biomass mineralized to CO_2 during the incubation (Jenkinson and Powlson, 1976). The k_c value, calculated from a range of microorganisms in controlled experiments, is assigned a general value of 0.45 (Jenkinson, 1988).

Jenkinson and Powlson (1976) relied on laboratory measurements of microbial cells added to soil. Voroney and Paul (1984) extended this work to include labile nitrogen, and measured both k_c and k_n (fraction of biomass nitrogen mineralized to inorganic nitrogen). A review of usage of ^{14}C to measure microbial biomass and turnover is given by Voroney *et al.* (1991), with step-by-step procedures for this research. They introduced carbon by labeling plants via photosynthetic pathways, and then followed the carbon into the microbial biomass via root exudates and turnover, and in turn into the soil organic matter.

A wide range of soils has been compared for biomass carbon calculated from biovolume (the measured volume of the cell), using the CFI method, and a ratio of biomass carbon from biovolume to biomass carbon, also from CFI, has been determined (Powlson, 1994) (Table 3.1). These ratios range from 0.86 to 1.25 from soils in arable lands and up to 6.47 from soils in deciduous woodland. Forest soils, including those with low pH, have proven more difficult to analyze for microbial biomass, and are considered next.

LIVERPOOL JOHN MOORES UNIVERSITY
LEARNING SERVICES

TABLE 3.1.   Comparison of Biomass Carbon as Calculated from Direct Microscopy and the Fumigation Incubation (FI) Method

Soil	Organic C (%)	pH	Biomass C calculated from biovolume[a] (μg C/g soil)	Biomass C calculated from FI method[b] (μg C/g soil)	Ratio of biomass C from biovolume to biomass C from FI
Arable[c]	2.81	7.6	550	547	1.01
Arable[c]	0.93	8.0	190	220	0.86
Deciduous woodland[c]	4.30	7.5	1540	1231	1.25
Arable[c]	2.73	6.4	390	360	1.08
Grassland[c]	9.91	6.3	3200	3711	0.86
Deciduous woodland[c]	2.95	3.9	330	51	6.47
Secondary rainforest[d]	1.46	7.1	430	540	0.80
Cleared forest[d,e]	1.23	6.2	260	282	0.92

[a]See Jenkinson et al. (1976) for method of calculation.
[b]Calculated using K_c of 0.45, not 0.5 as in the original paper.
[c]Temperate soils from the United Kingdom.
[d]Subhumid tropics, Nigeria.
[e]Arable cropping for 2 years after clearing secondary forest, Nigeria.
Source: From Powlson, 1994.
Adapted from Jenkinson et al., 1976.

The Chloroform Fumigation and Extraction (CFE) Technique

Vance et al. (1987) noted that low pH soils, particularly those in the range below pH 5.0, including many forest soils, were not well characterized for microbial biomass using the CFI procedure. They modified the CFI procedure (Jenkinson and Powlson, 1976) to the chloroform fumigation and extraction (CFE) procedure as follows (Vance et al., 1987): soil samples are fumigated with chloroform for 48 hours, the fumigated and nonfumigated control samples are extracted with $0.5 M$ K_2SO_4, and the resulting organic extracts are measured for carbon, nitrogen, and other elements. The difference between the total organic carbon from the chloroform-fumigated soils minus the nonfumigated controls, multiplied by the k_{ec} factor (see Chapter 9 for details) is the microbial biomass carbon. For soils with pH values less than 4, the k_{ec} values are usually lower, from 0.2 to 0.35 k_{ec} (Jenkinson, 1988). The CFE method has proven quite successful, and enables one to obtain microbial biomass values for carbon, nitrogen, phosphorus (Hedley and Stewart, 1982), and sulfur (Gupta and Germida, 1988).

A few authors have expressed concern about the extent of faunal contributions to the fumigation "flush." Protozoan biomass may be a significant contributor in some soils (Ingham and Horton, 1987), but usually constitutes less than 2% of total microbial carbon.

Some general comments on the methodology of the microbial biomass method are necessary. The CFI and CFE methods should be emp loyed within the context or intent of the methods originally described. Because they are bioassays, and not general chemical assays, they are not as robust as the latter. They can be misused, particularly if a great deal of organic matter substrate, waterlogging, or very low pH conditions are encountered (Powlson, 1994). However, the microbial biomass values are useful in the development and exercising of simulation models of labile carbon and nutrient turnover in a wide range of ecosystems (e.g., Parton *et al.*, 1987; Parton *et al.*, 1989a, 1989b; Jenkinson and Parry, 1989). Jenkinson *et al.* (2004) provide a helpful review of the chloroform fumigation techniques. Interestingly, they consider the fumigation and incubation (FI) technique to be obsolete, and urge caution in the usage of the k values, as had been noted by several investigators.

For reviews of biochemical methods to estimate microbial biomass, see Sparling and Ross (1993) and Alef and Nannipieri (1995a). For more specific details, see Chapter 9 (laboratory exercises) for comments on details of the microbial biomass estimation procedure. The ultimate "take home message" in studies of microbial biomass is the necessity to use more than one method to have some confidence in the numbers and hence biomass of the microorganisms measured. Although more time-intensive, it is advisable to compare biomass of microorganisms to direct counts made microscopically (Parkinson and Coleman, 1991).

Physiological Methods: SIR Technique

Additional methods for measuring microbial biomass include the substrate-induced-respiration (SIR) technique, first developed by Anderson and Domsch (1978). The SIR technique involves adding a substrate such as glucose to soil, and measuring the respiration resulting from the stimulated metabolic activity in the experimental soil sample, versus control treatments that received no carbon substrate. It is possible to measure the relative contributions of bacteria and fungi by using inhibitors (e.g., cycloheximide to inhibit fungal activity or streptomycin to inhibit bacterial activity). The assumption is that one measures only bacterial activity when fungi are inhibited, and vice versa. The technique requires some care, because soil texture may affect the apparent "resistance" to biocides. Further details of the technique are given by Beare *et al.* (1990, 1991), Insam (1990), Kjøller and Struwe (1994), and Alphei *et al.* (1995).

Additional Physiological Methods of Measuring Microbial Activity

There is a large body of literature dealing with the indirectly measured signs of metabolic activity, namely CO_2 output or oxygen uptake. The ratio of the two gases, in terms of either uptake or output, is very informative about the principal sources of carbonaceous compounds being metabolized. The ratio of CO_2 evolved to O_2 taken up, known as RQ, is lowest for carbohydrates, intermediate for proteins, and highest when lipids are the principal substrate being metabolized (Battley, 1987). In several studies, the microbial respiration per unit microbial biomass ($qCO_2 = \mu g\ CO_2$-$C/mgC_{mic}/h$) (Anderson and Domsch, 1978; Insam and Domsch, 1988; Anderson and Domsch, 1993; Anderson, 1994) was measured and found useful as an indicator of the overall metabolic status of a given microbial community. Additional metabolic quotients have been used to study influences of climate and temperature, soil management, heavy metals, and soil animals in ecosystems, notably the ratio of microbial carbon to organic carbon, expressed as a percentage of microbial carbon to total organic carbon, or C_{mic}/C_{org} (Table 3.2) (Anderson, 1994; Joergensen $et\ al.$, 1995). This follows from the assumption that terrestrial ecosystems in a near-steady state are characterized by a constant flow of nutrients and energy into and out of the ecosystem on a yearly basis, and entering and leaving the microbial biomass pool as well (Fig. 3.7) (Anderson, 1994). All of the foregoing is based on aerobic conditions. The extent of anaerobicity can be important at certain times, and needs to be carefully measured.

Enough data sets on microbial biomass carbon and nitrogen have accumulated by now that an extensive synthesis of temporal and latitudinal variation was carried out on data from more than 58 studies worldwide. For the entire data set, temporal variability was best predicted by

TABLE 3.2.   Examples of Studies in Soil Microbiology in Which Metabolic Quotients Have Been Applied

Field of study	Metabolic quotient[a]
Maintenance carbon requirement	m, qCO_2
Carbon turnover	qCO_2, μ, K_mGLUCOSE, Y, m, qD, C_{mic}/C_{org}
Soil management	qCO_2, qD, V_{max}, C_{mic}/C_{org}
Impact of climate and temperature	qCO_2, C_{mic}/C_{org}, qD
Impact of soil texture and soil compaction	qCO_2, qD
Impact of heavy metals	qCO_2
Ecosystems, ecosystem theory	qCO_2, qD, C_{mic}/C_{org}
Impact of soil animals	qCO_2

[a] m = maintenance coefficient; qCO_2 = metabolic quotient or specific respiration rate; μ = specific growth rate; K_m = Michaelis-Menten constant; V_{max} = maximum specific uptake rate; Y = growth yield; qD = specific death rate; C_{mic}/C_{org} = microbial carbon to organic carbon ratio expressed as a percentage of microbial carbon to total organic carbon.

See Anderson (1994) for specific references pertaining to usage of particular metabolic quotients.

Modified from Anderson, 1994.

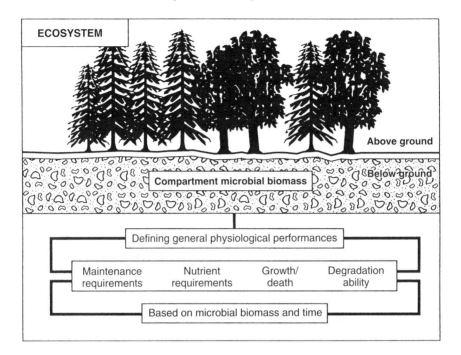

FIGURE 3.7. Working hypothesis for the application of metabolic quotients in ecosystem development at the synecological level (from Anderson, 1994).

a three-component model incorporating pH, soil carbon, and latitude (Wardle, 1998). The increasing latitude reflected higher interseasonal variations in temperature, causing greater interseasonal flux of the biomass. A majority of these studies provided data showing less than one turnover of the entire microbial biomass per year, reflecting the extreme scarcity of food for most of the microbial populations much of the time (reasons for this are discussed earlier in this chapter).

Enzyme Assays and Measures of Biological Activities in Soils

Numerous soil biologists/ecologists have used enzyme assays to measure soil biological activity (Coleman and Sasson, 1980; Nannipieri,

1994; Alef and Nannipieri, 1995b). Oxidoreductases, transferases, and hydrolases have been most studied. These assays have been considered of questionable value, mostly because of misapplication of the techniques and misinterpretation of the resulting data. The principal objection to soil enzyme assays is that the activities are substrate specific, and hence related to specific reactions and do not necessarily reflect organismal activities (Nannipieri *et al.*, 1990; Nannipieri, 1994). This concern is very well expressed by Nannipieri *et al.* (2002), who note that enzymes in soils can be in six different locations: (1) active and present intracellularly in living cells, (2) in resting or dead cells, (3) in cell debris, (4) extracellularly free in the soil solution, (5) adsorbed by inorganic colloids, or (6) associated in various ways with humic molecules. The preferred situation is being able to assay enzymes that are active and present intracellularly in living cells. The array of extracellular and intracellular distributions in the soil environment is expressed in Figure 3.8 (Nannipieri *et al.*, 2002), which depicts various aspects of overall enzyme diversity related to microbial functional diversity in soil.

It should be noted that enzymes related to particular target substrates, such as ligno-cellulases in leaf litter, may be relatively good predictors of mass loss. After early stages of mass loss caused by leaching and mineralization, the "middle stage" is often strongly correlated with enzyme activity. In the final stages, with less than about 25% of initial

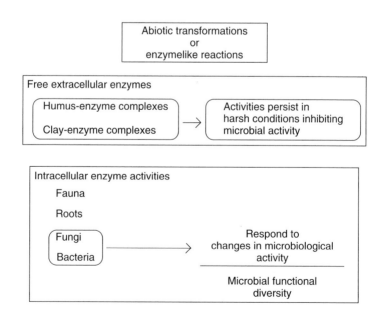

FIGURE 3.8. Various activities contributing to the overall enzyme activity measured in soil compared with those affecting the microbial functional diversity in soil (Nannipieri *et al.*, 2002).

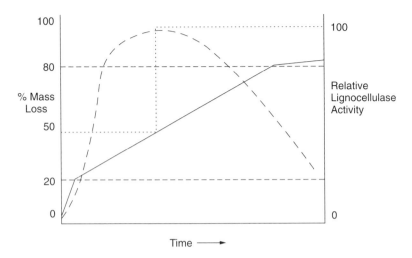

FIGURE 3.9. Idealized plot of lignocellulase activity (heavy dashed line) in relation to litter mass loss (solid line) through time. Lignocellulase activity traces a "bell-shaped" pattern over the course of litter decomposition, peaking (in this example) when cumulative mass loss reaches 45%. The dashed horizontal lines at 20% and 80% mass loss highlight breakpoints in the mass loss curve. During the early stages of litter decomposition rapid mass loss is often largely attributable to leaching and mineralization of soluble litter constituents. In the middle stage, lignocellulose degradation predominates. Throughout the late stages, the accumulation of humic condensates depresses microbial activity, stabilizing the remaining material (from Sinsabaugh *et al.*, 1994).

mass remaining, the accumulation of humic condensates depresses microbial activity, stabilizing the remaining material (Fig. 3.9) (Sinsabaugh *et al.*, 1994).

Many models of soil organic matter (SOM) decomposition are based on first order kinetics that assume that decomposition rate of a particular carbon pool is proportional to pool size and a simple decomposition constant ($dC/dt = kC$). In reality, SOM decomposition is catalyzed by extracellular enzymes that are produced by the microorganisms. A theoretical model to explore the behavior of a decomposition–microbial growth system that operates by exoenzyme catalysis used the following relationship: $D_C = K^*_d\ Enz_C$, where D_C = decomposition of polymeric material to produce available C; K^*_d is a single decomposition constant, and Enz_C = exoenzyme pool (Schimel and Weintraub, 2003). An enzyme kinetics analysis showed there must be some mechanisms to produce a nonlinear response of decomposition rates to enzyme concentration. This nonlinearity induces carbon limitation, regardless of the potential carbon supply. In a linked carbon and nitrogen version of the model, adding a pulse of carbon to a nitrogen-limited system increases respiration, while adding nitrogen decreases respiration (with carbon redirected from waste respiration to microbial growth). Previous conclusions drawn in the literature have assumed that the lack of a respiratory response by soil microbes to added nitrogen indicates that they are not nitrogen limited. This model of Schimel and Weintraub (2003) suggests that, while total carbon flow may be limited by the functioning of the exoenzyme system, in fact microbial growth may be nitrogen limited. This important finding should be the subject of several laboratory and field studies in the near future.

Direct Methods of Determining Soil Microbial Activity

Direct measurements of the activity of soil microorganisms have been a goal of soil biologists for a long time (Newman and Norman, 1943). This results from the basic thermodynamic fact that as organisms undergo metabolic activity, they emit heat from the enthalpy of reactions occurring in net catabolism (Battley, 1987). With the continuing trend toward miniaturization of circuitry and much better, more sensitive thermocouples, it is now possible to obtain direct measures of metabolic activities of organisms in small samples of soils with only a few milligrams of biomass (Sparling, 1981; Battley, 1987; Alef, 1995). Flow-microcalorimeters are now available that allow for the simultaneous measurements of CO_2 and N_2O production in soils (Albers *et al.*, 1995).

Another approach to direct measurement of microbial metabolic activity in a more fine-grained fashion is to measure microbial metabolic processes and identify the microorganisms responsible for particular biochemical reactions under field conditions. Radajewski *et al.* (2000) pioneered stable isotope probing of community-extracted DNA as a laboratory-based means of identifying microbial populations involved in ^{13}C-substrate metabolism. Padmanabhan *et al.* (2003) combined the approach of Radajewski *et al.* (2000) with an assay of a range of labile and recalcitrant organic compounds, linking the ^{13}C field release assay of respired $^{13}CO_2$ to DNA extraction analyses of the active microbial populations. Transient peaks of $^{13}CO_2$ released in excess of background were found in glucose- and phenol-treated soil within 8 hours of application. Across the 30-hour time span of the experiment, neither naphthalene nor caffeine additions stimulated $^{13}CO_2$ release above background. A total of 29 full sequences revealed that active populations included relatives of *Arthrobacter, Pseudomonas, Acinetobacter, Massilia, Flavobacterium*, and *Pedobacter* spp. for glucose; *Pseudomonas, Pantoea, Acinetobacter, Enterobacter, Stenotrophomonas*, and *Alcaligenes* spp. for phenol; *Pseudomonas, Acinetobacter* and *Variovorax* spp. for naphthalene; and *Acinetobacter, Enterobacter, Stenotrophomonas*, and *Pantoea* spp. for caffeine. All these genera belong to bacterial divisions or subdivisions that were recovered from soils in more than 25% of the studies surveyed in the review. This approach is a useful first step in taking powerful analytical tools to the field. However, Padmanabhan *et al.* (2003) note that the amendment-based approach used in this study would not be likely to identify less responsive, slow-growing (K-selected) members of the soil microbial community.

SOIL STERILIZATION AND PARTIAL STERILIZATION TECHNIQUES

A number of the techniques mentioned above involve drastic perturbations to soils, such as fumigation, for the purpose of determining numbers or biomasses of organisms residing within them. Huhta *et al.* (1989) examined the influence of microwave radiation on soil processes, noting that it seems to have a less drastic impact compared to autoclaving, gamma irradiation, or chloroform fumigation. Microwaving is particularly useful for removing various mesofaunal groups, leaving the microbial communities reasonably intact under moderate thermal energy inputs of 380 watts (W) for 3 minutes (Huhta *et al.*, 1989; Wright *et al.*, 1989). Unfortunately, some unwanted side effects were introduced with microwaving, principally a decreased water-holding capacity. Monz *et al.* (1991) found that microwaving was of limited use in their agricultural soils. We discuss methods to manipulate fauna, including group-specific chemical inhibitors or repellents, in Chapter 4.

CONCEPTUAL MODELS OF MICROBES IN SOIL SYSTEMS

Root–Rhizosphere Microbe Models and Experiments

As was noted in the discussion of primary production processes (Chapter 2), there are "hot spots" of activity, particularly of microbes in relationship to root surfaces and rhizospheres. When viewed as a transect through the rhizosphere, (e.g., 2 mm from the root surface or less,) there are arrays of rapidly growing bacteria and fungi, which have been called "fast" flora (Trofymow and Coleman, 1982) (Fig. 3.10). Moving up the root toward the shoot into older regions, one finds root hairs, then root cortical cells, which may be sloughing off into the surrounding soil. There are accompanying microbial and root grazers such as protozoa and nematodes, which are discussed in detail in Chapter 4. Out in the bulk soil, away from the rhizosphere (more than 4 mm from the root surface), occur some of the slower-growing, or "slow," bacteria and fungi, organic matter fragments, and some of the hyphae of either arbuscular mycorrhizal (AM) or ectotrophic mycorrhiza.

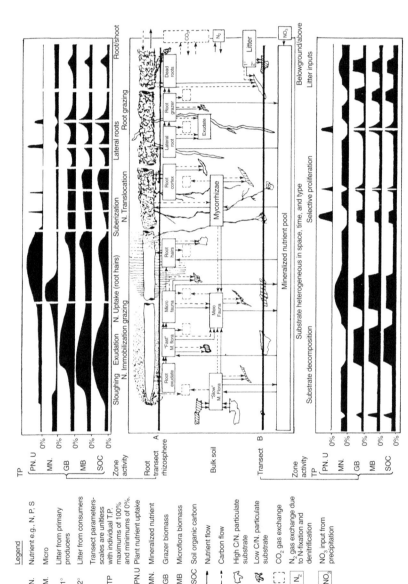

FIGURE 3.10. Conceptual diagram of a root–rhizosphere–soil system (from Trofymow and Coleman, 1982).

The field of mycorrhizosphere (Garbaye, 1991; Andrade *et al.*, 1998) research has taken a quantum leap forward with elegant microscopic methods, in conjunction with molecular tools to pinpoint organisms that are co-associates. Artursson and Jansson (2003) used bromodeoxyuridine (BrdU), as a thymidine analog, to identify active bacteria associated with AM hyphae. After adding BrdU to the soil and incubating for 2 days, DNA was extracted, and the newly synthesized DNA was isolated by immunocapture of the BrdU-containing DNA. The active bacteria in the community were identified by 16S rRNA gene PCR amplification and DNA sequence analysis. Based on gene sequence information, a selective medium was used to isolate the corresponding active bacteria. *Bacillus cereus* strain VA1, one of the bacteria identified by the BrdU method, was isolated from the soil and tagged with green fluorescent protein. By using confocal microscopy, this was shown to clearly attach to AM hyphae. This study by Artursson and Jansson (2003) is a pioneering attempt, using molecular and traditional approaches, to isolate, identify, and visualize (Fig. 3.11) a specific bacterium that is active in fallow soil and associates with AM hyphae.

Soil Aggregation Models

Soil aggregates, as noted in the section on soil structure in Chapter 1, play a central role in protecting pools of carbon and nitrogen, and are derived from a variety of sources. A mechanism particularly prevalent in many tropical soils is the physical aggregation process, which occurs abiotically as a physicochemical process (Oades and Waters, 1991). In both temperate and tropical soils, there are several biological processes

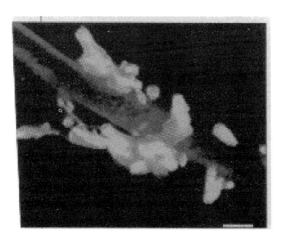

FIGURE 3.11. *Bacillus cereus* strain VA1 (*pnf8*) on a hyphal fragment from field soil (from Artursson and Jansson, 2003).

that result in the formation of "biological macroaggregates" (Fig. 3.12) (Six $et\ al.$, 2002b). These include the following three processes: (1) Fresh plant- and root-derived residues form the nucleation sites for the growth of fungi and bacteria. Macroaggregate formation is initiated by fungal hyphae enmeshing fine particles into macroaggregates. Exudates from both bacteria and fungi, produced as a consequence of decomposition of fresh residues, form binding agents that further stabilize macroaggregates ($t_{1,A}$). (2) Biological macroaggregates also form around growing roots in soils, with roots and their exudates enmeshing soil particles, thereby stimulating microbial activity ($t_{0,B}$ to $t_{1,B}$). (3) A third principal mechanism of biological macroaggregate formation in soils in all climates is via the action of soil fauna, particularly earthworms, termites, and ants. For example, earthworms often produce casts that are rich in organic matter ($t_{1,C}$) and are not stable when freshly formed and wet. During gut passage, the soil and organic materials are kneaded thoroughly and copious amounts of watery mucus are added as well. This molding process breaks bonds between soil particles, but can lead to casts that are quite stable upon drying. It is also worth noting that soil mesofauna, for example collembola and mites, are important in the SOM formation process through their production of copious amounts of fecal pellets. Effects of meso- and macrofauna on soil structure are discussed further in Chapter 4 on soil fauna.

The subsequent fate of macroaggregates follows a fascinating process over time, as noted by Six $et\ al.$ (2002b) summarizing from several literature sources. At first (time t_1), the young, freshly formed unstable macroaggregates (UA) are only stable when treated in very gentle fashion (that is, when the aggregates are taken from the field, brought to field capacity, subsequently immersed in water, and retained when gently sieved). The formation of water-stable aggregates (WSA) that can resist slaking (air drying and quick submersion in water before sieving) occurs by three processes (t_1 to t_2):

1. Under moist conditions ageing may increase stability by binding through microbial activity. Microbial activity is stimulated inside the biological macroaggregates, including worm casts because of their high organic matter content. In this process, substantial amounts of polysaccharides and other organics are deposited, serving to further stabilize the macroaggregates.

2. Dry–wet cycles can result in closer arrangements of primary particles, leading to stronger bonding and increased aggregate stability.

3. Biological and physicochemical macroaggregates, in the presence of active root growth, can become more stabilized by penetration of the aggregates by roots. This includes the roles of root exudates as

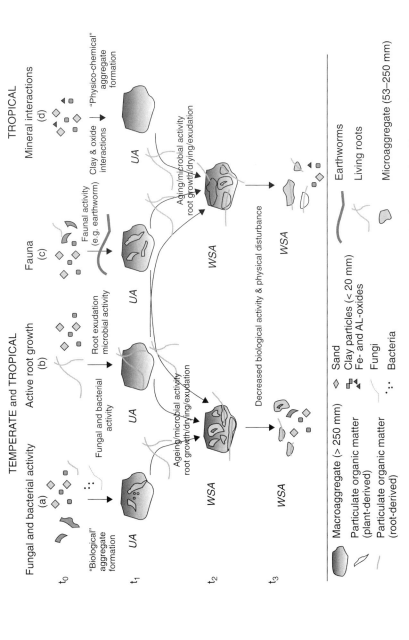

FIGURE 3.12. Aggregate formation and degradation mechanisms in temperate and tropical soils. Fungal and bacterial activity, active root growth, and earthworm activity are the biological aggregate formation agents in both temperate and tropical soils, whereas the mineral–mineral interactions in tropical soils are the physicochemical aggregate formation agents. UA = unstable aggregates; WSA = water-stable aggregates (from Six *et al.*, 2002b).

cementing agents, and the effects of the accompanying stimulated microbial activity.

In addition, roots influence aggregation physically both by exerting lateral pressures inducing compaction, and by continually removing water during plant transpiration, leading to drying of the soil and cohesion of soil particles around the roots. Note that this process is likely to be enhanced or intensified by mycorrhizal hyphae associated with the plant roots.

During macroaggregate stabilization (t_1 to t_2), the intra-aggregate particulate organic matter (POM) is further decomposed by microorganisms into finer POM (Six et al., 1998) (Fig. 3.13). This fine POM is increasingly encapsulated with minerals and microbial products, forming new microaggregates (53–250 µm) within the macroaggregates (Six et al., 1999). Similar processes may arise by stimulation from root exudation and mycorrhizal products, causing further encrustation of microbial products and mineral particles, forming microaggregates around the root-derived POM. Note that this microaggregate formation within macroaggregates is crucial for the long-term sequestration of carbon because microaggregates have a greater protective capacity to shield carbon against decomposition compared with macroaggregates.

The final phase of the aggregate turnover cycle (t_2 to t_3) occurs when the macroaggregates break down, releasing microaggregates and microbially processed soil organic matter (SOM) particles. The macroaggregates are more liable to break up over time, as the labile constituents of the coarse-sized SOM are consumed, microbial production of binding agents decreases, and the degree of association between the soil matrix and SOM decreases. Fortunately, microaggregates are still stable enough and not as sensitive to disruptive forces as the macroaggregates, and therefore survive (see Fig. 3.13). This is borne out by a table of mean residence time (MRT) (in years) of macro- and microaggregate–associated carbon (Table 3.3) (Six et al., 2002b). Note the essentially fivefold greater MRT for microaggregates (m) compared with macroaggregates (M).

The combined influences of physical, chemical, and biological factors in soil aggregate formation reach a peak when one includes the effects of arbuscular mycorrhizal fungi (AMF), both directly by physical binding, and indirectly by the production of the glycoprotein glomalin, as noted briefly in Chapter 2. In a controlled field plot study, five species (three grasses, one forb, and one legume) were grown in monocultures. Soil aggregate water stability (1–2 mm size class) was correlated with plant cover, root weight and length, AMF soil hyphal length, and glomalin concentrations (Rillig et al., 2002). Root length, soil glomalin, and percent cover contributed equally to water-stable aggregation using path

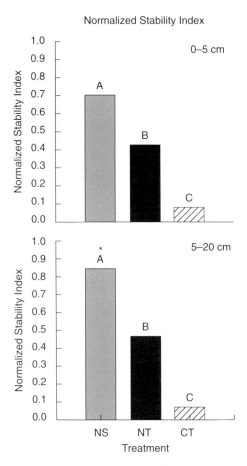

FIGURE 3.13.   Effect of management on the normalized stability index (NSI) at the 0-, 5-, and 5- to 20-cm depths. NS = native sod; NT = no-tillage; CT = conventional tillage. Values followed by a different uppercase letter within a depth are significantly different. Values followed by * for the 5- to 20-cm depth are significantly different from corresponding values in the 0- to 5-cm depth. Statistical significance determined at P > 0.05 according to Tukey's HSD mean separation test (from Six *et al.*, 1998).

analysis as the structural equation modeling approach. The direct effect of the glomalin was much stronger than the direct effect of the AMF hyphae alone, suggesting that this protein from the AMF is a very important hyphae-mediated mechanism of soil aggregate stabilization, at least for the larger macroaggregates of the 1–2 mm diameter size class.

For an extensive account of the many physical, chemical, and biological interactions involved in the dynamics of creation and dissolution of soil structure, refer to the masterful review by Baldock (2002).

TABLE 3.3. Mean Residence Time (MRT) (in years) of Macroaggregate– and Microaggregate–associated Carbon.

Ecosystem	Aggregate Size class[a]	(μm)	MRT
Tropical pasture	M	>200	60
	m	<200	75
Temperate pasture grasses	M	212–9500	140
	m	35–212	412
Soybean	M	250–2000	1.3
	m	100–250	7
Corn	M	>250	14
	m	50–250	61
Corn	M	>250	42
	m	50–250	691
Wheat-fallow, no-tillage	M	250–2000	27
	m	53–250	137
Wheat-fallow, conventional tillage	M	250–2000	8
	m	53–250	79
Average ± stder[b]	M		42 ± 18
	m		209 ± 95

[a]M = macroaggregate; m = microaggregate.
[b]stder = standard error.
From Six *et al*, 2002b.

Models: Organism and Process-Oriented

A recurrent theme that resonates throughout the field of soil ecology is the focus on organisms and population dynamics models used by community ecologists, and the use of process models at the ecosystem scale (Moore *et al.*, 1996; Smith *et al.*, 1998). Many ecosystem level models have included dynamics of the soil biota only implicitly, yet the intraseason dynamics of microbes and fauna, as is demonstrated in Chapters 4 and 5, often have a significant effect on nutrient availability and turnover. One of the more successful combinations of the organismal and process modeling approaches was by Paustian *et al.* (1990), who used four cropping systems (both annual and perennial crops) that varied in inorganic inputs and organic production in the growing season: (1) barley without fertilizer, (2) barley fertilized with $120\,kg\,N\,ha^{-1}\,yr^{-1}$, (3) a meadow fescue field with $200\,kg\,N\,ha^{-1}\,yr^{-1}$, and (4) a lucerne field with indigenous nitrogen fixation. The conceptual models of carbon and nitrogen flows in the four fields are presented in Figures 3.14 and 3.15 (Paustian *et al.*, 1990). Although there were no large differences in microbial biomass between treatments, the estimated microbial

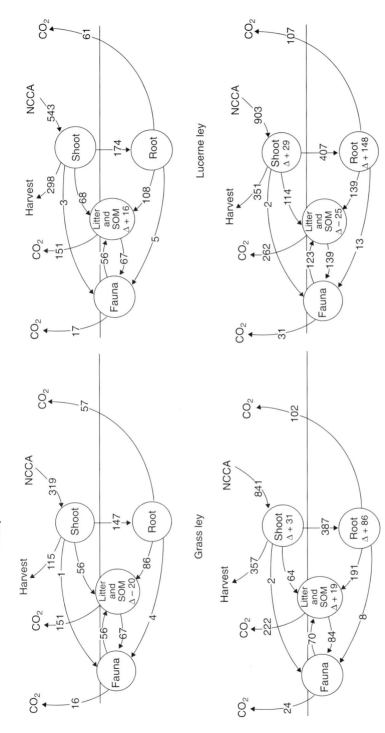

FIGURE 3.14. Budgets of annual carbon flows ($g\,C\,m^{-2}\,yr^{-1}$) for barley receiving no nitrogen fertilizer (B0), barley receiving 120 kg N ha^{-1} (B120), a grass ley receiving 200 kg N ha^{-1} (GL) and a N$_2$-fixing lucerne ley (LL). Budgets are based on data from 1982–1983 for the topsoil (0–27 cm). Compartment changes on an annual basis are denoted by delta (Δ) symbols; for the aboveground plant compartment this includes biomass, standing dead litter, and surface litter. Soil organic matter (SOM), including microbial biomass and soil litter, has been combined into a single compartment. NCCA is net canopy carbon assimilation (from Paustian et al., 1990).

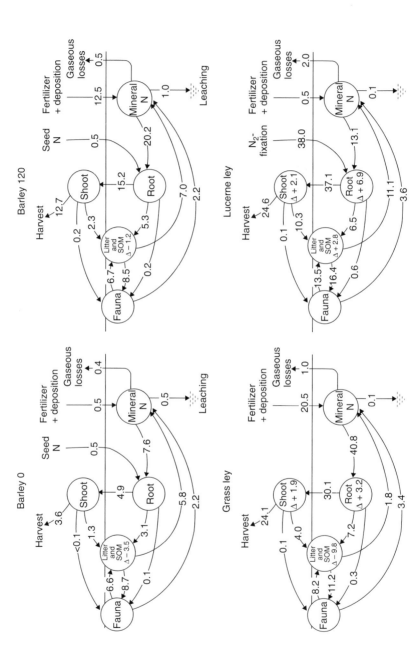

FIGURE 3.15. Budgets of annual nitrogen flows (g N m⁻² yr⁻¹) in the four cropping systems. See Figure 3.14 for further explanation of symbols (from Paustian *et al.*, 1990).

production was 50% greater in the perennial ley fields than in the barley treatments, with soil meso- and macrofauna biomass, consumption, and respiration being significantly higher as well. The main conclusions drawn from this impressive long-term (10-year) study is that microbial and faunal (biotic) interactions do occur, and they have a significant effect on nutrient turnover. As we note in Chapter 8, with current considerable concerns about the impacts of land use change and global change phenomena on key processes in soils, it is imperative to measure explicit changes in numbers and taxa of soil organisms that carry out these processes.

SUMMARY

The processes of consumption and decomposition are considered ecologically as system-level catabolism. The primary agents of decomposition are bacteria and fungi, often referred to as "microbial biomass."

Microbial production and turnover is measured in a number of indirect ways, both chemical and physiological, as well as in direct fashion, using high-magnification microscopy, or via energetics approaches such as calorimetry.

The microbial biomass, although relatively small (about 200–400 g · m^{-2} within 15 cm of the surface) relative to the total soil organic matter pool, has a rapid turnover time and serves as a principal food source for microbivorous fauna. It is also the source of labile nutrients, available for plant roots and other microbes. Hence the microbial community is indeed the "eye of the needle" through which virtually all of the decomposition carbon and nutrients must pass. In the course of microbial growth and turnover, there is a dynamic process of buildup of macroaggregates, which age and decay into more extensively protected microsites within microaggregates. Roles of fauna in the process of aggregate formation and alteration are also discussed further in Chapter 4.

Some very large strides forward in soil ecology have been taken by investigators who are linking microbial community structure and function. We discuss several examples in Chapter 8, regarding the effects of invasive plant species on microbial communities. It is apparent that a combined or synthetic approach using aspects of functional measures such as enzyme activities, when combined with the qualitative measures such as PLFA and the more quantitative measures of using 16S rRNA gene probes and cDNA analyses, along with the immunofluorescent approaches presented by Artursson and Jansson (2003), will yield considerable dividends in future studies.

4 Secondary Production: Activities of Heterotrophic Organisms—The Soil Fauna

INTRODUCTION

Animals, the other group of major heterotrophs in soil systems, exist in elaborate food webs containing several trophic levels. Some soil animals are true herbivores, because they feed directly on roots of living plants, but most subsist upon dead plant matter, microbes associated with dead plant matter, or a combination of the two. Still others are carnivores, parasites, or top predators. Actual heterotrophic production by the soil fauna is poorly known, because turnover of the faunal biomass, feeding rates, and assimilation efficiencies are difficult to assess. Estimates of biomass of soil animals are not common, and knowledge of the rates of energy or material transfer in food webs is fragmentary (Moore and de Ruiter, 1991; 2000). Analyses of food webs in the soil have emphasized numbers of the various organisms and their trophic resources. Analysis of the structure of these food webs reveals complex structures with many "missing links" poorly described (Walter *et al.*, 1991; Scheu and Setälä, 2002). Communities of soil fauna offer opportunities for studies of phenomena such as species interactions, resource utilization, or temporal and spatial distributions.

Animal members of the soil biota are numerous and diverse. The array of species is very large, including representatives of all terrestrial phyla. Many groups of species are poorly understood taxonomically, and details of their natural history and biology are unknown. For the microarthropods (discussed later in this chapter), only about 10% of populations have been explored, and perhaps only 10% of species described (André *et al.*, 2002). Protection of biodiversity in ecosystems clearly must include the rich pool of soil species.

Soil ecologists cannot hope to become experts in all animal groups. When research focuses at the level of the soil ecosystem, two things are required: the cooperation of zoologists and the lumping of animals into functional groups. These groups are often taxonomic, but species with similar biologies are grouped together for purposes of integration (Coleman *et al.*, 1983; 1993; Hendrix *et al.*, 1986).

The soil fauna also may be characterized by the degree of presence in the soil (Fig. 4.1) or microhabitat utilization by different life forms. There are transient species exemplified by the ladybird beetle, which hibernates in the soil but otherwise lives in the plant stratum of the garden. Gnats (Diptera) represent temporary residents of the soil, because the adult stages live aboveground. Their eggs are laid in the soil and their larvae feed on decomposing organic debris. In some soil situations, dipteran larvae are important scavengers. Cutworms also are temporary soil residents, whose larvae feed on seedlings by night and hide by day. Periodic residents spend their lives belowground, with adults such as the velvet mites emerging perhaps to reproduce. From this perspective, the soil food webs are linked to aboveground systems, making trophic analyses much more complicated. Even permanent residents of the soil may be adapted to life at various depths in the soil.

Among the microarthropods, collembolans are examples of permanent soil residents (see Fig. 4.1). The morphology of collembolans reveals their adaptations for life in different soil strata. Species that dwell on the soil surface or in the litter layer may be large, pigmented, and equipped with long antennae and a well-developed jumping appa-

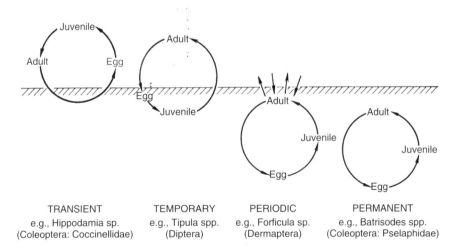

TRANSIENT	TEMPORARY	PERIODIC	PERMANENT
e.g., Hippodamia sp.	e.g., Tipula spp.	e.g., Forficula sp.	e.g., Batrisodes spp.
(Coleoptera: Coccinellidae)	(Diptera)	(Dermaptera)	(Coleoptera: Pselaphidae)

FIGURE 4.1. Categories of soil animals defined according to degree of presence in soil, as illustrated by some insect groups (from Wallwork, 1970).

ratus (furcula). Within the mineral soil, collembolans tend to be smaller with unpigmented, elongate bodies and much reduced furculae—there is no place to jump to.

Numerous researchers have marveled at the many and varied body-plans and size differences of the soil fauna. A generalized classification by length (Fig. 4.2) illustrates a commonly used device for separating the soil fauna into size classes: microfauna, mesofauna, macrofauna, and megafauna. This classification encompasses the range from smallest to largest (i.e., from about 1 to 2 micrometers [μm] of the microflagellates to several meters for giant Australian earthworms).

Body width of the fauna is related to their microhabitats (Fig. 4.3). The microfauna (protozoa, small nematodes) inhabit water films. The mesofauna inhabit air-filled pore spaces and are largely restricted to

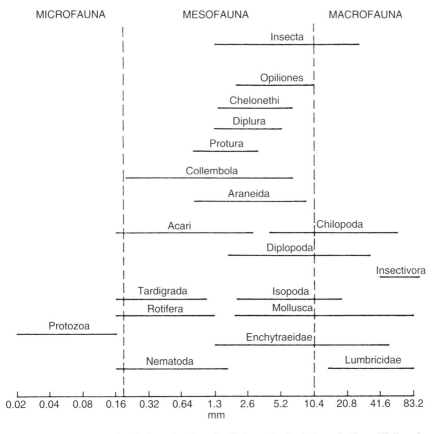

FIGURE 4.2.   A generalized classification of soil fauna by body length (from Wallwork, 1970).

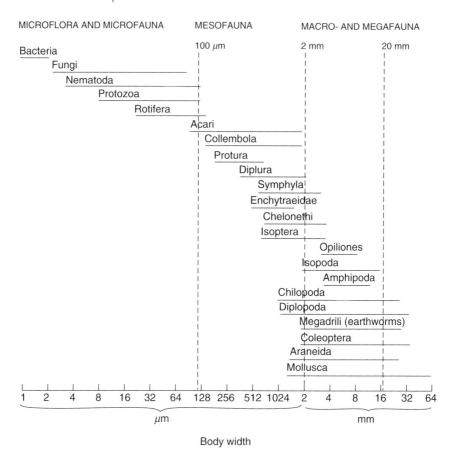

MICROFLORA AND MICROFAUNA   MESOFAUNA   MACRO- AND MEGAFAUNA

FIGURE 4.3.   Size classification of organisms in decomposer food webs by body width (from Swift *et al.*, 1979).

existing ones. The macrofauna, in contrast, have the ability to create their own spaces, through their burrowing activities, and like the megafauna, can have large influences on gross soil structure (Lee, 1985; Lavelle and Spain, 2001; van Vliet and Hendrix, 2003). Methods for studying these faunal groups are in large part size-dependent. Methods for studying the microfauna rely mainly upon techniques used for micro-biology. Mesofauna require microscopic techniques for study and specialized extraction procedures for collection. The macrofauna may be sampled as field collections, often by hand sorting, and populations of individuals are usually measured.

There is, of course, considerable gradation in the classification based on body width. The smaller mesofauna exhibit characteristics of the microfauna, and so forth. Nevertheless, the classification continues to

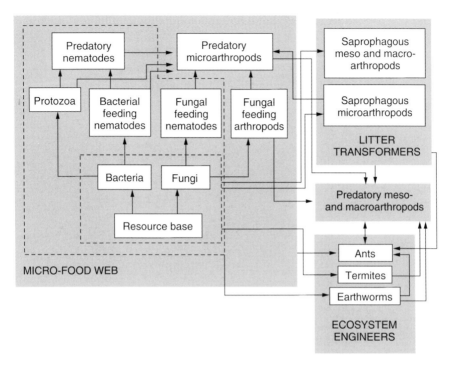

FIGURE 4.4. Organization of the soil food web into three categories—ecosystem engineers, litter transformers, and micro-food webs (after Wardle, 2002, and Lavelle *et al.*, 1995).

have considerable utility. Finally, the vast range of body sizes among the soil fauna suggests that their effects on soil processes take place at a range of spatial scales. Three levels of participation have been suggested (Lavelle *et al.*, 1995; Wardle, 2002). "Ecosystem engineers," such as earthworms, termites, or ants, alter the physical structure of the soil itself, influencing rates of nutrient and energy flow. "Litter transformers," microarthropods, fragment decomposing litter and improve its availability to microbes. "Micro-food webs" include the microbial groups and their direct microfaunal predators (nematodes and protozoans). These three levels operate on different size, spatial, and time scales (Fig. 4.4) (Wardle, 2002).

THE MICROFAUNA

The free-living protozoa of litter and soils belong to two Phyla: the Sarcomastigophora and the Ciliophora (Levine *et al.*, 1980). For practical purposes, we consider them in four ecological groups: the flagellates, naked amoebae, testacea, and ciliates (Lousier and Bamforth, 1990). A general comparison of body plans is given in Figure 4.5, showing representatives of the four major types. After a brief overview of the groups, we will consider aspects of their enumeration and identification.

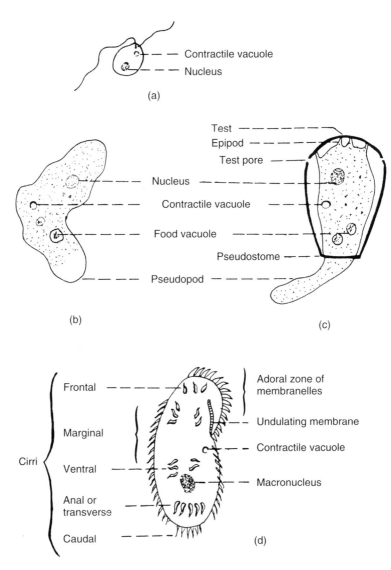

FIGURE 4.5.   Morphology of four types of soil protozoa: (a) flagellate (*Bodo*); (b) naked amoeba (*Naegleria*); (c) testacean (*Hyalosphenia*); (d) ciliate (*Oxytricha*) (from Lousier and Bamforth, 1990).

1. *Flagellates.* Named for their one or more flagella (whiplike propulsive organs), these are among the more numerous and active of the protozoa. They play a significant role in nutrient turnover by their often-intensive feeding activities, with bacteria as their principal prey items (Zwart and Darbyshire, 1991; Kuikman and Van Veen, 1989). Numbers have varied from 100 per gram in desert soils to more than 10^5 per gram in forest soils (Bamforth, 1980).

2. *Naked Amoebae.* These are among the more voracious of the soil protozoa, and are very numerous and active in a wide range of agricultural, grassland, and forested soils (Elliott and Coleman, 1977; Clarholm, 1981, 1985; Gupta and Germida, 1989). The dominant mode of feeding for the amoebae, as for the larger forms such as Ciliates, is phagotrophic (engulfing), with bacteria, fungi, algae, and other fine particulate organic matter being the majority of the ingested material (Bamforth, 1980; Bryant *et al.*, 1982). The highly plastic mode of existence of the naked amoebae is impressive; they have the ability to explore very small cavities or pores in soil aggregates and feed upon bacteria that would otherwise be considered inaccessible to predators (Foster and Dormaar, 1991).

3. *Testate Amoebae.* When compared with the naked amoebae, testate amoebae are often less numerous, except in moist, forested systems where they thrive. However, they are more easily censused by a range of direct filtration and staining procedures (Lousier and Parkinson, 1981). Detailed community production and biomass studies of testacea have been carried out in forested French sites by Coûteaux (1972, 1985) and in Canadian aspen forest lands (Lousier and Parkinson, 1984). For example, Lousier and Parkinson (1984) noted a mean annual biomass of 0.07 gram (g) dry weight$\cdot$m^{-2} of aspen woodland soil, much smaller than the average annual mass for bacteria or fungi, of 23 and 40 g, respectively. However, the testacean annual secondary production (new tissue per year) was 21 g dry weight$\cdot$m^{-2}, or essentially the entire average standing crop of the bacteria in that site.

Certain genera of testacea are also diagnostic of soil types. Foissner (1987a) notes that pioneer soil scientists, such as P. E. Mueller in the 1880s, were able to differentiate between mull and mor forms of humus by the kinds of testacea found (Table 4.1). They used ratios of abundance of forms, rather than exclusivity of presence or absence.

4. *Ciliates.* These protozoa, which have their own unusual life cycles and complex reproductive patterns, tend to be restricted to very moist or seasonally moist habitats. Their numbers are lower than other groups, with a general range of 10 to 500 per g of litter/soil. Ciliates can be very active in entering soil cavities and pores and exploiting bacterial food sources in them (Foissner, 1987a). In com-

TABLE 4.1.   Species Characteristic of the Testacean and Ciliate Communities in Mull and Mor Soils

	Testaceans		*Ciliophora*
Type of humus	*Characteristic species*	*Ratio of full and empty shells*	*Characteristic species*
Mull	*Centropyxis plagiostoma* *Centropyxis constricta* *Centropyxis elongata* *Plagiopyxis minuta* *Geopyxella sylvicola* *Paraquadrula* spp.	<1:2–5	*Urosomioda agilis* *Urosoma* spp. *Hemisincirra filiformis* *Engelmanniella mobilis* *Grossglockneria hyalina* *Colpoda elliotti*
Moder and Mor	*Trigonopyxis arcula* *Plagiopyxis labiata* *Assulina* spp. *Corython* spp. *Nebela* spp.	>1:2–5	*Frontonia depressa* *Bryometopus sphagni* *Dimacrocaryon amphileptoides*

From Foissner, 1987a.

mon with other protozoa, ciliates have resistant or encysted forms from which they can emerge when conditions become favorable for growth and reproduction, with the presence of suitable food sources (Foissner, 1987a). Ciliates, along with flagellates and naked and testate amoebae, can quickly reproduce asexually by fission. The flagellates, naked amoebae, and testacea can reproduce by syngamy, or fusion of two cells. For the ciliates, sexual reproduction occurs by conjugation, with the micronucleus undergoing meiosis in two individuals, and the two cells joining at the region of the cytostome and exchanging haploid "gametic" nuclei. Each cell then undergoes fission to produce individuals, which are genetically different from the preconjugant parents (Lousier and Bamforth, 1990).

As noted above for the testacea, some genera of ciliates are considered indicative of acid humus, and others more typical of higher-pH or "mild" humus. See Table 4.1 for species characteristic of mull and mor soils.

Methods for Extracting and Counting Protozoa

There is an extensive literature on approaches to extracting and counting protozoa. For much of the 20th century, researchers have been heavily influenced by the papers of Cutler (1920, 1923), Cutler *et al.* (1923), and Singh (1946). These authors favored the culture technique, in which small quantities of soil or soil suspensions from dilution series are incubated in small wells that are inoculated with a single species of bacteria as a food source. Based on presence or absence in each well, one can calculate the overall population density ("most probable number"). Other scientists, notably Coûteaux (1972) and Foissner (1987a), espouse the direct count approach, in which one examines soil samples, in water, to see what organisms are present in the subsample. The advantages of this approach are that it is possible to observe the organisms, which are immediately present, and not have to rely on the palatability of the bacterium used to inoculate the series of wells in the culture technique. The disadvantage of the direct count method, as noted by Foissner (1987b), is that one usually employs only 5–30 milligrams (mg) of soil, so as not to be overwhelmed with total numbers. Unfortunately, this discriminates against some of the more rare forms of testaceans or ciliates, which occur only infrequently, but may have a significant impact, if they happen to be very large. Given rather limited research budgets, it is seldom possible to employ a small army of staff to scan literally hundreds of slides of soil from a single sample site.

An additional complication is the fact that the culture technique attempts to differentiate between active (trophozoite forms) and inactive (cystic) forms by the treatment of replicate samples with 2% hydrochloric acid overnight. The acid kills off the trophic forms, and then, after washing in dilute NaCl, the counting continues. This assumes that all of the cysts will excyst after this drastic process; sometimes the assumption is met, but not always.

Distribution of Protozoa in Soil Profiles

Although protozoa are considered to be distributed principally in the upper few centimeters of a soil profile, they are also found at depths of more than 200 meters (m) in groundwater environments (Sinclair and Ghiorse, 1989). Small (2–3 μm cell size) microflagellates were found to decrease 10-fold in numbers during movement through 1 m in a sandy matrix under a trickling-filter facility (in dilute sewage), as compared to a 10-fold reduction in bacterial transport over a 10-m distance (Harvey et al., 1995).

Impacts of Protozoa on Ecosystem Function

Several investigators have noted the obvious parallel between the protozoan–microbe interaction in water films in soil, and on root surfaces, and in open-water aquatic systems (Stout, 1963; Coleman, 1976; Clarholm, 1994; Coleman, 1994a). The so-called "microbial loop" Pomeroy (1974) has proven to be a powerful conceptual tool; rapidly feeding protozoa may consume several standing crops of bacteria in soil (see Chapter 6) every year (Clarholm, 1985; Coleman, 1994b).

Darbyshire and Greaves (1967) noted that this tendency is particularly marked in the rhizosphere, which provides a ready food source for microbial prey. This was demonstrated impressively for protozoa in arable fields (Cutler *et al.*, 1923), and more recently for bacteria, naked amoebae, and flagellates in the humus layer of a pine forest in Sweden after a rain (Clarholm, 1994). Bacteria and flagellates began increasing immediately after a rainfall event and rose to a peak in 2–3 days; naked amoebae rose more slowly and peaked at days 4–5, and then tracked the bacterial decrease downward, as did the flagellates (Fig. 4.6).

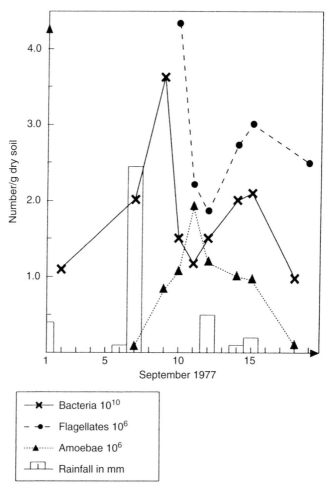

FIGURE 4.6.   Daily estimates of numbers of bacteria, amoebae, and flagellates in the humus layer of a pine forest after rain (from Clarholm, 1994).

Further information on protozoan feeding activities and their impacts on other organisms and ecosystem function are given in Griffiths (1994), Zwart *et al.* (1994), Poussard *et al.* (1994), and Bamforth (1997). Bonkowski *et al.* (2000) suggest that protozoa, and the bacteria they feed upon in the rhizosphere, produce plant-growth–promoting compounds that stimulate plant growth above and beyond the amounts of nitrogen mineralized in the rhizosphere. This is an interesting area of new process-oriented research, and we can expect further developments in the near future.

Protozoa and other microfauna are quite sensitive to environmental insults, and changes in the distribution and activities are diagnostic of changes in soil health (Gupta and Yeates, 1997). We address the issue of soil health more extensively in Chapter 8.

THE MESOFAUNA

Rotifera

Among the small fauna, rotifera are often found only when a significant proportion of water films exists in soils. They are usually considered to be aquatic organisms and may not be listed in major compendia of soil biota (Dindal, 1990); they are a genuine, albeit secondary, component of the soil fauna (Wallwork, 1976). While sampling for nematodes in the surface layers of agricultural fields near La Selva, in the Atlantic coastal forest of Costa Rica, one of the authors (Coleman) found virtually no nematodes, but large numbers of rotifers (tens of thousands per square meter), despite the soil being far from water saturation. The field was being maintained in a "bare fallow" regime with frequent weeding or denudation of vegetation, to deliberately reduce organic inputs. However, there seemed to be ample Cyanobacteria and perhaps other unicellular primary producers, which would have provided food for the rotifers. Some rotifers have been found in bagged leaf litter on forest floors in the southern Appalachian Mountains.

Features of Body Plan and General Ecology

More than 90% of soil rotifers are in the order Bdelloidea, or wormlike rotifers. In these creeping forms, the suctorial rostral cilia and the adhesive disc are employed for locomotion (Donner, 1966). Rotifers also form cysts to endure times of stress or lack of resources. Additional life history features of interest include the construction of shells from a body secretion, which may have particles of debris and/or fecal material adhering to it. Some rotifers will use the empty shells of Testacea, the thecate amoebae. The Bdelloidea are vortex feeders, creating currents of water that conduct food particles to the mouth for ingestion

(Wallwork, 1970). The importance of these organisms is largely unknown, although they may reach numbers exceeding 10^5 per square meter in moist, organic soils (Wallwork, 1970).

Rotifers are extracted from soil samples and enumerated using methods similar to those used for nematodes (see next section).

Nematoda

Nematodes, or roundworms, are among the most numerous of the multicellular organisms found in any ecosystem. As with the protozoa, they are primarily inhabiters of water films, or water-filled pore spaces in soils. Nematodes have a very early phylogenetic origin, but as with many other invertebrate groups, the fossil record is fragmentary. They are classified among the triploblastic pseudocoelomates (three body layers: ectoderm, mesoderm, and endoderm). In other words, nematodes have a body cavity for the gastrointestinal tract, but it is less well-differentiated than that for the true coelomates, such as annelids and arthropods.

The overall body shape is cylindrical, tapering at the ends (Fig. 4.7). In general, nematode body plans are characterized by a "tube within a tube" (alimentary tract/the body wall). The alimentary tract, beginning at the anterior end, consists of a stoma or stylet, pharynx (or esophagus), intestine, and rectum, which opens externally at the anus. The reproductive structures are quite complex, as shown in Figure 4.7. Some species are parthenogenetic, reproducing without sex. It is possible to view the internal structures of most nematodes because they have virtually transparent cuticles. The nematodes can be keyed out fairly readily to family and/or genus under a moderate magnification (about 100×) binocular microscope or in a Sedgwick–Rafter chamber on an inverted microscope (Wright, 1988), but species-specific characteristics must be determined under high magnification, using compound microscopes.

Nematode Feeding Habits

Nematodes feed on a wide range of foods. General trophic groupings include bacterial feeders, fungal feeders, plant feeders, and predators and omnivores. For the purposes of our general overview, one can use anterior (stomal or mouth) structures to differentiate general feeding, or trophic, groups (Fig. 4.8) (Yeates and Coleman, 1982; Yeates et al., 1993; Yeates, 1998). The feeding categories are a good introduction, but the feeding habits of many genera are either complex or poorly known. Thus immature forms of certain nematodes may be bacterial feeders, then become predators on other fauna once they have matured (Allen-Morley and Coleman, 1989). Some of the stylet-bearing nematodes (e.g.,

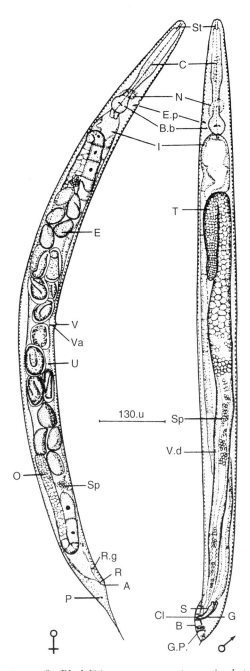

FIGURE 4.7. Structures of a *Rhabditis* sp., a secernentean microbotrophic nematode of the order Rhabditida. (Left) Female. (Right) Male. ST = stoma; C = corpus area of the pharynx; N = nerve ring; E.p = excretory pore; B.b = basal bulb of the pharynx; I = intestine; T = testis; E = eggs; V = vulva; Va = vagina; U = uterus; O = ovary; SP = sperm; V.d = vas deferens; R.g = rectal glands; R = rectum; A = anus; S = spicules; G = gubernaculum; B = bursa; P = phasmids; G.P = genital papillae; CL = cloaca (courtesy of *Proceedings of the Helminthological Society of Washington*) (from Poinar, 1983).

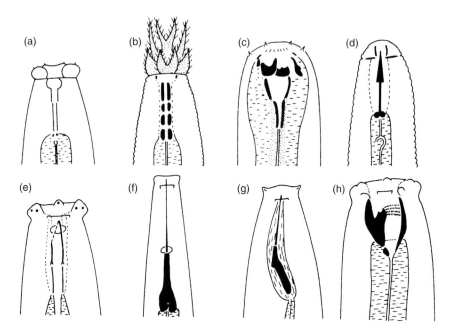

FIGURE 4.8.   Head structures of a range of soil nematodes. (a) *Rhabditis* (bacterial feeding); (b) *Acrobeles* (bacterial feeding); (c) *Diplogaster* (bacterial feeding, predator); (d) tylenchid (plant feeding, fungal feeding, predator); (e) *Dorylaimus* (feeding poorly known, omnivore); (f) *Xiphinema* (plant feeding); (g) *Trichodorus* (plant feeding); (h) *Mononchus* (predator) (from Yeates and Coleman, 1982).

the family Neotylenchidae), may feed on roots, root hairs, and fungal hyphae (Yeates and Coleman, 1982). Some bacterial feeders (e.g., *Alaimus*) may ingest 10-μm width cyanobacterial cells (*Oscillatoria*) despite the mouth of the nematode being 1-μm wide, indicating that the cyanobacterial cells can be compressed markedly by the nematode (Yeates, 1998). Recent laboratory studies (Venette and Ferris, 1998) have confirmed that population growth of bacterial-feeding nematodes is strongly dependent on the species of bacteria ingested. The six nematode species used in the study of Venette and Ferris (1998) reached maximal population growth rates when ingesting from 10^4 to 10^5 colony-forming units (CFUs) per nematode. Population growth rates (λ) under these controlled conditions ranged from 1.1 to greater than $12\,d^{-1}$, making these organisms ideal for detecting rapid changes in the soil environment.

Although specialized in nature, the feeding habits and impacts of entomopathogenic nematodes are quite marked in several soil environments worldwide (Hominick, 2002). The nonfeeding "infective juveniles," or third instar dauer larvae of nematodes in the family

Heterorhabditidae, live in the soil and search for hosts and disperse. An infective juvenile enters the insect host (which it senses along a CO_2 gradient) (Strong *et al.*, 1996) through a spiracle or other opening, punctures a membrane, then regurgitates the symbiotic bacterium *Photorhabdus luminescens*, which kills the host within 48 hours. A rapidly growing bacterial population then digests the insect cadaver and provides food for the exponentially growing adult nematode population inside. The symbiotic bacteria produce antibiotics and other antimicrobial substances that protect the host cadaver and adult nematodes inside from invasion by alien bacteria and fungi from the soil (Strong *et al.*, 1999). When the cadaver is exhausted of resources, reproduction shunts to infective juveniles, which break through the host integument and disperse into the soil. For example, as many as 410,000 *Heterorhabditis hepialus* infective juveniles are produced in a large ghost moth caterpillar (Strong *et al.*, 1996). In pot experiments, Strong *et al.* (1999) found that *Lupinus arboreus* seedlings, whose seedling survival decreased exponentially with increasing densities of root-feeding caterpillars, had virtually the entire negative effect of the herbivore cancelled upon the introduction of the entomopathogenic nematode into the system. For more information on dynamics of entomopathogenic nematodes in soil food webs, see Strong (2002).

For identifying fungal-based food chains, Ruess *et al.* (2002) have shown that the measurement of fatty acids specific to fungi can be traced to the body tissues of fungal-feeding nematodes. Although still in early stages of development, this technique shows considerable promise for more detailed biochemical delineation of food sources of specific feeding groups of nematodes.

Because of the wide range of feeding types and the fact that they seem to reflect ages of the systems in which they occur (i.e., annual versus perennial crops [Neher *et al.*, 1995], or old fields and pastures and more mature forests), nematodes have been used as indicators of overall ecological condition (Bongers, 1990; Ettema and Bongers, 1993; Yeates, 1999; Ferris *et al.*, 2001). This is a growing area of research in soil ecology; one in which the intersection between community analysis and ecosystem function could prove to be quite fruitful. We discuss some of these concepts further in Chapter 5 on decomposition and nutrient cycling.

Nematode Zones of Activity in Soil

As noted in Chapter 2, the rhizosphere is a zone of considerable metabolic activity for root-associated microbes. This extends also to the soil fauna, which may be concentrated in the rhizosphere. For example, Ingham *et al.* (1985) found up to 70% of the bacterial and fungal-feeding nematodes in the 4–5% of the total soil that was rhizosphere, namely the amount of soil 1–2 millimeters (mm) from the root surface (the rhizoplane). In comparison, Griffiths and Caul (1993) found that nematodes migrated to packets of decomposing grass residues, with considerable amounts of labile substrates therein, in pot experiments. They concluded that nematodes are seeking out these "hot spots" of concentrated organic matter, and that protozoa, also monitored in the experiment, do not.

Nematodes are very sensitive to available soil water in the soil matrix. Elliott *et al.* (1980) noted that the limiting factor for nematode survival often hinges on the availability of soil pore necks, which enable movement between soil pores. In recent studies, Yeates *et al.* (2002) measured the movements, growth, and survival of three genera of bacterial-feeding soil nematodes in undisturbed soil cores maintained on soil pressure plates. Interestingly, the nematodes showed significant reproduction even when diameters of water-filled pores were approximately 1 μm. This information should prove useful when determining biological interactions under field conditions, and indicates that soil nematodes may be more active over a wider range of soil moisture tensions than had been thought to be the case previously.

Nematode Extraction Techniques

Nematodes may be extracted by a variety of techniques, either active or passive in nature. For more accuracy in determination of populations, the passive or flotation techniques are generally preferred. The principal advantage of the oldest, active method, namely the Baermann funnel method, is that it is simple, requiring no fancy equipment or electricity. It is based on the fact that nematodes in soils will move about in the wetted soil and fall into the funnel itself. Thus samples are placed on coarse tissue paper, on a coarse mesh screen, and then placed in the cone of a funnel and immersed in water. Once they crawl through the moist soil and filter paper, the nematodes fall down into the neck of the funnel. Because nematodes have only circular and not longitudinal muscles, they do not stay in suspension in the water and fall to the bottom of the funnel stem, which was closed off with a screw clamp on a rubber hose. At the conclusion of the extraction (typically 48 hours), the nematodes in solution are drawn off into a tube and kept preserved for examination later. One drawback to the technique is that it allows dormant nematodes to become active and be extracted, so it may give a slightly inflated estimate of the true, "active" population at a given time. Other methods include filtration, or decanting and sieving, and flotation/centrifugation (Christie and Perry, 1951, Coleman *et al.*, 1999) to remove the nematodes from the soil suspension. When handling larger quantities of soil (up to 500 g) to recover large amounts of nematodes, various elutriation (extraction using streams of air bubbles in funnels) methods are employed. For details, see Gorny and Grüm (1993) (Fig. 4.9).

Tardigrada

These interesting little micrometazoans (ranging from 50 μm, the smallest juvenile, to 1200 μm, the largest adult) (Nelson and Higgins, 1990) are also called "water bears" because of their microursine appearance. They were named "Il Tardigrado," literally slow-stepper, because their slow movements resembled those of a tortoise, by the famous Italian abbot and natural history professor Lazzaró Spallanzani in 1776 (Nelson and Higgins, 1990).

Tardigrades are members of the monophyletic group known as Ecdysozoa, a clade of all molting animals that includes nematodes and arthropods (Garey, 2001). Tardigrades are bilaterally symmetrical with four pairs of legs, equipped with claws on the distal end, of various sizes and forms (Fig. 4.10). The sizes and shapes of the claws are used in keying genera and species. Perhaps their greatest notoriety in recent times has come from the marked recuperative powers that they show after having been kept dry in a state of "suspended animation" for many years or even decades. These studies (Crowe, 1975; Crowe and Cooper, 1971; Wright, 2001) have found that tardigrades recover well even after extreme environmental insults such as being plunged into liquid nitrogen. More generally, a series of five types of latency or virtual cessation of metabolism have been described: encystment, anoxybiosis, cryobiosis, osmobiosis, and anhydrobiosis (Crowe, 1975). All these are subsumed under the more general term "cryptobiosis" (Keilin, 1959), or hidden life, which was first described by Antonie van Leeuwenhoek in a Royal Society lecture in 1702 (Wright, 2001). Being highly resistant, or resilient, to various environmental insults, tardigrades exemplify a recurrent thread throughout biology in general, and soil biology in particular: the selective advantages to "waiting out" a spell of bad microclimate and being able to reactivate and become active in a given patch in the soil, years or decades later.

Tardigrades occur predominantly in the surface 1–3 cm of many grassland soils, but certain genera (e.g., the *Macrobiotus*-group species), are quite numerous at depths up to 10 cm in subalpine coniferous forest (Ito and Abe, 2001). They may serve as "early-warning devices" for environmental stress. Tardigrades were found to be the most sensitive organism measured in a several-year study of the effects of dry-deposition of SO_2 on litter and soil of a mixed-grass prairie ecosystem (Grodzinski and Yorks, 1981; Leetham *et al.*, 1982). They are thought to feed on algal cells and debris in the interstices of moss thalli and probably have a rather broad diet of various microbial-rich bits of

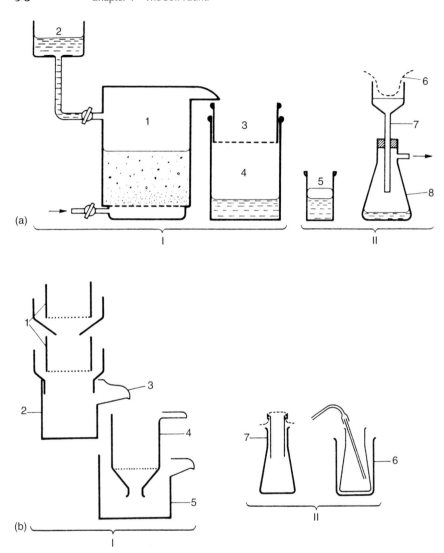

FIGURE 4.9.   Flotation apparatus for extraction of soil invertebrates (Ladell's modified flotation apparatus). (a) Edwards and Denis, 1962; (b) Salt, 1953. I: Separation of organic matter from the mineral components of the soil; II: Separation of soil invertebrates and organic residue. The process uses these materials: (a) 1 = flotation vessel with air supply from below, 2 = container with salt solution, 3 = sieve, 4 = sedimentation tank, 5 = glass oil separator, 6 = gauze with the material under study, 7 = glass funnel, 8 = flask connected to suction pump. (b) 1 = vessels with coarse and fine sieves, 2 = container collecting the residue, 3 = outlet, 4 = final collecting sieve, 5 = water tank, 6 = water bath, 7 = glass for oil-water flotation (from Kaczmarek, 1993).

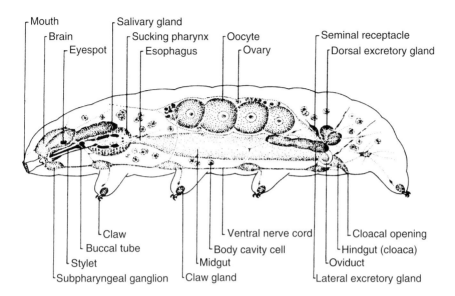

Mouth
Brain
Eyespot

Salivary gland
Sucking pharynx
Esophagus

Oocyte
Ovary

Seminal receptacle
Dorsal excretory gland

Claw
Buccal tube
Stylet
Subpharyngeal ganglion

Ventral nerve cord
Body cavity cell
Midgut
Claw gland

Cloacal opening
Hindgut (cloaca)
Oviduct
Lateral excretory gland

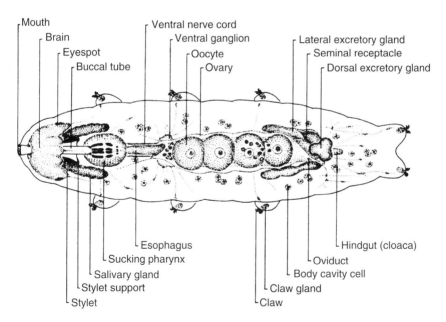

Mouth
Brain
Eyespot
Buccal tube

Ventral nerve cord
Ventral ganglion
Oocyte
Ovary

Lateral excretory gland
Seminal receptacle
Dorsal excretory gland

Esophagus
Sucking pharynx
Salivary gland
Stylet support
Stylet

Oocyte
Ovary
Claw gland
Claw

Hindgut (cloaca)
Oviduct
Body cavity cell

FIGURE 4.10. Eutardigrade internal anatomy (*Macrobiotus hufelandii*, female); lateral view (above) and dorsal view (below) (*Soil Biology Guide,* Dindal, D. L., ©1990, John Wiley & Sons, New York. Reprinted by permission of John Wiley & Sons, Inc.).

soil organic matter. Tardigrades have also been observed to feed voraciously on nematodes when in culture (G. W. Yeates, personal communication). Tardigrades have been found in large numbers (up to 2000 per 10 square meters of soil surface), and are particularly associated with lichens, mosses, liverworts, and rosette angiosperms (Nelson and Higgins, 1990; Nelson and Adkins, 2001). They are found also in very cold, dry habitats, such as the Antarctic dry valleys, where they feed on the particulate organic detritus brought in by windward movement of algal cells from lake ice at one end of the long valleys (D. H. Wall, personal communication).

Tardigrades may be extracted from soils and various substrates by Baermann funnels (Petersen and Luxton, 1982), or flotation and sieving through a 44-μm sieve, or by the sucrose flotation and centrifugation technique used for extracting nematodes (Christie and Perry, 1951).

Microarthropods

Large numbers of the microarthropod group (mainly mites and collembolans) are found in most types of soils. A square meter of forest floor may contain hundreds of thousands of individuals representing thousands of different species (Fig. 4.11). Microarthropods have a significant impact on the decomposition processes in the forest floor (see Chapter 5) and are important reservoirs of biodiversity in forest ecosystems.

Microarthropods also form an important set of linkages in food webs. Many microarthropods feed on fungi and nematodes, thereby linking the microfauna and microbes with the mesofauna. Microarthropods in turn are prey for macroarthropods such as spiders, beetles, ants, and centipedes, thus bridging a connection to the macrofauna. Even some of the smaller megafauna (toads, salamanders) feed upon microarthropods. We emphasize, again, the need to study soil as an ecosystem. Analysis of one part of the food web, the microarthropods for example, falls short if other components are ignored.

In the size spectrum of soil fauna (see Figs. 4.2 and 4.3), the mites and collembolans are found among the mesofauna. Members of the microarthropod group are unique, not so much by their body size as by the methods used for sampling them. Microarthropods are too small and numerous to be sampled as individuals. Instead, small pieces of habitat (soil, leaf litter, or similar materials) are collected and the microarthropods extracted from them in the laboratory. In this manner they resemble certain of the microfauna such as nematodes, rotifers, or tardigrades. Most of the methods used for microarthropod extraction are either variations of the Tullgren funnel ("Berlese funnel"), which uses heat to desiccate the sample and force the arthropods into a collec-

FIGURE 4.11.   An example of the thousands of species of soil microarthropods (D. A. Crossley, Jr. photo).

tion fluid, or flotation in solvents or saturated sugar solutions followed by filtration (see Chapter 9). Edwards (1991) gives an extensive review of these procedures. Both approaches to sampling microarthropods have their proponents. Generally, flotation methods work well in low organic, sandy soils whereas Tullgren funnels perform best in soils with high organic matter content. Flotation procedures are much more laborious than is Tullgren extraction.

Choice of method also depends upon the objectives of the sampling program. If numbers of individuals are to be measured, a large set of small samples may be needed. Estimations of species number may be better served by fewer, larger samples. In any case, extraction methods are never completely efficient and, indeed, efficiency of sampling is seldom estimated (André et al., 2002). Consequently, Walter and Proctor (1999) concluded that enumeration of microarthropods was an "intellectually vacuous" exercise. However, valid comparisons of microarthropod abundance in different habitats may be obtained even if extraction efficiencies, though unknown, are similar.

Microarthropod densities vary during seasons within and between different ecosystems (Table 4.2). Generally, temperate forest floors with large accumulations of organic matter support high numbers, whereas tropical forests where the organic layer is thin contain lesser numbers of microarthropods (Seastedt, 1984b). Disturbance or perturbation of soils usually depresses microarthropod numbers. Tillage, fire, and pesticide applications typically reduce populations but recovery may be rapid and microarthropod groups respond differently.

Soil mites usually outnumber collembolans but these become more abundant in some situations. In the springtime, forest leaf litter may develop large populations of "snow fleas" (*Hypogastrura nivicola* and related species). Among the mites themselves the oribatids usually dominate but the delicate Prostigmata may develop large populations in cultivated soils with a surface crust of algae. Immediately following cultivation, numbers of astigmatic mites have been seen to increase dramatically (Perdue, 1987).

TABLE 4.2.   Abundance of Microarthropods in Soils from Various Ecosystems

Ecosystem	Microarthropods (10^3 per m^2)	Reference
Fallow crop fields, Nigeria	40–68	Adejuyigbe et al., 1999
Corn tillage plots, Guelph, Canada	16–17	Winter and Voroney, 1990
No-tillage plots, North Carolina	1–30	House and Worsham, 1987
Cedar plantation, Nagoya, Japan	48–149	Hijii, 1987
Deciduous forest, Tennessee	36.9	Reichle et al., 1975
Deciduous forest, North Carolina	88	Lamoncha and Crossley, 1998
Burned tallgrass prairie, Kansas	35–50	Seastedt, 1984
Unburned tallgrass prairie, Kansas	63–77	Seastedt, 1984
Mediterranean desert, Negev	1–2	Steinberger and Wallwork, 1985
North American desert, American Southwest	1–8	Steinberger and Wallwork, 1985
Phryganic ecosystem, Greece	20–60	Sgardelis et al., 1981

Soil microarthropods are significant reservoirs of biodiversity but it is not clear exactly how diverse they may be. Estimation of species richness is a difficult problem for many types of soil organisms (fungi, bacteria and nematodes, for example, as well as microarthropods). In an extensive review, André *et al.* (2002) report that at most 10% of soil microarthropod populations have been explored and 10% of species described. Thus, according to those authors, the contribution of soil fauna to global biodiversity remains an enigma. Consequently, the mechanisms underlying the large species diversity of the microarthropods continue to elude us. The decline in numbers of taxonomic specialists for these groups has been noted (Behan-Pelletier and Bissett, 1993) as a contributing factor to our inadequate information base for microarthropods.

Unlike the macroarthropods, the mites and collembolans have little or no effect on soil structure. Their dimensions allow them to use existing spaces in soil structure. Even the large, soft-bodied members of the mite group Prostigmata do not seem to create their own passageways. Some litter-feeding species do burrow into substrates such as petioles of decaying leaves and create tunnels, but these have no direct effect on soil structure per se. The microarthropods resemble the microfauna in this characteristic.

Collembola

Among microarthropods, collembolans are often equal to soil mites in numerical abundance. They are worldwide in distribution and occur in all biomes, from tropic to arctic and from forest to grassland and desert and throughout the soil profile. Collembolans (Fig. 4.12) have the common name of "springtails" from the fact that many of the species are able to jump by means of a lever attached to the bottom of the abdomen. They also have a unique ventral tube (*collophore*), which seems to function in osmoregulation, and a springing apparatus (*furcula* and *tenaculum*) ventrally on the abdomen (absent in some groups). Most species are small, at most a few millimeters long, but may be brightly colored. They are ubiquitous members of the soil fauna, often reaching abundances of 100,000 or more per square meter. They occur throughout the upper soil profile, where their major diet appears to be fungi associated with decaying vegetation. In the rhizosphere, they are often the most numerous of the microarthropods. Surface-dwelling forms, inhabiting the litter layer, are usually well equipped with furculas (see Fig. 4.12). Residents of the deeper soil layers generally have no furcula, or only a rudimentary one, and typically lack pigmentation and eyes (Petersen, 2002).

The position of the Collembola in the world of arthropods continues to puzzle specialists. Classically these small, wingless arthropods have

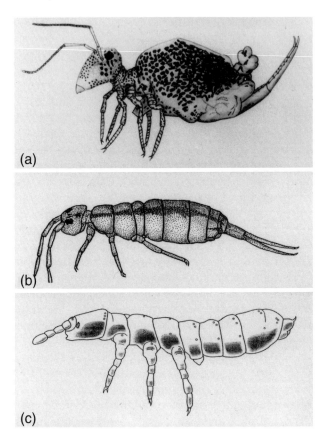

FIGURE 4.12. (a) A symphypleonid collembolan (*Sminthurus burtcheri*) (Snider, 1969). (b) An arthropleonid collembolan (*Isotomurus palustris*) (Snider, 1967). (c) An onychiurid collembolan (Onychiuridae: *Protaphorura* sp.). Note the absence of the furcula (jumping apparatus) on the eyeless, soil-dwelling onychiurid, in contrast to the other litter-dwelling forms.

been listed among the class Insecta (e.g., Boudreaux, 1979) but some authors suggest that they deserve a class of their own (Manton, 1970). Their mouthparts are held in a unique cone-shaped structure. Collembolans lack such features as compound eyes or wings, but do resemble insects by having three body regions. The head bears a pair of antennae. The thorax is three-segmented and bears three pairs of legs. The collembolan abdomen consists of only six segments, less than the insect model. The collembolan ventral tube, the collophore, is not found in other groups of arthropods.

The classification of the Collembola is relatively stable at the generic level, although many species remain unnamed and taxonomy of the group is based almost entirely on external morphology (Hopkin, 1997).

Resolution of the higher taxonomic categories will require a close examination of the fauna of the entire world (Christiansen and Bellinger, 1998). In North America, the taxonomic analysis by Christiansen and Bellinger (1998) is the standard reference and provides keys to genera and known species. On a worldwide basis, the literature is scattered over a wide variety of journals and other publications. Hopkin (1997) offers regional checklists of the collembolan fauna. The publications of Gisin (1962, 1963, 1964, and others cited therein) are essential for the study of European Collembola.

With the development of computerized access to the World Wide Web, another large array of resources has recently become available. A good search engine will locate several thousand Web sites referencing collembolans and offering keys, check lists, lists of specialists, and other valuable information. These resources include such assets as a world list of collembola, interactive keys to species in some of the genera, a list of references to Collembola beginning in 1995, and a catalog of the neotropical species. We anticipate that the World Wide Web will become even more valuable as a source of information about microarthropods. (We do not offer specific Web addresses because they are often subject to change.) A cautionary note: Web pages are seldom peer reviewed.

Identification of collembolans requires use of a microscope and magnifications as high as 400×. Preliminary sorting of samples to family levels can be performed with a dissecting microscope, once some familiarity with the group has been gained. Recognition of genera and species will require slide mounts (see Chapter 9). Collembolans will float on the surface of many collection fluids, due to their very hydrophobic cuticle, and special collection fluids are recommended (see Chapter 9).

Families of Collembola

In the current system of classification, a dozen or so families of collembolans are arranged in three major groups (Christiansen and Bellinger, 1998). The suborder Arthropleona contains the so-called "linear" collembola, the great majority of species, in two sections: the Poduromorpha and the Entomobryomorpha.

Poduromorph collembolans [Fig. 4.12(c)] include the important families Hypogastruridae and Onychiuridae, whose species are dwellers in mineral soil layers, and the family Poduridae with a single darkly pigmented species, *Podura aquatica*, whose natural habitat is standing water.

Onychiurid collembolans [Fig. 4.12(c)] almost always have no furcula; eyes are reduced or absent and, if present, are unpigmented. They possess pseudocelli, cuticular organs which have nothing to do with vision but which can extrude a defensive oil when disturbed, an alarm pheromone (Usher and Balogun, 1966). Onychiurids feed in the rhizosphere. Curl and Truelove (1986) argue persuasively that these collembolans are attracted to plant roots and are important in rhizosphere dynamics. In experiments, collembolans protected cotton plants from the root pathogen *Rhizoctonia solani* by selectively grazing that fungus from the plant roots. These rhizosphere inhabitants may prove to be effective biological control agents (Fig. 4.13, Curl and Truelove, 1986). Onychiurids are not well sampled with Tullgren funnels; they do not appear to respond to the heating and drying process in Tullgren extractions. Estimates of numbers of Onychiurids are best made with flotation methods (see Chapter 9) (Edwards, 1991).

The family Hypogastruridae includes several common species whose populations may build up to huge numbers. These include the "snow flea," *Hypogastrura nivicola*. That species multiplies under winter snows and, on warm days, appears to boil out onto the surface (Christiansen, 1992). Another related species, *H. armata*, is common in the litter layer of hardwood forests during the winter months (Snider, 1967). We have found *H. armata* (Fig. 4.14) to be the predominant winter microarthropod in hardwood litterbags in the southern Appalachians (Crossley and Coleman, unpublished data).

The Entomobryomorpha [Fig. 4.12(b)] includes the large family Entomobryidae in which the furcula is well developed. The collembolans are primarily dwellers of surficial soil layers, in forest canopies or on tree trunks. Laboratory cultures of one species, *Sinella curviseta*, have found a valuable role as prey for cultures of spiders (Draney, 1997). Species in the family Tomoceridae (Fig. 4.15) include large forms with long antennae, found in upper litter layers of forest floors throughout the Holarctic region.

Members of the family Isotomidae are a highly variable set of species; the group is in need of serious taxonomic revision (Christiansen and Bellinger, 1998). This family includes *Folsomia candida* (Fig. 4.16), a species widely used in laboratory experiments and in the assessment of the effects of toxic substances. Its reproductive biology has been thoroughly explored and culture methods well developed (Snider, 1973). In fact, the microbial gut flora of *F. candida* has been extensively explored using DNA probing methods. It was found to be a frequently changeable but selective habitat, possibly indicating that soil microarthropods could modify the species makeup of soil microbial communities (Thimm *et al.*, 1998). The European "glacier flea," *Isotoma saltans*, is active on ice at temperatures below freezing and feeds on pollen grains trapped on the glacier surface (Christiansen, 1992).

The third major group of Collembola, the suborder Symphypleona, includes the spherical or globular collembolans. It is a smaller group than the Arthropleona and much more uniform in habits (Christiansen

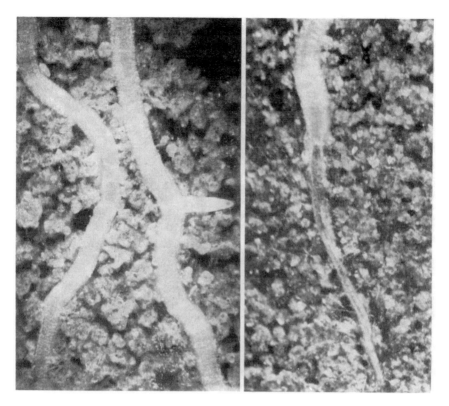

FIGURE 4.13.   Collembolan protection of roots from infection by *Rhizoctonia solani*. (Left) Roots from pathogen-infested soil with mycophagous collembola; (right) diseased root from pathogen-infested soil without collembola (from Curl and Truelove, 1986).

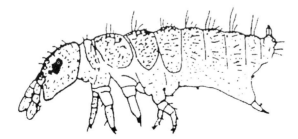

FIGURE 4.14.   Drawing of *Hypogastrura armata*, common in wintertime in Coweeta forest floors, in western North Carolina in the United States (from K. Christiansen, with permission).

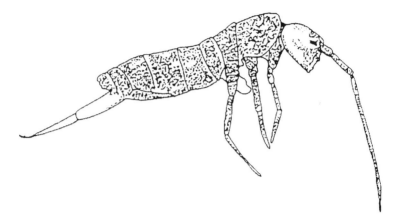

FIGURE 4.15. *Tomocerus dubius* (from K. Christiansen, with permission).

FIGURE 4.16. *Folsomia candida* adults and juveniles. The largest individual is 2 mm in length (from Hopkin, 1997).

and Bellinger, 1998). The family Sminthuridae [Fig. 4.12(a)] is a large and cosmopolitan one, being active jumpers, dwellers in surficial litter layers, on vegetation, and in the canopies of tropical humid forests. Often brightly colored, these collembolans are readily collected with Tullgren funnels or pitfall traps (see Chapter 9), but may also be collected by sweeping through grassy vegetation with a white enamel pan. The family Neelidae consists of tiny globular forms lacking eyes and with short antennae. The family is cosmopolitan but poorly studied (Hopkin 1997).

Population Growth and Reproduction

Many collembolans are opportunistic species, capable of rapid population growth under suitable conditions. They often respond to disturbances of the soil environment. In agricultural systems, spurts of growth may follow plowing or cultivation (Hopkin, 1997). In forests, fire may stimulate collembolan abundances, as may the broadscale application of pesticides (Butcher et al., 1971). Collembolans occur in aggregations. In samples of soils, they are not found at random, but occur in groups. Aside from the statistical problems of assessment of population size, aggregations pose ecological questions as well. In laboratory investigations, Christiansen (1970) and Barra and Christiansen (1975) analyzed collembolan responses to habitat variables (i.e., moisture and substrate) and food resources. Although these were important, the major variable seemed to be a behavioral one. Collembolans possess aggregation pheromones (Krool and Bauer, 1987), which probably function in bringing the sexes together for reproductive purposes. Earlier, reproductive pheromones were identified by Waldorf (1974). Many collembolan species are capable of rapid, even explosive, population growth under ideal conditions. Gist et al. (1974) analyzed life tables for *Sinella curviseta* under laboratory conditions for 170 days; they found an intrinsic rate of increase of 0.036 per day and a replacement rate (R_o) of 515 per female.

Sperm transfer is by means of spermatophores, either actively passed or deposited on pedicels and located by the females. Eggs are laid in groups. Development is continuous; the number of instars ranges between 2 and 50 or more (Christiansen and Bellinger, 1998). Collembolans become sexually active with the fifth or sixth instar but continue to molt throughout life, in contrast with the Insecta, which do not molt after reproduction.

Parthenogenic reproduction is common in many collembolan species, including the commonly cultured *Folsomia candida*. Many species are bisexual, especially those in the Entomobryidae (e.g., *Sinella curviseta*).

Collembolan Feeding Habits

Collembolans are generally considered to be fungivores, with occasional ingestion of other animals, decomposing plant or animal residue, or fecal material. As noted previously, they are often considered to be nonspecific feeders but this conclusion is controversial (Petersen, 2002). Gut content analysis of field-collected specimens or field observation in rhizotrons (Gunn and Cherrett, 1993) often reveals a wide variety of materials, including fungi, plant debris, and animal remains. Laboratory choice studies, in contrast, have found that collembolans have specific food preferences, choosing one fungal species over others (similar discrepancies in feeding analysis have been noted for oribatid mites; see later section on mites). Bengtsson *et al.* (1994), in laboratory experiments, reported that the collembolan *Protaphorura armata* showed an increased dispersal rate if a favored fungus was present as far away as 40 cm (cited in Petersen, 2002).

Models of soil food webs usually place collembolans as fungivores (e.g., Coleman, 1985; Hunt *et al.*, 1987; Moore and de Ruiter, 2000). However, like many of the soil fauna, collembolans in general defy such exact placement into trophic groups. Living plant tissue may be consumed and even dead animal material or feces in cultures. Many collembolan species will eat nematodes when those are abundant (Gilmore and Potter, 1993). Some species may be significant in the biological control of nematode populations (Gilmore, 1972). Feeding on nematodes does not seem to be selective; collembolans do not distinguish between saprophytic and plant parasitic nematodes. In the words of Hopkin (1997), "Indeed the opportunistic nature of the feeding behaviour of many species of Collembola may be one reason for their success."

Collembolan Impacts on Soil Ecosystems

The direct effect of collembolans on ecosystem processes such as energy flow appears to be quite small. Their biomass is relatively tiny, their respiration rates are but a small fraction of total soil CO_2 efflux, and their feeding rates account for only a small amount of microbial activity. They share these characteristics with other soil microarthropods (Gjelstrup and Petersen, 1987). These conclusions have lead Andrén *et al.* (1999) to a sardonic statement, to wit: "Soil animals exist. I like soil animals. They respire too little. Ergo, they must CONTROL something!" Those authors caution us to avoid an overly enthusiastic appraisal of the importance of microarthropods in soil ecosystems. Nevertheless, manipulation experiments have shown important impacts of collembola on nitrogen mineralization, soil respiration, leaching of dissolved organic carbon, and plant growth (Filser, 2002). These system responses may be viewed as indirect effects. Assessing the importance of Collembola in soil ecosystems needs to be done in the context of the intact system and may be expected to vary with temperature, moisture, season, and interactions with other biota.

Grazing upon fungal hyphae appears to be the major contribution of Collembola in the decomposition process. Such grazing on fungal hyphae may be selective, thus influencing the fungal community (Table 4.3). Indirectly, such direct effects on the fungal community may have indirect effects on nutrient cycling (Moore *et al.*, 1987). Selective grazing by the collembolan *Onychiurus latus* changed the outcome of competition between two basidiomycete decomposer fungi (Newell, 1984a, b), allowing an inferior competitor to prosper. Grazing upon fungi may actually increase general fungal activity in soils and stimulate fungal growth. The relationship between fungal and collembolan population dynamics is not straightforward, however, because some collembolan species may actually reproduce more successfully on least favored foods (Walsh and Bolger, 1990). Collembola have been demonstrated to have complex interactions with several fungal species simultaneously. Cotton was grown in a greenhouse with four fungal species, the pathogen *Rhizoctonia solani* and three known biocontrol fungi (including two sporulating Hyphomycetes), and the rhizosphere-inhabiting collembolan *Proisotoma minuta*. The collembolan preferentially fed on the pathogenic fungus and avoided the biocontrol fungi (Lartey *et al.*, 1994).

Acari (Mites)

The soil mites, Acari, chelicerate arthropods related to the spiders, are the most abundant microarthropods in many types of soils. In rich forest soil, a 100-g sample extracted on a Tullgren funnel may contain as many as 500 mites representing almost 100 genera (Table 4.4). This much diversity includes participants in three or more trophic levels and varied strategies for feeding, reproduction, and dispersal. Often, ecologists analyze samples by a preliminary sorting of mites into suborders. Identification of mites to the family level is a skill readily learned under

TABLE 4.3. Compensatory Growth of Fungi in Response to Collembolan Grazing

Fungal species	Collembolan species	Growth relative to controls
Botrytis cinerea	*Folsomia fimetaria*	$-^a$
Coriolus versicolor	*Folsomia candida*	$-/+^b$
Mortierella isabellina	*Onychiurus armatus*	$-^a$
Verticillium bulbillosum	*Onychiurus armatus*	$+$
Penicillium spinulosum	*Onychiurus armatus*	$+$
Field Soil Dilution	*Hypogastrura tullbergi*	$+^a$
	Folsomia regularis	$+^a$

[a] Bacteria were present.
[b] Increase with fungi grown on high nutrient medium, otherwise decrease.
After Lussenhop, 1992.

TABLE 4.4.   Densities (Number per m²) of Soil Microarthropods in Four Forest Types

Taxa	Mixed hardwood	Aspen woodland	Spruce forest	Scots pine forest
Oribatei	56,000	123,000	212,000	425,000
Prostigmata[a]		25,000	96,000	250,000
Mesostigmata	1500	7400	14,000	8600
Collembola	7500	71,000	46,000	60,000

[a]Not estimated.
Modified from Wallwork, 1983.

the tutelage of an acarologist. Expert assistance is necessary for identifications of soil mites to genus or species. By combining slide mounts with examination of specimens in alcohol, reasonably accurate sorting of samples can be performed.

Four suborders of mites occur frequently in soils: the Oribatei, the Prostigmata, the Mesostigmata, and the Astigmata. Specimens in alcohol can usually be assigned to one of these suborders. Slide mounts are required for placement of dubious specimens. Techniques for slide preparations are given in Chapter 9. Krantz (1978) provides keys to all families of the Acari. Keys to families or superfamilies of soil mites themselves may be found in Dindal (1990). As is the case for the collembola, taxonomic aids may be found in sites on the World Wide Web. Interactive keys have also recently become available (Walter and Proctor, 2001).

The soil mites are a subset of the Acari. Occasionally, mites from other habitats are extracted from soil samples. During autumnal leaf drop in deciduous forests, foliage-inhabiting mites may enter the soil food webs and may be found in Tullgren extractions. This group includes, for example, plant-feeding mites (red spider and false spider mites) and their predators. Occasionally mites parasitic upon vertebrates may be seen in soils from the vicinity of mammal nests, or host-seeking larval forms may occur in Tullgren samples. These stragglers doubtless enter into the soil food webs. But the more numerous species are the true soil mites.

Among the four mite groups, the oribatids are the characteristic mites of the soil and are usually fungivorous, detritivorous, or both. Mesostigmatid mites are nearly all predators on other small fauna, although some few species are fungivores and these may become numerous in some situations. Acarid mites are found associated with rich, decomposing nitrogen sources and are seldom abundant except in agricultural soils or stored products. The Prostigmata contains a broad diversity of mites with a variety of feeding habits and strategies. Very little is known of the niches or ecological requirements of most soil mite species, but some tantalizing information is emerging from field research (Walter *et al.*, 1987; Hansen, 2000). As noted previously for Collembola, some mites may play significant roles as consumers of plant pathogenic fungi. In ecosystems where most primary productivity occurs below ground (e.g., grasslands), where nematode biomass is high in root rhizospheres, nematophagous arthropods (including mites) could be significant predators of plant-feeding nematodes (Gerson *et al.*, 2003).

Oribatid Mites

Oribatids (Figs. 4.17–4.20) are an ancient group. They have the richest fossil record of any mite group, dating back to the Devonian period, or 350–400 million years ago (Labandeira *et al.*, 1997). Specimens from a Devonian site near Gilboa, NY, contain organic matter visible in their guts, attesting to a long relationship between oribatid mites and decomposing vegetable matter (Norton *et al.*, 1987). Some oribatid species are Holarctic in distribution, being widely distributed in forest floors of Europe and North America. A general catalog of oribatids of North America was published by Marshall *et al.* (1987). Oribatid mites occur in all terrestrial ecosystems, from arctic (Behan-Pelletier and Norton, 1983) to tropics (Balogh and Balogh, 1988, 1990; González *et al.*, 2001) and from deciduous forest (Hansen, 2000; Lamoncha and Crossley, 1988) to desert (Santos and Whitford, 1981).

A combination of three factors makes oribatid mites unique among the soil fauna. First, their sheer numbers are impressive. They are the most numerous of the microarthropods (Travé *et al.*, 1996). Second, they often possess juvenile polymorphism. Many immature stadia do not resemble their adults. Unlike most other mites, the immatures of some of the Oribatei (the "higher" oribatids) (Norton, 1984) are morphologically so dissimilar from the adult stadia that it is frequently impossible to correlate the two based on morphology alone (compare Figs. 4.17 and 4.18). Despite the differences in morphology, immature stages and adults can usually be cultured on the same resources. Third, oribatids reproduce relatively slowly, in contrast to the other microarthropods. One or two generations per year are usual, and females do not lay many eggs. Indeed, some common species are parthenogenic. Oribatids are sometimes considered to be "K" specialists, and, in contrast, collembolans would be "r" specialists, or opportunistic species (MacArthur, 1972). However, orbatids' K-style traits may be a constraint of low secondary production, not an adaptation or specialization.

Oribatids differ from other microarthropods by having a sclerotized, often-calcareous exoskeleton. In this they resemble the millipedes,

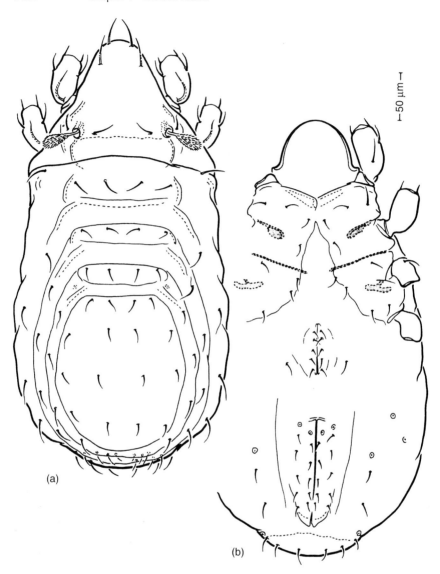

FIGURE 4.17. *Eueremaeus columbianus* (Berlese), tritonymph; (a) dorsal aspect; (b) ventral aspect (from Behan-Pelletier, 1993).

snails, and isopods. Most oribatids are brown to tan in color, although some primitive species are nearly colorless. The exoskeleton contains high calcium levels even in the primitive, lightly colored species (Fig. 4.19) (Todd *et al.*, 1974). When present, the chemical form of the mineral deposits is principally calcium carbonate, calcium oxalate, or calcium

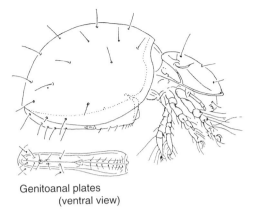

Genitoanal plates
(ventral view)

FIGURE 4.18. A "box" mite of the oribatid family Phthiracaridae. For protection, the legs can be withdrawn beneath the hinged prodorsum (after Baker *et al.*, 1958).

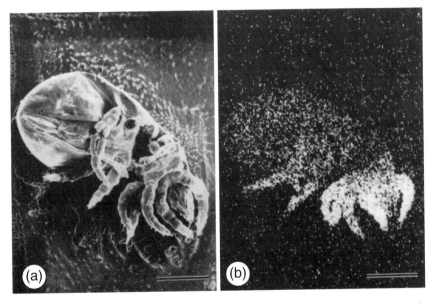

FIGURE 4.19. Oribatid scanning electron micrograph (SEM) with calcium in exoskeleton. (a) and (b) are micrographs of an oribatid mite (*Hypochthonius* sp.). A 100-μm marker is indicated at lower right. (a) Image made by secondary electrons. (b) Map of calcium distribution made from X-ray image (from *Todd et al.*, 1974).

phosphate. Presumably, oribatids are able to sequester calcium by feeding on fungi. Senescent fungal hyphae contain crystals of calcium oxalate, which may be metabolized by the mites (Norton and Behan-Pelletier, 1991). Oribatids often possess a cerotegument, a secretion

FIGURE 4.20. Scanning electron micrographs of oribatid mites, showing modified setae (a–f) (from Valerie Behan-Pelletier, personal communication).

layer of the integument that may be highly sculptured (Norton *et al.*, 1997).

Abundance and Diversity of Oribatid Mites

Oribatid densities in forest soils are in the range of 50,000–500,000 per square meter (Table 4.4). Coniferous forests typically support high numbers of oribatid mites, followed by deciduous hardwood forest, grassland, desert, and tundra. In arctic tundra and in some grassland or savanna habitats, oribatids may be outnumbered by prostigmatid mites (see later section). Cultivation of agricultural fields reduces oribatid populations to an average of about 25,000 per square meter. Population cycles in agroecosystems are often initiated by harvest and cultivation procedures, which change patterns of residue input into soils.

The diversity of oribatid mite species is large in soils from many different localities. In a southern Appalachian hardwood forest, Hansen (1997) reported 170 species of adult oribatid mites from litter and soil, with as many as 40 species in a single 20-cubic-centimeter sample core through the soil profile. Many oribatid mite species are widely distributed across a variety of habitats in Europe and North America, including such common species as *Oppiella nova, Tectocepheus velatus,* and *Scheloribates laevigatus.* Tropical soils also contain a diverse community of oribatid mites. Noti *et al.* (2003) examined three ecosystems in a sere in the Democratic Republic of Congo and reported a total of 149 species of oribatids. They found that the number of species dropped regularly from forest to savanna, where sampling revealed 105 oribatid species. This high species diversity has caused some authors (i.e., Anderson, 1975; André *et al.*, 2001) to consider "the enigma of the oribatids" (as a comparison with "the paradox of the plankton"), in view of the seeming uniformity of forest floor habitats. Indeed, explaining this prodigious diversity of apparently similar species with similar requirements is a long-standing question (Hansen, 2000; Bolger, 2001).

Tropical and temperate forest canopies may support large numbers of oribatid mites, in such abundance that canopy-inhabiting mites have been dubbed "arboreal plankton" (Walter and Proctor, 1999). Tree canopies have long been known to support phytophagous spider mites and their mite predators (Phytoseiidae and relatives). Oribatid mites in forest canopies have been ignored until recently (Winchester, 1997; Behan-Pelletier and Walter, 2000). Forest canopies contain a variety of microhabitats, most of which have been exploited by oribatids. As yet there appears to be no standardization of collection methods such as have been developed for forest floors, and density estimates are hard to come by. Behan-Pelletier and Walter (2000) suggest that canopy oribatid species assemblages may be distinct from those of the adjacent forest floor, although some species occur in both types of habitat. Canopy oribatids exhibit some behavioral and morphological modifications suiting them to the peculiarities of that habitat. Individuals may tend to form small clusters during daytime hours, and sexual dimorphism is better developed than in forest soil species (Behan-Pelletier and Walter, 2000). Species generally have short, clubbed bothridial setae, perhaps an adaptation to the stronger air currents in the forest canopy. Gut analyses of canopy oribatids suggests that they utilize a similar broad spectrum of material as do forest floor oribatids; they appear to be primarily mycophagous but their guts include a mixture of dead vascular plant material, lichens, and fungal hyphae.

Population Growth

Populations of oribatid mites in forest floors show peaks of activity beginning in spring and continuing through the summer months, and again in mid-autumn (for those species producing two generations per year). Peaks of abundances of immatures show a gradual progression during the summer (Fig. 4.21) (Reeves, 1967). Population densities for many species are markedly higher during these months. Is there a sequence of oribatid species—a succession—in decomposing forest leaf litter, corresponding to a succession of fungal species? Crossley and Hoglund (1962) found such a general relationship. In a detailed study, Anderson (1975) concluded that the dominant oribatid species rapidly colonized litterbags containing beech and chestnut leaf disks. The mites fed upon the succession of fungi, but no succession of the mites themselves was demonstrated.

Oribatid Feeding Habits

Information about feeding habits and nutrition of oribatid mites remains elusive, despite numerous detailed studies (e.g., Luxton, 1972, 1975, 1979; Mueller *et al.*, 1990; Walter and Proctor, 1999). Oribatids in culture will eat a variety of substrates, but may feed differently under field conditions. In the simplest classification, oribatids are separated into feeders on fungi (microphytophages), on decomposing vegetable matter (macrophytophages), or on both (panphytophages) (Schuster, 1956). This classification has some utility, but more specific feeding habits can be identified. Some species, the phthiracarids or box-mites, are largely macrophytophages. Many oribatids appear to be indiscriminate fungal feeders, ingesting fungal hyphae or fruiting bodies of a variety of species (Mitchell and Parkinson, 1976; Siepel and de Ruiter-Dijkman, 1993). Others may be selective. Adults of *Liacarus cidarus*, when offered a variety of foods, preferred the mold *Cladosporium* (Arlian and Woolley, 1970) (Table 4.5). Anderson (1975) studied competition between two generalist feeders, *Hermanniella granulata* and *Nothrus sylvestris*. When isolated, these two species used similar foodstuffs (based on gut analyses). But when kept together in soil-litter microcosms, the two species changed their feeding and their utilization

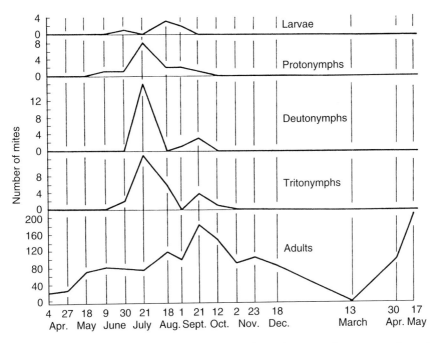

FIGURE 4.21.   Oribatid immatures and adult abundances by season (larvae, nymphs, and adults of *Oppia subpectinata* [Oudemans]) (from Reeves, 1967).

TABLE 4.5.   Feeding Behavior of Adult *Liacarus cidarus* When Offered Different Resources in Laboratory Cultures

Resource	Heavy feeding	Light feeding	Intermittent feeding	Failed to feed
Lichens	—	X	—	—
Yeast and sugar	—	—	X	—
Pine cone scales	—	X	—	—
Cladosporium	X	—	—	—
Aspergillus	—	—	—	X
"Mushrooms"	—	—	X	—
Potato dextrose agar and *Cladosporium*	X	—	—	—
Potato dextrose agar	—	X	—	—
Pine litter	—	—	X	—
Yeast	—	—	—	X
Trichoderma	—	—	—	X

After Arlian and Woolley, 1970.

of habitat space. *Hermanniella* moved into the litter (A_o) layers while the *Nothrus* population increased in the F (A_i) layer.

Several lines of evidence—gut analyses, feeding trials in cultures, chemical considerations—suggest that oribatid mites, as a group, are fungal feeders. Some exceptions exist, such as the phthiracarid group, which may tunnel into coniferous needles or twigs and ingest the spongy mesophyll. Others bore into the pedicels of oak leaves (Hansen, 1999). These species may contain a gut flora of bacteria that allow them to digest decomposing woody substrates. Possibly, their nutrition is derived from bacteria or fungi embedded in the woody tissue. Examination of gut contents for most other oribatids shows that they feed primarily on fungi. A survey of 25 species from the North American arctic found that more than 50% were panphytophagous (feeding on microbes and plant debris), but nearly all contained fungal hyphae or spores. Similarly, a study of Irish species found that 15 of 16 species were generalist feeders, having both fungi and plant remains in their guts (Behan and Hill, 1978; Behan-Pelletier and Hill, 1983). Occasional fragments of collembolans were discovered in some of the guts. Morphology of the chelicerae seems related to feeding type (Kaneko, 1988). Xylophagous (wood-feeding) species have large, robust chelicerae. Fragment feeders are generally smaller and have smaller chelae.

Further refinement of guild designations for oribatid mites has been made, based on their digestive capabilities as evidenced by their cellulase, chitinase, and trehalase activity in field populations (Siepel and de Ruiter-Dijkman, 1993). Using these three enzymes, it was possible to recognize five major feeding guilds: herbivorous grazers, fungivorous grazers, herbo-fungivorous grazers, fungivorous browsers, and opportunistic herbo-fungivores. In this classification, grazers are species which can digest both cell walls and cell contents; browsers can digest only cell contents. Siepel and de Ruiter-Dijkman (1993) also recognized two minor guilds: herbivorous browsers and omnivores. Further work with other groups of gut enzymes may provide additional information concerning resource utilization, and fungal feeding in particular, by guilds of oribatid mites. This is a most rewarding area for research in biodiversity, considering the multitude of species and niche dimensions for oribatids in forest floor habitats (Anderson, 1975; Hansen, 2000).

The oribatids themselves are prey for some insects such as scydmaenid beetles (Molleman and Walter, 2000), pselaphid beetles (Park, 1947), or ants (Matsuko, 1994). Some vertebrates such as salamanders have been found to contain oribatids in their guts. In general, the hard exoskeleton of oribatid mites protects them from smaller predators such as prostigmatic and mesostigmatic mites (see later section).

Oribatid Impacts on Soil Ecosystems

Oribatid mites can affect organic litter decomposition and nutrient dynamics in forest floors, but they appear to do so indirectly by grazing on microbial populations or fragmenting plant detritus, and thus influencing the decomposition process (Petersen and Luxton, 1982; Seastedt, 1984b). Calcium dynamics may be an exception, because oribatid mites can store and process a significant portion of the calcium input in forest litterfall (Gist and Crossley, 1975). Several mechanisms of microarthropod–microfloral interaction have been proposed (Lussenhop, 1992), including consequences of selected grazing, dispersal of fungal spores or inocula, and stimulation of fungal growth or bacterial activity. The importance of oribatid mites, as with other fauna, cannot be interpreted outside the context of the entire suite of soil biota. The milieu of decomposer organisms, vegetable matter in various stages of decomposition, and localized "hot spots" of activity requires careful analysis and is a focus of important current research (Edwards, 2000; Hansen, 2000; Moore and de Ruiter, 2000; Bolger *et al.*, 2000).

Prostigmatic Mites

The Prostigmata contains a large array of soil species (Fig. 4.22). Many of these species are predators, but some families contain fungal-feeding mites and these may become numerous. Like the oribatids, prostigmatic mites are an ancient group with fossil representatives from the Devonian era. Keys to families of Prostigmata (also known as Actinedida) are provided by Krantz (1978) and Kethley (1990).

Some families of prostigmatic mites include species which are predators, microbial feeders, plant feeders, or parasites (Kethley, 1990). The fungal feeding species (such as members of the family Eupodidae) are opportunistic, able to reproduce rapidly following a disturbance or a sudden shift in resources. Small species of Prostigmata are the common mites of Antarctic soil surfaces, of drained lake beds with algal blooms, of plowed and fertilized agricultural fields, of tidal marshlands or burned prairie soils, and so forth (Lussenhop, 1976; Luxton, 1967, 1981; Perdue and Crossley, 1989; Seastedt, 1984; Strandtmann, 1967; Tevis and Newell, 1962). Species in the families Eupodidae, Tarsonemidae, Nanorchestidae, and some of their relatives feed on algae or fungi; their populations may grow rapidly to large sizes. In these situations, the Prostigmata may become more numerous than the oribatid mites. In general they are more numerous in temperate than in tropical or subtropical habitats (Luxton, 1981b).

Nematodes are an important part of the diet of the smaller Prostigmata. Whitford and Santos (1980) found that mites in the family Tydei-

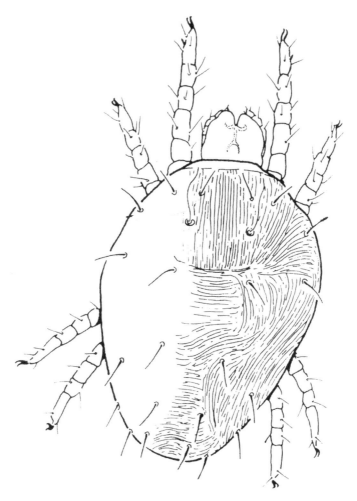

FIGURE 4.22.   Prostigmatid mite: family Tydeidae (*Lorryia* sp.) (after Krantz, 1978).

dae were important regulators of nematode populations in desert soils of the southwestern United States. Some of the smaller predatory species may utilize fungi on occasion. Walter (1988) observed predation on nematodes by prostigmatic mites in the families Bimichaelidae and Alicorhagiidae; members of the latter family also ingested fungal hyphae but with lower reproductive success. Many types of fungal-feeding Prostigmata have small, stylet chelicerae and may simply pierce fungal hyphae (members of the Nanorchestidae and Nemataly-chidae families) (Walter, 1988). That author also examined gut contents in slide mounts of more than 500 small prostigmatic mites collected in various localities in North America and elsewhere. Many of these specimens contained a fungal food bolus (Table 4.6). Both fungal feeding and predation on nematodes appear to be widespread among the tiny Prostigmata. Kethley (1990) lists the known feeding habits of 30 families or superfamilies of soil-inhabiting Prostigmata, and summarizes their biology, ecology, and type of habitat.

In general, the larger predaceous Prostigmata feed upon other arthropods or their eggs; the smaller species are nematophagous. Some Prostigmata have well-defined patterns of predation. The "grasshopper mite," *Allothrombium trigonum*, feeds exclusively upon grasshopper eggs; the larval stages of the mite are parasitic on grasshoppers. The large red "velvet mites" (*Dolicothrombium* species), which erupt in numbers following desert rains, are predaceous on termites. The pestiferous "chiggers" are the larval stages of mites in the family Trombiculidae; the adults are predaceous on collembolans and their eggs. Collembolans may be an important diet item for the larger Prostigmata such as members of the families Bdellidae, Cunaxidae, and the trombidioid families.

TABLE 4.6. Percentages of Field-Collected Prostigmatid Mites with Boluses or a Central Mass of Particulate Fungal Hyphae in Their Guts

Taxon	Individuals Examined (no.)	With fungal bolus (no.)	Other inclusions
Terpnacaridae	38	26	—
Alicorhagiidae			
Alicorhagia	98	23	nematode stylets (3) tardigrade claws (1)
Stigmalycus	32	10	nematode (1)
Grandjeanicidae	20	17	diatoms
Oehserchestidae	38	10	—
Lordalychidae	49	0	fungal mass
Bimichaelidae			
Alycus	52	0	—
Bimichaelia	36	0	—
Pachygnathus	1	0	—
Petralycus	13	0	—
Nanorchestidae			
Nanorchestes	24	0	spore like bodies (1)
Speleorchestes	25	0	caecal masses (11)
Nematalycidae			
Cunliffia	24	0	crystals? (21)
Gordialycus	31	0	—
Psammolycus	13	0	—
Sphaeroluchidae	18	0	—

After Walter, 1988.

Eggs of collembola were used successfully to culture pest chiggers (*Eutrombicula alfreddugesi* and relatives) (Crossley, 1960). It is difficult to assess the importance of the soil-dwelling Prostigmata, even in those cases when their numbers escalate. Those species predaceous on nematodes may have an impact (Whitford and Santos, 1980), but it is seldom quantified. The biomass of prostigmatic mites is generally small, only a fraction of the total acarine mass (Kethley, 1990), and their total respiration is comparatively small (Luxton, 1981b). When populations of fungal or algal feeders reach high population sizes, we suspect that they may have some impact on their food base, but the magnitude of the effect is unknown.

Mesostigmatic Mites

The Mesostigmata (Fig. 4.23) contains fewer soil inhabiting species than do Oribatida or Prostigmata. Krantz and Ainscough (1990) include keys to families and genera of the soil inhabiting species of Mesostigmata. Many of the Mesostigmata are parasitic on vertebrates or invertebrates (Krantz 1978), and some of these may be captured in soil samples. The true soil species are almost all predators. A few species (in the Uropodidae, for example) are polyphagous, feeding on fungi, nematodes, and juvenile insects (Gerson *et al.*, 2003), and may become somewhat numerous in agroecosystems (Mueller *et al.*, 1990). Mesostigmatic mites are not as numerous as oribatids or prostigmatid mites, but are universally present in soils and may be important predators. As with the Prostigmata, the larger species tend to feed on small arthropods or their eggs; the smaller species are mainly nematophagous. Some of the species in the family Laelapidae are voracious predators on red spider mites and false spider mites feeding on aboveground vegetation. In the soil itself, members of the genus *Hypoaspis* (Fig. 4.24) are important predators of small insect larvae.

Walter and Ikonen (1989) found that mesostigmatic mites were the most important predators of nematodes in grasslands of the western United States. In contrast, they found that the larger Prostigmata were predators on arthropods or their eggs. Of 63 species of mesostigmatic mites tested, only 6 did not readily feed on nematode prey. The mites each consumed 3 to 8 nematodes per day. Western grassland soils have little surface plant litter.

Forest floors, with abundant surface litter, contain a larger spectrum of mesostigmatic mites. The forest litter inhabitants (families Veigaiidae and Macrochelidae, for example) are bigger species and are predaceous on arthropods or their eggs. Mesostigmatic mites of the mineral soil layers are the smaller, colorless Rhodacaridae and relatives, and are nematophagous. The size of soil Mesostigmata diminishes with increas-

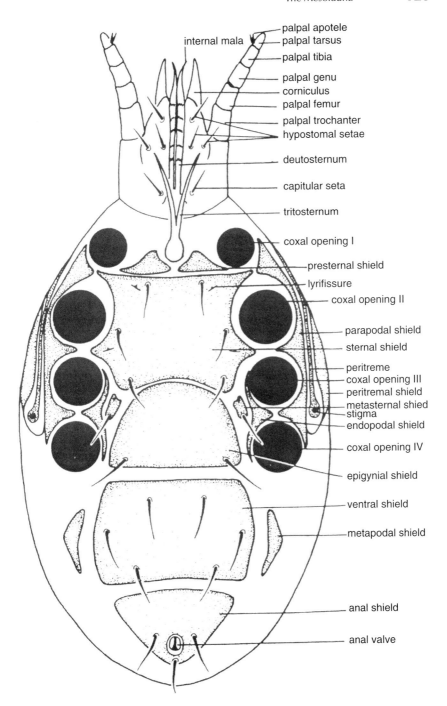

FIGURE 4.23. Macrochelid mite (from Krantz, 1978).

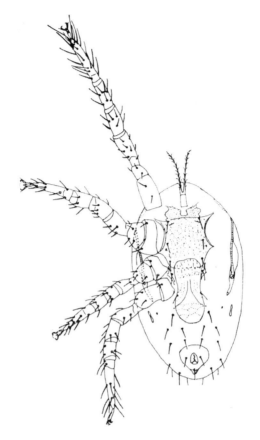

FIGURE 4.24.   *Hypoaspis marksi* (from Strandtmann and Crossley, 1962).

ing depth in the soil (Coineau, 1974). In forest floor habitats, members of the genus *Veigaia* (Fig. 4.25) are inhabitants of litter and humus layers; smaller species (*Dendrolaelaps*) occurs in the humus–soil interface, and the minute *Rhodacarellus* (Fig. 4.26) is found in mineral soil (Krantz and Ainscough, 1990). Many species of Mesostigmata have a close relationship with various insect species, a relationship that often includes the soil environment (Hunter and Rosario, 1988). Several genera in the cohort Gamasina are also considered useful as bioindicators of habitat and soil conditions (Karg, 1982).

Astigmatic Mites

The Astigmata (Fig. 4.27) are the least common of the soil mites, although they may become abundant in some habitats (Luxton, 1981a). The free-living Astigmata favor moist environments high in organic

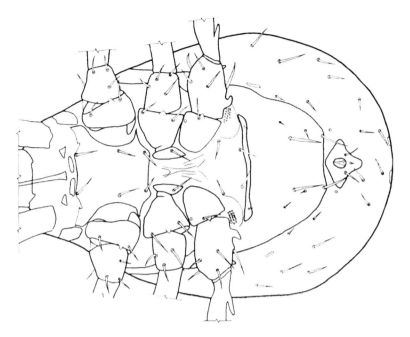

FIGURE 4.25.   *Veigaia uncata*, ventral view of female (Krantz, 1978).

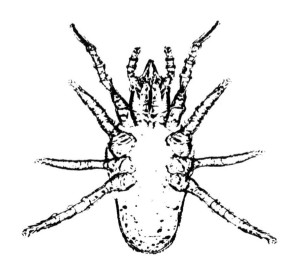

FIGURE 4.26.   *Rhodacarellus* sp.

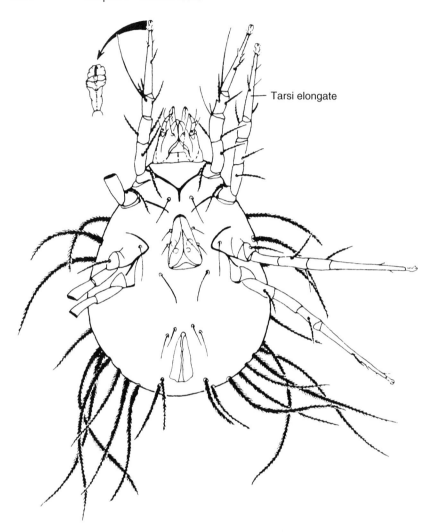

Tarsi elongate

FIGURE 4.27. Astigmatid mite (*Glycyphagus domesticus* [DeGeer]) (Oregon, United States), venter of female with detail of pretarsus I (from Krantz, 1978).

matter. The Astigmata contain some important pests of stored grain. They become abundant in some agroecosystems following harvest, or after application of rich manures. In agroecosystems of the Piedmont region of Georgia in the United States, Perdue (1987) found a marked increase in Astigmata following autumnal harvest and tillage (Fig. 4.28). Incorporation of residues and winter rains produced moist, organic residues suitable for the mites. The springtime plowing, under drier conditions, did not lead to increases of astigmatic mites. Tomlin (1977)

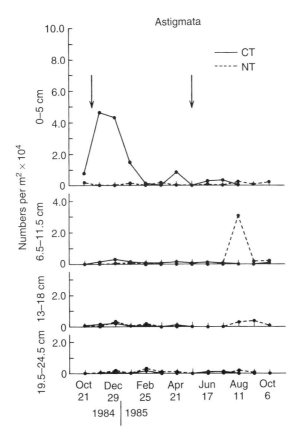

FIGURE 4.28. Vertical distribution of astigmatic mites in conventional and no-tillage agroecosystems. Arrows indicate autumn and spring dates for mowing, tillage, and planting. Numbers increased under conventional tillage following autumn tillage, but not following spring tillage (from Perdue and Crossley, 1990).

described a buildup of astigmatic mites following pipeline construction in Ontario, Canada. The mites were associated with accumulations of residue under moist conditions. Philips (1990) provided keys to families and genera of soil inhabiting Astigmata.

Most of the soil Astigmata are microbial feeders. Those with chelate chelicerae are able to chew vegetable material, fungi, and algae (Philips, 1990). Members of the Anoetidae have reduced chelae; their palpi are highly modified strainers for filter feeding on microbial colonies (Philips, 1990).

Occasionally, species in the family Acaridae become pests in microbiology laboratories, where they reproduce rapidly on agar plates. They are readily cultured on Baker's yeast; a few grams of yeast left untended in a collembolan culture will soon become infested with acarids. We have found astigmatic mites as contaminants in some Tullgren extractions. When fresh agricultural products are stored in the laboratory, large populations of Astigmata may develop, and may wander into Tullgren funnels during extractions. Similar population excursions of prostigmatid mites (family Cheyletidae) have also yielded extensive contamination of Tullgren samples. It is good practice to operate some empty, "control" funnels to check for the possibility of wandering microarthropods in the funnel room.

Other Microarthropods

In addition to mites and collembolans, Tullgren extractions contain a diverse group of other small arthropods. Although not numerous in comparison to mites and collembolans, Tullgren extractions may have abundances of several thousand per square meter. Collectively, the "other" microarthropods have relatively small biomasses and probably have no major impact on soil ecology. Such a judgment may be premature, in view of the general lack of information about their ecology.

Small spiders and centipedes, occasional small millipedes, insect larvae, and adult insects occur in soil cores extracted on Tullgren funnels. Most of these are better sampled as macroarthropods using hand sorting or trapping methods. Some insects (small larvae of carabid and elaterid beetles, thrips, pselaphid beetles, tiny wasps) are sometimes numerous enough to be effectively sampled from soil cores. Social insects, such as ants and termites, require special sampling considerations.

Protura

Proturans (Fig. 4.29) are small, wingless, primitive insects readily recognized by their lack of antennae and eyes (Bernard, 1985). Seldom as numerous as the other microarthropods, proturans occur in a variety of soils worldwide, often associated with plant roots and litter. Keys to families and genera of the Protura were published by Copeland and Imadaté (1990) and by Nosek (1973).

Numbers reported in the literature range between 1000 and 7000 per square meter at best (Petersen and Luxton, 1982). They penetrate the soil to surprising depths (25 cm), considering that they do not appear to be adapted for burrowing (Price, 1975; Copeland and Imadaté, 1990). Their feeding habits remain unknown. Observations that they feed on mycorrhizae (Sturm, 1959) have not been verified, but their occurrence in the rhizosphere of trees with mycorrhizae would support Sturm's observations.

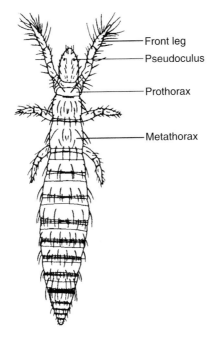

Front leg
Pseudoculus
Prothorax
Metathorax

FIGURE 4.29. Proturan.

Diplura

Diplurans are small, elongate, delicate, primitive insects. They have long antennae and two abdominal cerci. Most diplurans are euedaphic, but some are nocturnal cryptozoans, hiding under stones or under bark during the day. They occur in tropical and temperate soils in low densities. In Georgia Piedmont agroecosystems, the authors have sampled dipluran populations with Tullgren extractions, finding populations of approximately 50 per square meter.

Two common families of diplurans are found in soils, readily separable by their abdominal cerci. Campodeidae (Fig. 4.30) have filiform cerci; Japygidae (Fig. 4.31) have cerci modified as pinchers. Keys to families and subfamilies were provided by Ferguson (1990a). The japygids are predators on small arthropods (such as collembolans), nematodes, and enchytraeids. The cerci are used in capturing prey. Campodeids are predators on mites and other small arthropods, but also ingest fungal mycelia and detritus (Ferguson, 1990a). These animals are adapted for life in the soil by their elongate narrow form, sensory antennae, and sensory cerci.

LIVERPOOL
JOHN MOORES UNIVERSITY
AVRIL ROBARTS LRC
TEL. 0151 231 4022

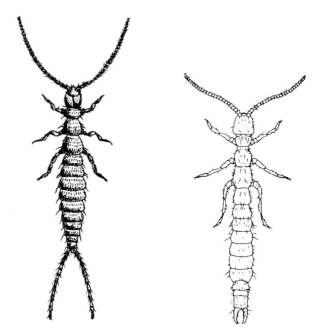

FIGURE 4.30.   Campodeidae.   **FIGURE 4.31.**   Japygidae.

Microcoryphia

The jumping bristletails (family Machilidae) were formerly included in the order Thysanura with silverfish and relatives, but now are placed in the separate order Microcoryphia. They are closely related to one another (Ferguson, 1990b). Machilids, when disturbed, can leap a distance of 10 cm (Denis, 1949). They feed on a wide variety of primitive plant materials such as lichens and algae. We have observed Machilids (presumably *Machilis* sp.) on rocky cliff faces in Georgia. They emerge at dusk from cracks in the rock surface on Stone Mountain and other granite domes on the Georgia Piedmont, and may reach densities of 50 per square meter. They return to their crevices at dawn. They fall prey to spiders (*Pardosa lapidocina*) on these outcrops (Nabholz et al., 1977).

Pseudoscorpionida

Pseudoscorpions (Fig. 4.32) are minute copies of their more familiar relatives, the scorpions, except that they lack tails and stingers. They occur throughout the terrestrial world except for Arctic and Antarctic regions. Pseudoscorpions are small cryptozoans, hiding under rocks and bark of trees, but they are occasionally extracted from leaf litter samples

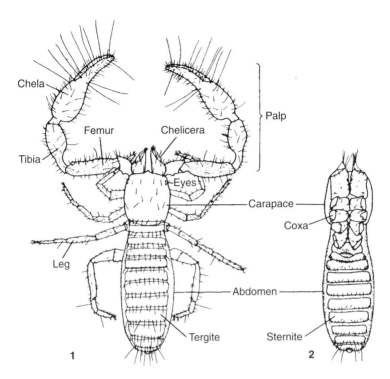

FIGURE 4.32.   Pseudoscorpion.

in Tullgren funnels. They are predaceous on small arthropods, nematodes, and enchytraeids. Keys to families and genera were provided by Muchmore (1990).

False scorpions are found in a number of habitats but not in large numbers. They can move readily through small spaces and crevices. Wallwork (1976) notes that two important habitat features for pseudoscorpions are high humidity and the availability of small crevices. Forest leaf litter provides both of these features, as do bark of decomposing logs, caves, nests of small mammals, and similar habitats.

Hand collecting is successful but Tullgren extraction is the usual means of sampling pseudoscorpions (Hoff, 1949; Muchmore, 1990).

Symphyla

Symphylids (Fig. 4.33) are small, white, eyeless, elongate, many-legged invertebrates that resemble tiny centipedes. They differ from centipedes in several characteristics, but superficially symphylids have but 12 body segments and 12 pairs of legs, whereas centipedes have at least 15 pairs of legs, the first pair modified as fangs. Edwards (1990)

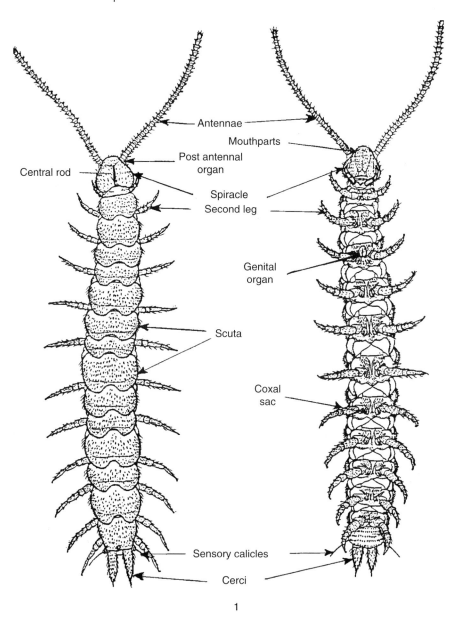

Antennae

Mouthparts

Post antennal organ

Central rod

Spiracle

Second leg

Genital organ

Scuta

Coxal sac

Sensory calicles

Cerci

1

FIGURE 4.33.   Symphylid (Dindal, 1990).

provides a partial key to genera, and notes that the North American fauna of Symphylids is badly in need of further revision. Symphylids are part of the true eudaphic fauna, occurring in forest, grassland, and cultivated soils. They are omnivorous and can feed on the soft tissues of plants or animals (Edwards, 1959). Some species reach pest status in greenhouse soils where they feed on roots of seedlings (Edwards, 1990). Symphylids have silk glands near the end of the abdomen. The function of silk strands for these soil dwellers is obscure.

Pauropoda

Pauropods (Fig. 4.34) are tiny (1.0–1.5 mm long) terrestrial myriapods with 8–11 pairs of legs and a distinctive morphological feature—branched antennae (Scheller, 1988). They are white to color-less and blind; these characteristics make them members of the true eudaphic fauna. Pauropods occur in soils worldwide but are not well known. They are commonly collected in Tullgren extractions but are sel-dom numerous, usually fewer than 100 per square meter. In forests, they inhabit the lower litter layers, F-layers, and mineral soil; they also occur in agricultural soils. It is generally assumed that pauropods are fungus feeders, but they may also be predaceous. Little information has been accumulated about their biology or ecology (Scheller, 1990). The taxonomy of the group is in need of revision. Although considered to be poor in species, probably less than 20% of extant species have been described. For example, Scheller (2002), working in the Great Smoky Mountains National Park, has found more than 30 species of Pauropods previously undescribed in that region, with seven or eight of them new to science.

Enchytraeidae

In addition to earthworms (discussed in the next section), another important family of terrestrial oligochaeta is the Enchytraeidae. This group of small, unpigmented worms (Fig. 4.35), also known as "pot-worms," is classified within the "microdrile" oligochaetes and consists of some 600 species in 28 genera. Species from 19 of these genera are found in soil, the remainder occurring primarily in marine and freshwater

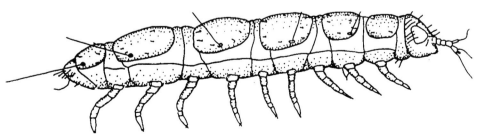

FIGURE 4.34. Pauropoda (Dindal, 1990).

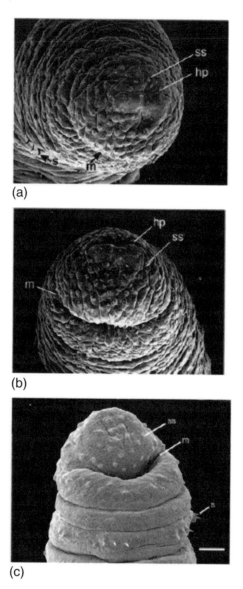

(a)

(b)

(c)

FIGURE 4.35. Anterior end of *Mesenchytraeus solifugus* and *Enchytraeus albidus*. (a) End view of *M. solifugus* displaying head pore (hp), and sensory structures (ss) at tip of prostomium. The mouth (m) and one cluster of setae (s) are visible in the lower region of the image. (b) Ventral view of *M. solifugus* displaying mouth, head pore, and sensory structures. (c) Ventral view of *E. albidus* displaying mouth, sensory structures, and four setal clusters. Scale bar = 50 μm (from Shain *et al.*, 2000).

habitats (Table 4.7) (Brinkhurst and Cook, 1980; Dash, 1990; van Vliet, 2000). The Enchytraeidae are thought to have arisen in cool temperate climates where they are commonly found in moist forest soils rich in organic matter; interestingly, Tynen (1972) described the occurrence of "ice worms," which were enchytraeids that emerged onto snow- and ice-covered ground in British Columbia. Various species of enchytraeids are now distributed globally from subarctic to tropical regions.

Taxonomic organization of the European Enchytraeidae was defini- tively treated by Nielsen and Christensen (1959, 1961, and 1963); much less work has been done in other parts of the world. More recently, keys to the common genera were presented by Dash (1990). Identification of enchytraeid species is difficult, but genera may be identified by observ- ing internal structures through the transparent body wall of specimens mounted on slides (Fig. 4.36).

The Enchytraeidae are typically 10–20 mm in length and they are anatomically similar to the earthworms, except for the miniaturization and rearrangement of features overall. They possess setae (with the exception of one genus), and a clitellum in segments XII and XIII, which contains both male and female pores. Sexual reproduction in enchy- traeids is hermaphroditic and functions similarly to that in earth- worms. Cocoons may contain one or more eggs and maturation of newly

TABLE 4.7.   Enchytraeid Genera and Their Occurrence in Various Environments

Genera occurring in soil	Other genera	Environment
Achaeta	Aspidodrilus	epizoic on earthworms
Bryodrilus	Barbidrilus	fresh water
Buchholzia	Enchylea	only found in Enchytraeid culture
Cernosvitoviella	Enchytraeina	marine
Cognettia	Grania	marine
Enchytraeus	Pelmatodrilus	epizoic on earthworms
Enchytronia	Propappus	fresh water
Fridericia	Randidrilus	marine
Guaranidrilus	Stephensoniella	marine
Hemienchytraeus		
Hemifridericia		
Henlea		
Isosetosa		
Lumbricillus		
Marionina		
Mesenchytraeus		
Oconnorella		
Stercutus		
Tupidrilus		

From van Vliet, 2000.

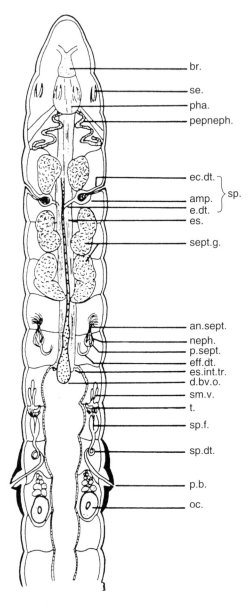

FIGURE 4.36. Morphological characters of an enchytraeid worm. *amp.*, ampulla: *an. sept.*, ante-septal; *br.*, brain; *d.bv.o.*, dorsal blood vessel origin; *ec.g.*, ectal gland; *eff.dt.*, efferent duct; *e.op.*, ental opening; *es.*, esophagus; *es.int.tr.*, esophageal intestinal transition; *m.pha.*, muscular pharynx; *neph.*, nephridia; *oc.*, oocyte; *pha.*, pharynx; *p.b.*, penial bulb; *pepneph.*, peptonephridia; *p.sept.*, postseptal; *se.*, setae; *sept.g.*, septal gland; *sm.v.*, seminal vesicle; *sp.*, spermatheca; *sp.dt.*, sperm duct; *sp.f.*, sperm funnel; *t.*, testes (*Soil Biology Guide,* Dindal, D. L., ©1990, John Wiley & Sons, New York. Reprinted by permission of John Wiley & Sons, Inc.).

hatched individuals ranges from 65 to 120 days, depending on species and environmental temperature (van Vliet, 2000). Enchytraeids also display asexual strategies of parthenogenesis and fragmentation, which enhance their probability of colonization of new habitats (Dósza-Farkas, 1996).

Enchytraeids ingest both mineral and organic particles in the soil, although typically of smaller size ranges than those ingested by earthworms. Numerous investigators have noted that finely divided plant materials, often enriched with fungal hyphae and bacteria, are a principal portion of the diet of enchytraeids; microbial tissues are probably the fraction most readily assimilated, because enchytraeids lack the gut enzymes to digest more recalcitrant soil organic matter (Brockmeyer, 1990; van Vliet 2000). Didden (1990, 1993) suggested that enchytraeids feed predominantly upon fungi, at least in arable soils, and classified a community as 80% microbivorous and 20% saprovorous. As with several other members of the soil mesofauna, the mixed microbiota that occur on decaying organic matter, either litter or roots, are probably an important part of the diet of these creatures. The remaining portions of the soil organic matter, after the processes of ingestion, digestion, and assimilation, enter the slow-turnover pool of soil organic matter. Zachariae (1963, 1964) studied the nature of enchytraeid feces and found that they had no identifiable cellulose residues. In addition, Zachariae suggested that so-called "collembolan soil," said to be dominated by collembolan feces (particularly low-pH mor soils) were really formed by Enchytraeidae. Mycorrhizal hyphae have been found in the fecal pellets of enchytraeids from pine litter (Fig. 4.37) (Ponge, 1991). There is also the strong likelihood that enchytraeids consume and further process larger fecal pellets and castings of soil fauna such as collembolans and earthworms (Zachariae, 1964; Rusek, 1985).

Enchytraeid densities range from less than 1,000 to more than 140,000 individuals per square meter in intensively cultivated agricultural soil in Japan and a peat moor in the United Kingdom, respectively (Table 4.8). In a subtropical climate, Coleman *et al.* (1994a) reported enchytraeid densities of 4,000 to 14,000 per square meter in agricultural plots in the Piedmont of Georgia, whereas van Vliet *et al.* (1995) found higher densities (20,000 to 30,000 individuals per square meter) in surface layers of deciduous forest soils in the southern Appalachian Mountains of North Carolina. Although enchytraeid densities are typically highest in acid soils with high organic content, Didden (1995) found no statistical relationship between average density and annual precipitation, annual temperature, or soil pH over a broad range of data; local variability may be at least as great as variation on a wider scale. Enchytraeid densities show both spatial and seasonal variations. Vertical distributions of enchytraeids in soil are related to organic matter hori-

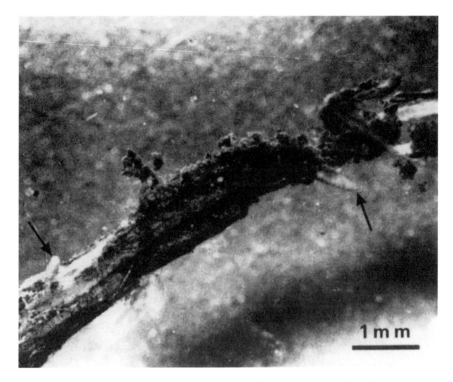

FIGURE 4.37. Two enchytraeid worms, indicated by arrows, tunneling through a pine needle (fecal pellets have been deposited on the outside) in the F_1 layer (modified from Ponge, 1991).

zonation; up to 90% of populations may occur in the upper layers in forest and no-tillage agricultural soils, but densities may be higher in the Ah horizon of grasslands (Davidson *et al.*, 2002). Seasonal trends in enchytraeid population densities appear to be associated with moisture and temperature regimes (van Vliet, 2000).

Enchytraeids have been shown to have significant effects on organic matter dynamics in soil and on soil physical structure. Litter decomposition and nutrient mineralization are influenced primarily by interactions with soil microbial communities. Enchytraeid feeding on fungi and bacteria can increase microbial metabolic activity and turnover, accelerate release of nutrients from microbial biomass, and change species composition of the microbial community through selective grazing. However, Wolters (1988) found that enchytraeids decreased mineralization rates by reducing microbial populations and possibly by occluding organic substrates in their feces. Thus the influence of enchytraeids on soil organic matter dynamics is the net result of both enhancement and

TABLE 4.8. Enchytraeid Abundances (annual average number/m^2) in Different Ecosystems and Locations

Ecosystem	Location	Density (no. m^{-2})
Forest		
Douglas fir	Wales	134,300
Pinus radiata 50 stems/ha	New Zealand	64,002
Pacific silver fir, mature stand	WA, United States	49,400
Pinus radiata 200 stems/ha	New Zealand	39,270
Spruce	Norway	34,700
Rhododendron-Oak 1160 m altitude	NC, United States	32,630
Rhododendron-Oak 750 m altitude	NC, United States	26,811
Pine	Norway	22,900
Pinus radiata 100 stems/ha	New Zealand	21,391
Scots pine forest	Sweden	16,200
Deciduous forest	United Kingdom	14,590
Spruce	South Finland	13,400
Pacific silver fir, young stand	WA, United States	11,400
Pinus radiata 0 stems/ha	New Zealand	10,647
Spruce	South Finland	8200
Spruce	North Finland	4000
Arable land		
Sugarbeet	The Netherlands	30,000
Winterwheat	The Netherlands	19,437
NT corn-clover	GA, United States	16,830
CT corn-clover	GA, United States	15,270
Potato field	Poland	13,200
Barley, no N	Sweden	10,000
Rye field	Poland	9800
Barley, 120 kg N	Sweden	8100
Rice/wheat/barley (organic)	Japan	4940
Rice/wheat/barley (conven.)	Japan	525
Moor		
Juncus peat	United Kingdom	145,000
Nardus	United Kingdom	71,000
Blanket bog	United Kingdom	40,000
Fen	Canada	5600
Grassland		
Grassland soil	Sweden	24,000
Lucerne ley	Sweden	9900
Grassland 10 sheep/ha	Australia (NSW)	6000
Grassland 30 sheep/ha	Australia (NSW)	2300

Modified from van Vliet, 2000.

inhibition of microbial activity, depending on soil texture and population densities of the animals (Wolters, 1988; van Vliet, 2000).

Enchytraeids affect soil structure by producing fecal pellets which, depending on the animal size distribution, may enhance aggregate stability in the 600–1000 μm fraction (Didden, 1990). In forest floors, these pellets are composed mainly of fine humus particles, but in mineral soils, organic matter and mineral particles may be mixed into fecal pellets with a loamy structure (Kasprzak, 1982). Davidson *et al.* (2002) estimated that enchytraeid fecal pellets constituted nearly 30% of the volume of the Ah horizon in a Scottish grassland soil (Fig. 4.38). Encapsulation or occlusion of organic matter into these structures may reduce decomposition rates. Burrowing activities of enchytraeids have not been well studied, but there is evidence that soil porosity and pore continuity can increase in proportion to enchytraeid body size (50–200 μm diameter) (Rusek, 1985; Didden, 1990). Van Vliet *et al.* (1993, 1997) observed that enchytraeids in small microcosms increased soil porosity and hydraulic conductivity, depending on the distribution of organic matter and enchytraeid population densities.

Enchytraeids are typically sampled in the field using cylindrical soil cores of 5–7.5 cm diameter; large numbers of replicates may be needed for a sufficient sampling due to the clustered distribution of enchytraeid populations (van Vliet, 2000). Extractions are often done with a wet-funnel technique (O'Connor, 1955), similar to the Baerman funnel extraction used for nematodes. In this case, soil cores are submerged in water on the funnel and exposed for several hours to a heat and light source from above; enchytraeids move downward and are collected in the water below (see van Vliet, 2000, for a comparison of modifications of this technique).

(a)   (b)

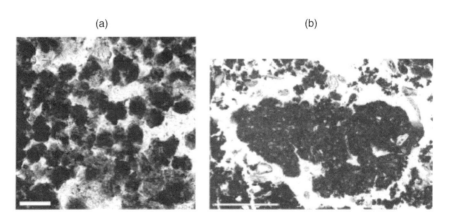

FIGURE 4.38.   Thin section micrographs of fecal pellets in a grassland soil. (a) Derived from enchytraeids (scale bar = 0.5 mm); (b) Derived from earthworms (scale bar = 1.0 mm) (from Davidson *et al.*, 2002).

THE MACROFAUNA

Macroarthropods

Larger insects, spiders, myriapods, and others are considered together under the appellation "macroarthropods." Typical body lengths range from about 10 mm to as much as 15 cm (Scolopendromorph centipedes) (Shelley, 2002). The group includes an artificial mix of various arthropod classes, orders, and families. Like the microarthropods, the macroarthropods are defined more by the methods used to sample them rather than by measurements of body size.

Large soil cores (10 cm in diameter or greater) may be appropriate for euedaphic species; arthropods can be recovered from them using flotation techniques (Edwards, 1991). Mechanical or hand sorting of soils and litter is more time-consuming but yields better estimates of population size. In rare instances, capture–mark–recapture methods have been used to estimate population sizes of selected macroarthropod species, but the assumptions for this procedure are violated more often than not (Southwood, 1978).

Pitfall traps have been widely used to sample litter- and surface-dwelling macroarthropods (Banerjee, 1970; Greenslade, 1964; Michail, 1993) (see Chapter 9). This method catches arthropods that blunder into cups filled with preservative. Absolute population estimates are difficult to obtain with pitfall traps (Gist and Crossley, 1973) but the method yields comparative estimates when used with caution.

Many of the macroarthropods are members of the group termed "cryptozoa," a group consisting of animals that dwell beneath stones, logs, under bark, or in cracks and crevices (Cole, 1946). Cryptozoans typically emerge at night to forage; some are attracted to artificial lights. The cryptozoa fauna is poorly defined and is not an ecological community in the usual sense of the term. The concept remains useful, however, for identifying a group of invertebrate species with similar patterns of habitat utilization.

Importance of the Macroarthropods

However they are sampled, the macroarthropods are a significant component of soil ecosystems and their food webs. Macroarthropods differ from their smaller relatives in that they may have direct effects on soil structure. Termites and ants in particular are important movers of soil, depositing parts of lower strata on top of the litter layer (Fig. 4.39). Emerging nymphal stages of cicadas may be numerous enough to disturb soil structure. Larval stages of soil-dwelling scarabaeid beetles sometimes churn the soil in grasslands. These and other macroarthro-

LIVERPOOL JOHN MOORES UNIVERSITY
LEARNING SERVICES

FIGURE 4.39. Fire ants *(Solenopsis invicta)* as soil movers (D. A. Crossley, Jr. photo).

pods are part of the complex that has been termed "ecological engineers."

Some macroarthropods participate in both above- and belowground parts of terrestrial ecosystems. Many macroarthropods are transient or temporary soil residents (see Fig. 4.1), and thus form a connection between food chains in the "green world" of foliage and the "brown world" of the soil. Caterpillars descending to the soil to pupate or migrating armyworm caterpillars are prey to ground-dwelling spiders and beetles. A ground beetle species, *Calosoma sycophanta* ("the searcher"), was imported from Europe for biological control of the gypsy moth (Kulman, 1974).

Macroarthropods may have a major influence on the microarthropod portion of belowground food webs. Collembola, among other microarthropods, are important food items for spiders, especially immature stadia, thus providing a macro-to-micro connection. Other macroarthropods such as cicadas emerging from soil may serve as prey for some vertebrate animals (Lloyd and Dybas, 1966), thus providing a link to the larger megafauna.

Among the macroarthropods there are many litter feeding species, such as the millipedes, that are important consumers of leaf, grass, and wood litter. These arthropods have major influences on the decomposition process, thereby impacting rates of nutrient cycling in soil systems.

And, the reduction of vertebrate carrion is largely accomplished through the actions of soil-dwelling insects (Payne, 1964). The vast array of macroarthropod species in soil systems constitutes a major reservoir of biodiversity. As with the mites and collembolans, the functional significance of this diversity is not evident. Intuitively, it would seem that the large number of species participating in belowground food webs should increase their stability and enhance the recovery following disturbance, but the concept has remained elusive.

Isopoda

Terrestrial isopods (Fig. 4.40) are crustaceans, but are typical crypto-zoa, occurring under rocks and in similar habitats. Although they are distributed in a variety of habitats, including deserts, they are susceptible to desiccation. Adaptations to resist desiccation include nocturnal habits, the ability to roll up into a ball, low basal respiration rates, and restriction of respiratory surfaces to specialized areas. Considered to be general saprovores, isopods can feed upon roots or foliage of seedlings. Isopods possess heavy, sclerotized mandibles and are capable of considerable fragmentation of decaying vegetable matter. They display some selectivity in preferences for different leaf species. Digestive processes in the terrestrial isopods encompass a wide extent of biochemical complexity, with detoxification of ingested phenolics in the foregut, digestion by endogenous and bacterial enzymes in the anterior hindgut, absorption of nutrients, and microbial proliferation in the posterior hindgut (Fig. 4.41) (Zimmer, 2002). In the laboratory, terrestrial isopods feed upon fecal pellets dropped by themselves or by any other isopod (Zimmer, 2002). There is some doubt about how common this trait is

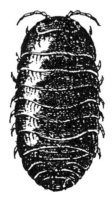

FIGURE 4.40.   A terrestrial isopod, *Armadillidium vulgare*. Left, extended; right, rolled into a ball (from Metcalf and Flint, 1939).

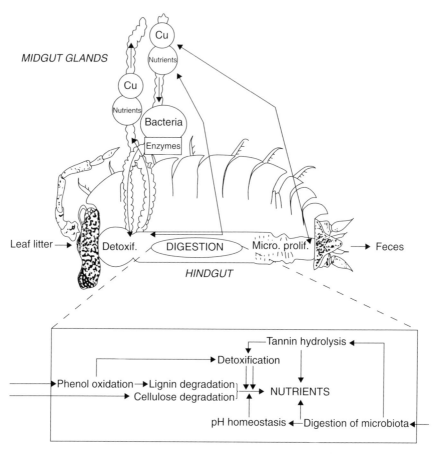

FIGURE 4.41. Digestive processes in the hindgut of *Porcellio scaber* (Porcellionidae), including the ingestion of leaf litter, detoxification of ingestion phenolics (Detoxif.) in the foregut, digestion in the anterior hindgut through the activity of endogenous and bacterial enzymes, adsorption of nutrients and copper, microbial proliferation (Microb. Prolif.) in the posterior hindgut, and egestion of feces (from Zimmer, 2002).

expressed in the field due to the difficulty of finding feces beneath the litter layer. However, microbially inoculated feces represent a microbial "hot spot" generating microbial metabolites that might allow the isopod to "home in" on a desirable food source (Zimmer *et al.*, 1996).

The isopod *Porcellio* has an excretory system that exposes its products to the external environment. Urine from the nephridia is channeled into a water-conducting system on the ventral surface. Ammonia is lost to the atmosphere and oxygen is absorbed during this flow. The ammonia-free water is then reabsorbed in the rectum (Eisenbeis and Wichard, 1987).

Diplopoda

Millipedes (Diplopoda) (Fig. 4.42) are a group of widely distributed saprophages. They are major consumers of organic debris in temperate and tropical hardwood forests, where they feed on dead vegetable matter. Millipedes are also inhabitants of arid and semiarid regions, despite their dependence on moisture. Millipedes lack a waxy layer on their epicuticle and are subject to rapid desiccation in environments with low relative humidity. Some are true soil forms, others seem restricted to leaf litter or to cryptozoan habitats. They can be loosely grouped into (1) tubular, round-backed forms such as the familiar *Narceus*; (2) flat-backed forms (many Polydesmid millipedes); and (3) pillbug types, which roll into a ball. Millipedes range widely in length. Typical North American forms are 5–6 cm in length; tropical ones may reach nearly 20 cm in length. Hoffman (1999) has published a checklist of millipedes of North and Middle America. Keys to North American families of millipedes were published by Hoffman (1990). For an account of millipede biology and ecology, see Hopkin and Read (1992).

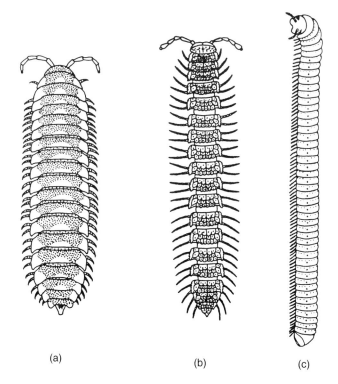

(a) (b) (c)

FIGURE 4.42. Representatives of three families of millipedes: (a) Xylodesmidae; (b) Polydesmidae; (c) Spirobolidae (from Kevan and Scudder, 1989).

Millipedes become abundant in calcium-rich, high rainfall areas in tropical and temperate zones. The southern Appalachian Mountains of the eastern United States support a large millipede population. Millipedes can be important in calcium cycling. They have a calcareous exoskeleton, and because of their high densities they can be a significant sink for calcium. Millipedes are major consumers of fallen leaf litter, and may process some 15–25% of calcium input into hardwood forest floors. In desert areas, millipedes are active following rains, especially in desert shrub communities. They avoid hot, dry conditions by concealment under vegetation or debris (Crawford, 1981). Millipedes are vulnerable to desiccation, because their cuticle generally lacks a waterproof layer, their gas exchange system is not closed, and they lose a considerable amount of water through the mouth, in defecation, and during reproduction (Wolters and Ekschmitt, 1997).

Millipedes appear to be selective feeders, avoiding leaf litter high in polyphenols and favoring litter with high calcium content (Neuhauser and Hartenstein, 1978). Freshly fallen leaves are generally avoided, even though assimilation efficiency is much higher from that source (David and Gillon, 2002). Some millipedes are obligate coprophages. When McBrayer (1973) cultured millipedes in containers, which excluded their feces, the millipedes lost weight. When a small tray containing feces was added to the cultures, the millipedes consumed it and prospered. Such obligate coprophagy indicates a close relationship with bacteria necessary for digestion of vegetable material. It is not known whether millipedes possess a unique gut flora of microbes.

Chilopoda

Centipedes (Chilopoda) are common predators in soil, litter, and cryptozoan habitats (Fig. 4.43). They are all elongate, flattened, active forms. Centipedes occur in biomes ranging from forest to desert. The large desert centipedes (*Scolopendromorpha*) are some 15-cm long; tropical centipedes may exceed 30 cm (Shelley, 2002). Lithobiids are the common brown, flat centipedes of litter in hardwood forests. The elongate, slim geophilomorph centipedes are euedaphic in forest habitats, where they prey on earthworms, enchytraeids, and Diptera larvae (Lock and Dekoninck, 2001). Like the millipedes, centipedes lose water through their cuticles at low relative humidities. They avoid desiccation by seeking moist habitats, and by adjusting their diurnal activities to humid periods in desert and sand dune habitats.

Centipedes are distinguished from superficially similar organisms by the presence of forcipules, the modified first segment upon which the head rests (Mundel, 1990). This segment bears the pincerlike fangs, which have poison ducts opening at their tips. Five orders of centipedes

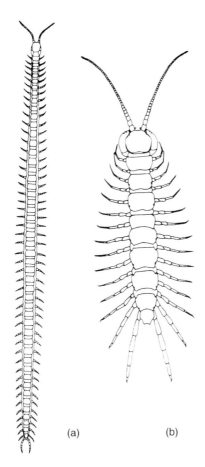

(a)   (b)

FIGURE 4.43. Representatives of two centipede families: (a) Geophilidae (*Geophilus proximus* Koch); (b) Lithobiidae (*Lithobius forficatus* [L.]).

are recognized, and Mundel (1990) provides keys to orders and families of centipedes of the world.

All centipedes are predators but may ingest some leaf litter on occasion—it can sometimes be seen in their guts. Centipedes are fast runners and actively pursue and capture small prey such as collembolans.

Scorpionida

The scorpion, the archetypal generalized arachnid with its long, segmented, stinger-bearing abdomen and chelate palpi, needs no description. It was obvious to the ancients—the only zodiacal sign bearing the name of a soil organism. Scorpions (Fig. 4.44) are inhabitants of warm, dry, tropical, and temperate regions but reach their greatest diversity in deserts. They are highly mobile predators of other arthropods and occasionally even small vertebrates. The selection of prey items utilized is large; for one species of scorpion more than 100 different prey were recorded (Crawford, 1990). Relatively few individual scorpions forage at any one time. Different species demonstrate different patterns of predation, some being "sit and wait" predators and others acting as mobile hunters (Crawford, 1990). Scorpions are also cannibalistic to an unusual extent (Williams, 1987).

Typical cryptozoans, scorpions hide under rocks or logs, or in crevices, during the day and emerge at night to feed. In the southeastern United States, scorpions may be trapped by placing wet cloth on the ground at night; the dampness will attract them. The scorpion cuticle will fluoresce under black light, offering a nighttime survey procedure.

Scorpions' stings are painful, about the same as a honeybee sting but of shorter duration. In the southeastern United States, the tiny scorpion *Vejovis carolinus*, commonly found under bark of pine stumps, invades houses on occasion. A species with similar habits, *Centruroides vittatus*, occurs from east Texas to the southwestern United States (Shelley and Sisson, 1995). Only a few species of scorpions are deadly. Of

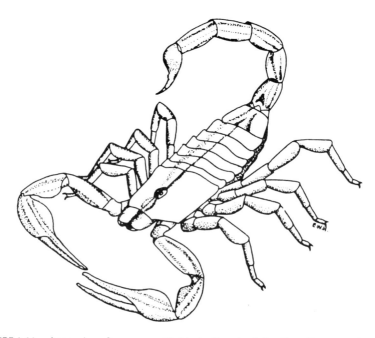

FIGURE 4.44.   A scorpion, shown magnified twice its actual size (from Borror *et al.*, 1981).

the 1500 species worldwide, only about 20–25 are dangerous, all in the family Buthidae. Where dangerous species occur, antivenom is usually available (Jackman, 1997).

The impact of scorpions on their ecosystems is unknown. They are not numerous, but in desert ecosystems they may be dominant predators (Polis, 1991).

Araneae

Spiders (Araneae) (Fig. 4.45) are another familiar group of carnivores. They are solitary hunters, exhibiting a range of strategies from "sit and wait" with silken webs to active pursuit of prey. They are found in all terrestrial environments except truly polar (Arctic/Antarctic) regions. Many species are found in aboveground habitats, but some are cryptozoans in litter and on the soil surface. Some small spiders are euedaphic (Fig. 4.46). Some of the small litter-inhabiting spiders could be considered microarthropods. Spiders may be active hunters or "sit and wait" dwellers in retreats. Wolf spiders (Lycosidae) (Fig. 4.47) are common wandering predators in leaf litter and on soil surfaces, and are often captured in pitfall traps. They are conspicuous ground-dwelling predators in agroecosystems (Draney, 1997).

Spider taxonomy is a dynamic discipline. There are about 100 families in the order, arranged in several suborders (or infraorders). Kaston's (1978) guide, *How To Know The Spiders,* is an excellent introduction for the novice arachnologist. And Roth's (1993) *Spider Genera of North America* is invaluable for workers in the United States. Regional works

FIGURE 4.45. A typical wolf spider, *Lycosa communis* (Lycosidae) (from Wise, 1993).

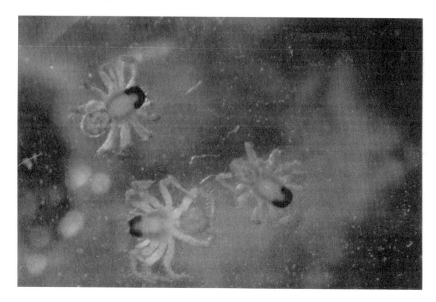

FIGURE 4.46. Small spiders (D. A. Crossley, Jr., photo).

FIGURE 4.47. *Rabidosa rabida* (Lycosidae).

such as Jackman (1997) should not be overlooked, nor should information currently available on the World Wide Web.

Sampling methods for spiders run the gamut from Tullgren extraction to hand collection, sorting of litter samples, and pitfall trapping. Spiders have complex behavioral patterns including mating rituals and defense of territories. Wolf spiders may wander some distance between forest, meadow, and agroecosystem (Draney, 1997), so that assessment of population size may become complicated.

Little is known about the ecology of the smaller soil- and litter-dwelling spiders (Linyphiidae and relatives). Most information about spiders comes from studies of web-spinning species or the jumping spiders (Attidae) in vegetation (Foelix, 1996). There are large numbers of small spiders in deciduous forest leaf litter (about 100 per square meter). Their habitat usage and prey selection are not well known. Spiders in laboratory microcosms show some prey selectivity, but leaf litter spiders doubtless feed more opportunistically.

A number of spider species are ant mimics (Foelix, 1996). These species copy the body shapes and coloring of ants, and move in an antlike manner. The spider's front legs are elongated and thin, and mimic the antennae of the ant. The significance of this mimicry is not known, but it is suspected to confer some protection from predation by birds. Spiders that live together with ants seldom prey upon the ants (Jackson and Willey, 1994).

The impact of spiders on their ecosystems is not well known. Their effectiveness as biological control agents has been discounted because of their slow reproduction. As noted above, spiders are strongly territorial, with complicated mating rituals—adaptations that tend to hold down population sizes even when prey is abundant (Wise, 1993). In woodland forest floors, the large number of spiders argues that they must have an impact on the insect population there. In contrast, numbers of spiders in agricultural soils seem lower (Draney, 1997). In an experimental study, Lawrence and Wise (2000) found a "top-down" effect of spiders on litter decomposition rates. When spiders were removed from experimental areas of a forest floor, collembolan populations increased. Subsequently, straw in litterbags decomposed more rapidly in those areas. These results suggest that spider predation may reduce collembola populations enough to lower rates of litter disappearance on the forest floor.

Opiliones

Harvestmen (Opiliones) are delicate, shy forms that are among the largest arachnids in woodlands. Their bodies are small but their legs may be unusually long, suggesting that their habitat is litter surface or exposed areas. Smaller, shorter-legged forms inhabit loose leaf litter or small spaces (Edgar, 1990). Others are inhabitants of caves, and have reduced eyes and reduced pigmentation (Goodnight and Goodnight, 1960). Opilionids have no venom glands yet are considered to be largely predaceous. Some species occur high in foliage, others in subcanopy, some on soil surface, and some (smaller forms) in litter layers. Opilionids are slow reproducers, usually with one generation per year. They are active predators in the daylight but seem to be primarily crepuscular (active dawn and dusk). They possess repugnatorial glands, the secretion of which is offensive to predators (Blum and Edgar, 1971).

Opilionids (*Opilio* means "a shepherd" in Latin) resemble the mites in that the cephalothorax (prosoma) and abdomen (opisthosoma) are broadly fused, so that the body is oval. They differ superficially from the mites in that opilionids have 6–10 segments in their abdomen; mites have none. Of course, they are larger than mites as well. Edgar (1990) provides a key to genera and species of North American Opiliones (exclusive of Mexico).

Solifugae

Solifugae, or solpugids, are desert arachnids with large distinctive curved chelicerae, often as long as the cephalothorax (Fig. 4.48). They are ferocious predators capable of rapid movement. Common names include sun-spiders, false-spiders, wind-spiders, or wind-scorpions (in recognition of their rapid movement), or camel-spiders, among other names. The term "camel spider" refers to a prominent arch-shaped plate on the prosoma (Punzo, 1998). (They do not run down camels and eat their stomachs, a rumor circulated among troops during the recent war in Iraq.) Solifugae occur in tropical and temperate deserts worldwide.

Most species of Solifugae are nocturnal predators, emerging from relatively permanent burrows to feed upon a variety of arthropods. They do not have poison glands. They are generalist predators and attack a wide variety of arthropods, as well as small lizards, birds, and mammals (Punzo, 1998). In North American deserts, immature stages of solpugids feed extensively on termites (Muma, 1966).

Uropygi

This order of arachnids contains some of the largest species, up to 10 cm in length. The North American form, *Mastigoproctus giganteus* (Fig. 4.49), occurs from Florida to Arizona (Jackman, 1999). The distinc-

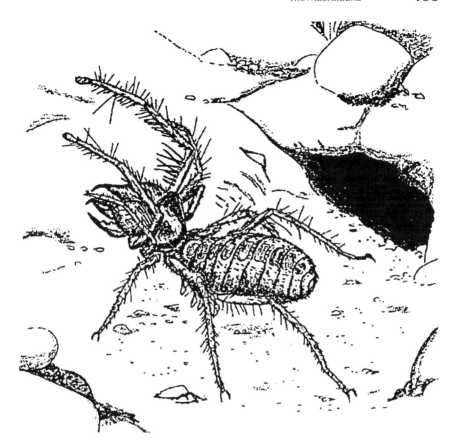

FIGURE 4.48.   Solpugid, or camel spider (from Punzo, 1998).

tive long, whiplike tail has no stinger. When disturbed, the arachnid emits acetic acid from a gland at the base of the tail, giving rise to the common name "vinegaroon." Uropygids are nocturnal predators, utilizing natural retreats or burrowing into sand. They have poor vision and depend upon vibrations to locate their prey.

The Pterygote Insects

Many of the higher, winged insects (Pterygota) are residents of soils and participate in food webs there. Some are permanent soil inhabitants, and all stages of the life history are found in soil or on the soil. Immature stages of other species are true soil dwellers—white grubs, wireworms, and cutworms, for example—whereas their flying adult forms live in vegetation and feed in aboveground food chains. All of the

FIGURE 4.49. Uropygi—"Vinegaroon."

major winged insect orders—the Coleoptera (beetles), Lepidoptera (butterflies and moths), Hymenoptera (bees, wasps and ants), and Diptera (flies)—include soil-dwelling species. The Isoptera (termites) are essentially soil insects and are saprophages. The Homoptera (aphids, cicadas), Orthoptera (grasshoppers and crickets), and minor orders such as the Dermaptera (earwigs) contain soil-dwelling species or life history stages. Indeed, of the 26 pterygote insect orders, all but seven contain at least some species that are involved in soil food webs in one way or another (Greenslade, 1985). Space does not permit us to review thoroughly these extensive and important groups, or to discuss all species groups that impact soils or soil food webs. We refer the reader to a textbook of entomology, such as Rosomer and Stoffolano (1994) or Borror *et al.* (1989), and to field guides such as that of White (1983) for aids in identification and basic biology of these groups. We can offer only a very superficial treatment of the higher insects. Nevertheless, this group includes important species that are root feeders, predators, and modifiers of soil structure—animals that the soil ecologist can hardly ignore.

Coleoptera

Beetles, the largest order of insects, have soil species that are predatory, phytophagous, or saprovores. Some are permanent residents, others are temporary, and many are transient members of soil food webs. Beetles are particularly abundant in tropical ecosystems, where many species remain to be named. For identification of beetles to family see White's (1983) guide. Dillon and Dillon (1961) is an invaluable, well-illustrated manual for eastern North America.

The Carabidae (Fig. 4.50), the ground beetles, are among the more familiar insects caught in pitfall traps or active on the soil surface of agroecosystems (Purvis and Fadl, 2002). *Harpalus pennsylvanicus* is frequently caught in pitfall traps. Some members of the genus ingest seeds but most are predators. Larval stages are euedaphic and may be sampled with Tullgren extraction. Adults and larval stages of *Calosoma sycophonta*, the searcher, climb trees in search of prey. Bell (1990) provides a taxonomic key to adults and larvae of Carabidae.

Darkling beetles, family Tenebrionidae (Fig. 4.51), are abundant in desert ecosystems; their habits are similar to those of the carabid beetles. Most of them are scavengers or saprophytic on decaying vegetation (White, 1983). Adults are surface active whereas larvae are euedaphic.

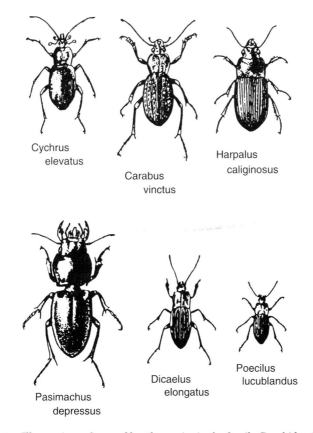

Cychrus
elevatus

Carabus
vinctus

Harpalus
caliginosus

Pasimachus
depressus

Dicaelus
elongatus

Poecilus
lucublandus

FIGURE 4.50. Illustrations of ground beetle species in the family Carabidae (Lutz, 1948).

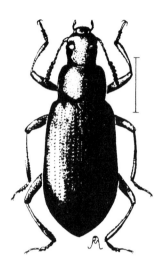

FIGURE 4.51. Tenebrionidae dorsal view of *Alobates* sp. (false mealworm) 20–23 mm (from Arnett, 1993).

Pitfall traps catch large numbers of desert tenebrionids in the springtime, when adults emerge from the pupal stage. The larvae resemble wireworms (larvae of elaterid beetles) and are called "false wireworms." They are considered to be saprovores (Crawford, 1990).

Rove beetles (Staphylinidae) are a large family of common, distinctive species (Fig. 4.52), often caught in pitfall traps. Most species appear to be predaceous (both larvae and adults) but a few are saprophagous. The adults are agile runners on the soil surface. Frequently, the tip of the abdomen is turned up as they scurry along the ground. They are attracted to decaying vegetation or carrion (Dillon and Dillon, 1961). Most species have well-developed wings and can fly, but wing reduction is usual in euedaphic species. Keys to adults and larval stages of soil inhabiting genera of Staphylinidae are provided by Newton (1990).

Scarab beetles (Scarabaeidae) (Fig. 4.53) are members of a large family of beetles, some colorful or metallic green, sometimes multicolored, often brown or black. Males may have horns on the head or pronotum.

Scarab beetles may be separated into two groups based on their feeding habits (Dillon and Dillon, 1961). One contains species that feed upon dung or carrion. Species in the other group feed upon leaves, flowers, and pollen as adults and on plant roots or decaying wood as larvae. Some species excavate burrows under pats of dung and provision these with dung for their larvae; other species live on the surface of dung pats (Curry, 1994). "Tumble bugs" chew off a piece of dung, work it into a ball,

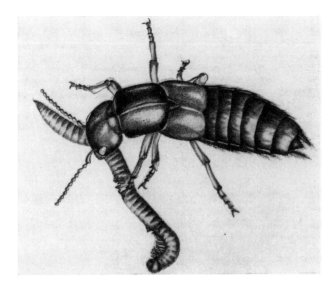

FIGURE 4.52. A predaceous staphylinid beetle (*Staphylinus badipes*) attacking a milli-pede (*Ophyiulus pilosus*) (Snider, 1984).

roll it to a burial site, and deposit an egg in it. The sacred scarab of ancient Egypt is a member of this group (Tashiro, 1990). Some scarab beetles are important due to their role in hastening the decay of dung of large animals. In Australia, where there was no native coprophagous fauna, dung from domestic animals accumulated, fouling pastures and immobilizing nutrients (Gillard, 1967).

Tiger beetles (Cicindellidae) are predators whose larvae dig pits where they sit and wait for prey (Fig. 4.54). Adults are rapid runners and fliers, often pouncing suddenly on their prey. Conspicuous on the soil surface in open, sunlit areas, adult tiger beetles are usually irides-cent green or blue.

Wireworms (larvae of the family Elateridae) are significant root feeders in forests and agroecosystems, where they can be destructive to certain crops. Tan to brown, wireworms are slender and have a hard cov-ering on their bodies. Adult elaterids are called "click beetles" because of their ability to snap the hinge between pro- and mesothorax. If the insect is on its back, it can right itself by snapping and projecting itself into the air, turning over repeatedly in the process (Dillon and Dillon, 1961). Adult elaterids are occasionally captured in pitfall traps.

Beetles, as a group, bridge the gap between mesofauna and macro-fauna. They are of more economic importance for their phytophagous activities above ground than for their participation in soil food webs.

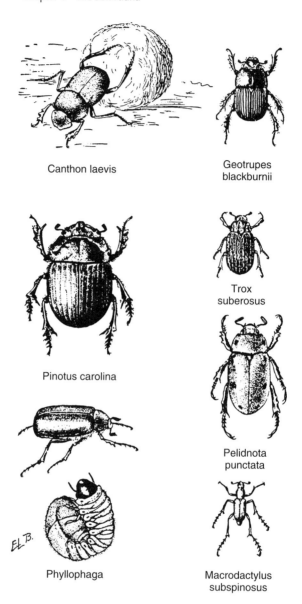

Canthon laevis

Geotrupes
blackburnii

Trox
suberosus

Pinotus carolina

Pelidnota
punctata

Phyllophaga

Macrodactylus
subspinosus

FIGURE 4.53.   Some scarabaeid beetles: *Canthos laevis*, a tumble-bug with a ball of dung in which an egg is laid; *Geotrupes blackburnii* and *Pinotus carolina*, also dung beetles. *Trox suberosus* lays eggs in carrion. Larvae of *Pelidnota punctata* live in decaying oak or hickory stumps. The J-shaped larvae of *Phyllophaga* species feed upon roots of plants (Lutz, 1948).

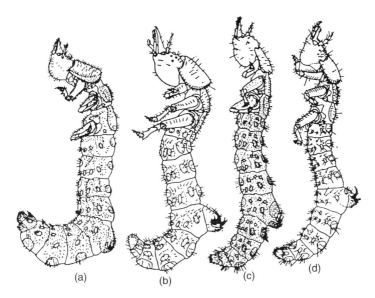

(a) (b) (c) (d)

FIGURE 4.54. Larvae of tiger beetles, family Cicindelidae. (a) *Amblychila cylindriformis*; (b) *Omus californicus*; (c) *Tetracha carolina*; (d) *Cicindela limbalis*. The predaceous larvae lie in wait in vertical burrows with their heads flush with the soil surface, and held in place by hooks on the hump protruding from the fifth abdominal segment (from Frost, 1942).

The predatory activities of beetles are especially significant in agricultural systems because they prey on pest species of insects. Beetles are important agents in the reduction of dung and animal carcasses, and in the early stages of wood decomposition on the forest floor (Wallwork, 1982; Hanula, 1995).

Hymenoptera

The order Hymenoptera is one of the largest orders of insects. It contains two groups of soil insects of large importance: ants (Fig. 4.55) and ground-dwelling wasps (Fig. 4.56). The Formicidae, the ants, are probably the most significant family of soil insects, due to the very large influence they have on soil structure. Bees and wasps, other hymenopteran insects, also impact soils because they may nest there. The ants are in a category by themselves.

Ants are widely distributed (from arctic to tropics), numerous, and diverse. Ant communities contain many species, even in desert areas (Whitford, 2000). Local species diversity is large, especially in tropical areas (Kempf, 1964). Populations of ants are equally large. About one-

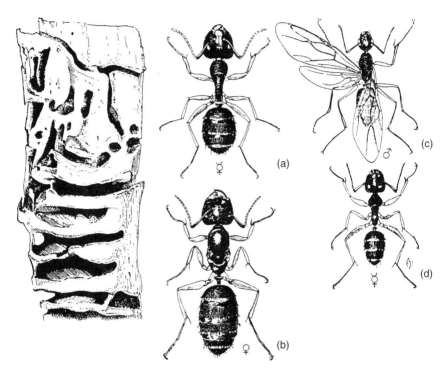

FIGURE 4.55. Carpenter ants and their galleries in deadwood. Shown are a large neuter worker (a), a winged male (b), a wingless female (c), and a small neuter worker (d) (from Henderson, 1952).

third of the animal biomass of the Amazonian rain forest is composed entirely of ants and termites, with each hectare containing in excess of 8 million ants and 1 million termites (Hölldobler and Wilson, 1990). Furthermore, ants are social insects, living in colonies with several castes (Fig. 4.55).

Ants have a large impact on their ecosystems. They are major predators of small invertebrates (including oribatid mites) (E. O. Wilson, personal communication). Their activities reduce the abundance of other predators such as spiders and carabid beetles (Wilson, 1987). Ants are "ecosystem engineers," moving large volumes of soil as much as earthworms do (Hölldobler and Wilson, 1990). Ant influences on soil structure are particularly important in deserts (Table 4.9) (Whitford, 2000), where earthworm densities are low.

Given the large diversity of ants, identification to species is problematic for any but the taxonomist skilled in the group. Wheeler and Wheeler (1990) offer keys to subfamilies and genera of the Nearctic ant fauna. Bolton's (1994) identification guide to the ant genera of the world is very well illustrated. For a review of the ants, their biology, ecol-

TABLE 4.9. Estimated Quantities of Soil[a] Brought to the Surface by Desert Ants and Termites During the Construction of Feeding Galleries and Nest Chambers, and in Nest Repair.

Ant species	Location	Turnover rate
Ant community	***Atriplex vesicaria* shrub steppe**	**350–420**
Aphaenogaster barbigula	*Cailtris-Eucalyptus* open woodland	3360
(funnel ants)		
Ant community	Heath, Western Australia	310
Ant community	Wandoo woodland, Western Australia	200
Ant communities	Variety of Chihuahuan Desert	21.3–85.8
	shrublands and grasslands	
Termite species		
Heterotermes aureus	Sonoran Desert, Arizona, USA	750
Gnathamitermes perplexus		
Macrotermes subhyalinus	Senegal	675–950
Gnathamitermes tubiformans	Chihuahuan Desert, USA	
	Mixed grassland-shrubland	4095
	Creosotebush shrubland	801
	Black-grama grassland	981
	Watershed	2600

[a]kg ha^{-1} year^{-1}.
From Whitford, 2000.

ogy, and social structure, the work by Hölldobler and Wilson (1990) is unsurpassed.

Most of the solitary wasps in the superfamily Vespoidea construct nests in the soil. (Fig. 4.56) The adult female wasp first constructs a small nest cavity. Then a suitable prey item (another insect or a spider) is located, which the wasp then stings to paralyze it and hauls it to the nest. An egg is laid on the paralyzed victim, and it is then entombed. Some of the social wasps, especially *Vespula* spp., nest in the ground. Often natural cavities such as abandoned rodent burrows are used as nesting sites. Vespids are carnivorous, feeding their larvae on captured prey (insects or spiders), although adult wasps generally feed on nectar, sap, or similar juices (Michener and Michener, 1951).

Diptera

Many of the true flies can be considered soil insects, at least in some stage of their life histories. At least 75 of the 108 dipteran families in North America have some contact with soil ecosystems (McAlpine, 1990). This listing excludes strictly aquatic families, aboveground herbivores, and some parasitic species. Many species that live in aboveground habitats pupate in the soil, thus participating, involuntarily,

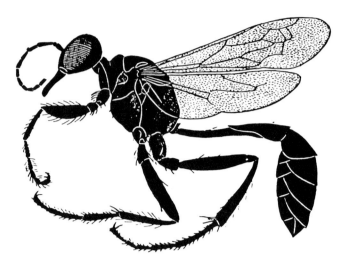

FIGURE 4.56. A digger-wasp (family Sphecidae). These solitary wasps usually prepare nests in the soil, which they provision with arthropod prey before depositing eggs (Pratt and Stojanovich, 1967).

in soil food webs. McAlpine (1990) provides a well-illustrated key to families of Diptera that have relations with soil systems.

Many species of fly larvae are important saprovores in soils. They are restricted to moist situations rich in organic matter. Some larvae are predatory and these have adaptations to reduce moisture loss; they occur in drier situations (Teskey, 1990). Fly larvae have a major impact on decomposition rates of carrion. Together with some beetle species, maggots of various types hasten the decomposition rate significantly. When Payne (1965) used window screen to exclude insects from decaying corpses of baby pigs, the bodies became mummified and decomposed slowly compared with corpses exposed to insect attack. Fly larvae are also important in forensic entomology, where their identification has been helpful in determining time of death of human corpses (Catts and Haskell, 1990).

Isoptera

The Isoptera (Fig. 4.57), the termites, are among the most important of soil fauna, in terms of their impact on soil structure and on decomposition processes. Termites are social insects with a well-developed caste system. Through their ability to digest wood they have become economic pests of major importance in some regions of the world (Lee and Wood, 1971; Bignell and Eggleton, 2000). Termites are highly successful,

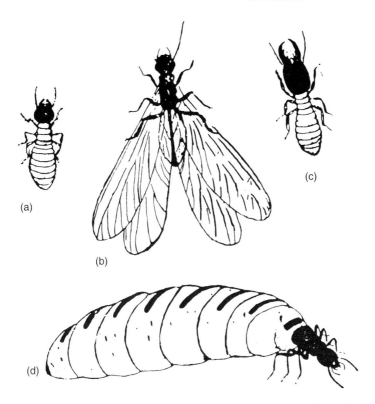

FIGURE 4.57. Isoptera (Castes of termites: (a) worker, (b) winged reproductive, (c) soldier, (d) queen (courtesy of Banks and Snyder and the U.S. National Museum) (from Borror *et al.*, 1981).

constituting up to 75% of the insect biomass and 10% of all terrestrial animal biomass in the tropics (Wilson, 1993; Bignell, 2000).

Termites in the primitive families, such as Kalotermitidae, possess a gut flora of protozoans, which enables them to digest cellulose. Their normal food is wood that has come into contact with soil. Most species of termites construct runways of soil and some are builders of spectacular mounds (Fig. 4.58). Members of the phylogenetically advanced family Termitidae do not have protozoan symbionts, but possess a formidable array of microbial symbionts (bacteria and fungi) that enable them to process and digest the humified organic matter in tropical soils, and to grow and thrive on such a diet (Breznak, 1984; Bignell, 1984; Pearce, 1997). Interestingly no one adaptive feature or mechanism appears to distinguish the guts of soil-feeding termites. As a result, approximately 67% of the genera in the family Termitidae now consist of these forms. A

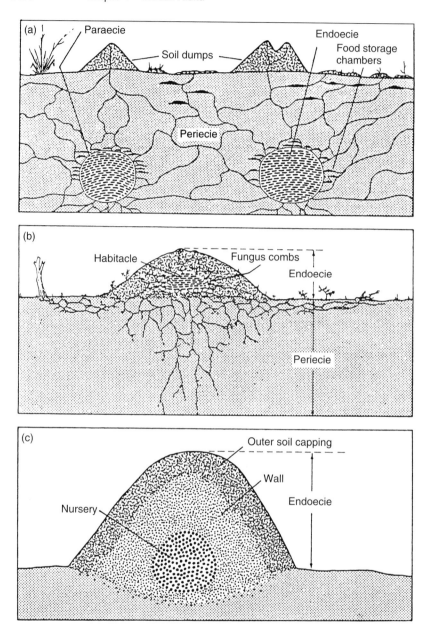

FIGURE 4.58. Termite Mounds: Diagrammatic representation of different types of concentrated nest systems. (a) *Hodotermes mossambicu*. (b) *Macrotermes subhyalinus*. (c) *Nasutitermes exitiosus* (from Lee and Wood, 1971).

speculative and generalized sequence of events in a typical Termitinae soil-feeder gut is given in Fig. 4.59 (Brauman *et al.*, 2000).

A number of inquilines (organisms existing in, and sharing, common space) occur in termite nests—ants, collembolans, mites, centipedes, and beetles that have become morphologically specialized for that habitat.

Although termites are mainly tropical in distribution, they occur in temperate zones and deserts as well. Termites have been called the tropical analogs of earthworms, because they reach a large abundance in the tropics and process large amounts of litter. Three nutritional categories include wood-feeding species, plant- and humus-feeding species, and fungus growers. This latter group lacks intestinal symbionts and depends upon cultured fungus for nutrition. Termites have an abundance of unique microbes living in their guts. Using the criterion of 97% sequence identity, one recent study of bacterial microbiota in the gut of the wood-feeding termite *Reticulitermes speratus* found 268 phylotypes of bacteria (16S rRNA genes, amplified by polymerase chain reaction [PCR]), including 100 clostridial, 61 spirochaetal, and 31 *Bacteroides*-related phylotypes (Hongoh *et al.*, 2003). More than 90% of the phylo-

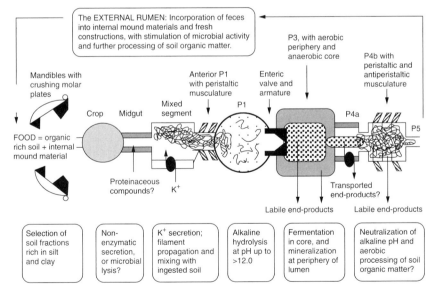

FIGURE 4.59. Hypothesis of gut organization and sequential processing in soil-feeding *Cubitermes*-clade termites. The model emphasizes the role of filamentous prokaryotes, the extremely high pH reached in the P1, and the existence of both aerobic and anaerobic zones within the hindgut. Major uncertainties have question marks. Not to scale (from Brauman *et al.*, 2000).

types were found for the first time. Others were monophyletic clusters with sequences recovered from the gut of other termite species. It should be noted that cellulose digestion in termites, which was once considered to be solely due to the activities of fungi, protists, and occasionally bacteria, has now been convincingly demonstrated to be endogenous to termites. Endogenous cellulose-degrading enzymes occur in the midguts of two species of higher termites in the genus *Nasutitermes*, and in the Macrotermitinae (which cultivate basidiomycete fungi in elaborately constructed gardens) as well (Bignell, 2000).

In contrast to the carbon degradation situation, only prokaryotes are capable of producing nitrogenase to "fix" N_2, or dinitrogen. This process occurs in the organic-matter–rich, microaerophilic milieu of termite guts. Some genera have bacteria that fix relatively small amounts of nitrogen, but others, including *Mastotermes* and *Nasutitermes*, have from 0.7 to greater than 21 micrograms (μg) nitrogen fixed·g fresh weight. This equals 20–61 μg of nitrogen per colony per day, which would double the nitrogen content if N_2 fixation was the sole source of nitrogen and the rate per termite remained constant (and the nitrogen content of termites is assumed to be 11% on a dry weight basis) (Breznak, 2000).

Termites are one of the three major earth-moving groups of invertebrates (the other two are earthworms and ants). Mound-building termite species have a major impact on the distribution and composition of soil mineral and organic matter. Where there is rich, well-drained grassland (e.g., in the Ivory Coast), humivorous and fungus-growing termites are common. In areas of poor drainage, these species are absent and grass feeders such as *Trinervitermes* and some *Macrotermes* spp. are common. In some cases, farmers grow crops on mounds. This is advantageous in areas that are flooded, such as paddy fields in Southeast Asia (e.g., Thailand) (Pearce, 1997). In desert regions of North America (e.g., the Chihuahuan desert of southern New Mexico), termites are considered "keystone arthropods," removing and processing large amounts of dead and dying net primary production every year (see Table 4.9) (Whitford, 2000). For a masterful exposition of the role of termites in ecosystems worldwide, refer to Bignell and Eggleton (2000).

Other Pterygota

As we noted above, most terrestrial insect orders have members that participate in soil systems, either by burrowing, pupating, or even feeding there. At times they may be present in some numbers, or exert an unusual influence on food webs.

The Orthoptera, grasshoppers and crickets, lay eggs in soils and some are active on the soil surface. Crickets, Gryllidae, may be abundant in pitfall traps set in meadows or agroecosystems (Blumberg and Crossley, 1983).

The Psocoptera, psocids, are a small order of insects that occasionally become abundant in leaf litter. They feed on organic detritus, algae, lichens, and fungus (Aldrete, 1990).

The order Homoptera, cicadas, aphids, and others, has members important as belowground herbivores and as soil movers. Cicadas are noisy, active flyers as adults. The immature stages feed upon the roots of perennial plants until mature, a time period that may last 13–17 years for periodical cicadas (*Magnicicada* spp.). In tallgrass prairie soils, cicadas are abundant insects; their annual emergence can result in a significant flux of nutrients from belowground to aboveground (Callaham *et al.*, 2000).

Gastropoda

Terrestrial gastropods (snails and slugs) (Fig. 4.60) (Burch and Pearce, 1990) are major players among herbivores and detritivores in many ecosystems, particularly agroecosystems (Byers *et al.*, 1989).

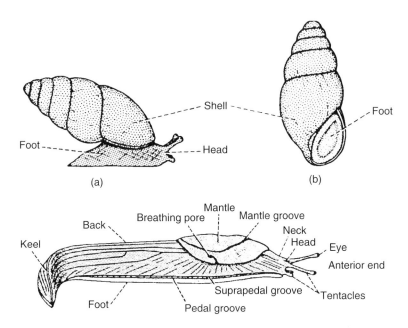

FIGURE 4.60.   Terrestrial gastropods (snails and slugs): At top, a nonoperculated (pulmonate) snail. It does not have a protective operculum to seal the shell aperture when the animal has withdrawn into its shell. (a) Active snail; (b) Inactive snail withdrawn into its shell, with only the surface of its foot showing. At bottom, slug body terminology (from Burch and Pearce, 1990).

They have been studied much less than the arthropod fauna in forests. They tend to require moist conditions and the presence of significant amounts of calcium for their metabolic needs, but some gastropods exist successfully in low pH and low calcium environments (Burch and Pearce, 1990). The terrestrial gastropod faunas have become rich and diversified as they have invaded many habitats. Nearly a thousand species occur in North America north of Mexico, and similar rich faunas occur throughout Europe (South, 1992). Land gastropods are quite speciose and ubiquitous in the eastern United States, with at least 500 species, including both snails and slugs, being described from the eastern United States (Hubricht, 1985, cited in Hotopp, 2002). Snails seem to key in on structural attributes of their environment. In an extensive study of community patterns of 108 snail species in the Great Lakes region, Nekola (2003) found that soil surface architecture—deep organic horizon soils (deeper than 4 cm, "duff") versus thin horizon soils subtended by live roots (less than 1 cm deep, "turf")—accounted for 43% of the variability of snail distributions.

In Danish beech forests, Petersen and Luxton (1982) measured $822 \, mg \cdot m^{-2}$ gastropod biomass, an amount that was exceeded by only diplopods and earthworms in the fauna of that forest floor. Terrestrial gastropods feed primarily on plants, but may prefer decaying or senescent tissues. Numerous basidiomycetes are consumed as well, including some that are highly toxic to mammals. Only a few gastropods feed on animals, but several may feed on carrion. The feces of gastropods retrieved from the wild include soil particles, which may be due to humic acids as a required substrate in their diet (this is particularly true for helicid snails grown in culture) (Speiser, 2001). Feeding rates on leaf litter range from 9.3 to $28.1 \, mg \cdot g^{-1}$ live weight of slugs in European forest floors (Jennings, 1975). Assimilation rates of *Agriolimax reticulatus* feeding on fresh *Ranunculus repens* (lotus) leaves were greater than 78%, and snails' assimilation rates ranged between 40 and 70%, feeding on either fresh leaf or leaf litter material (Mason, 1974). Pallant (1974) estimated rates of assimilation for several slug species to average 161.3 Joules $100 \, mg^{-1}$ dry weight in grasslands and $141.5 \, J \, 100 \, mg^{-1}$ dry weight in woodland, in Europe.

For a review of extensive phylogenies of terrestrial gastropods, using 28S rDNA and morphological data, see Barker (2001).

Sampling Techniques for Gastropods

Two different sampling techniques were employed by Hotopp (2002), working in the forests of the central Appalachian Mountains: (1) timed searching, and (2) sieving litter. The former approach constitutes a 10-

minute search of leaf litter surface, rocks, woody debris, and live plant stems across a 200 square meter sample plot. It tends to be more efficient in finding large snails and slugs. The latter method consists of placing litter from Oi and Oe horizons on a 10-mm sieve placed to a depth of about 10 cm, shaking it 50 times, turning it over, and shaking it 50 more times. This method is more efficient in retrieving small specimens including the ecological dominant, *Punctum minutissimum*, which is about 1mm in width.

Oligochaeta—Earthworms

Earthworms are the most familiar and, with respect to soil processes, often the most important of the soil fauna. As observed by many farmers and gardeners and reported in the popular literature, the importance of earthworms arises from their influence on soil structure (e.g., aggregate or crumb formation, soil pore formation) and on the breakdown of organic matter applied to soil (e.g., fragmentation, burial, and mixing of plant residues). These observations have led to numerous studies of the potential benefits of earthworms in agriculture, waste management, and land remediation (Edwards, 2004).

While the scientific literature on earthworms officially began with Linnaeus's taxonomic description of *Lumbricus terrestris* more than 200 years ago, the modern era of earthworm research began with Darwin's (1881) last book, *The Formation of Vegetable Mould Through the Actions of Worms, with Observations of Their Habits,* which called attention to the beneficial effects of earthworms: *"It may be doubted whether there are many other animals which have played so important a part in the history of the world, as have these lowly organized creatures."* Since then, a vast literature has established the importance of earthworms as biological agents in soil formation, organic litter decomposition, and redistribution of organic matter in the soil (see Lee, 1985; Hendrix, 1995; Edwards and Bohlen, 1996; and Lavelle *et al.*, 1999).

Despite the common reference in the popular literature to "the earthworm," there is great diversity and a wide range of adaptations to environmental conditions among the earthworm fauna. More than 3500 earthworm species have been described and it is estimated that considerably more await discovery and description (Fragoso *et al.*, 1999).

Earthworms are classified within the phylum Annelida, class Oligochaeta, and order Opisthophora. Although there is not universal agreement on taxonomic classification, recent analyses suggest 16 families, 6 comprising aquatic or semiaquatic worms (cohort Aquamegadrili plus suborder Alluroidina), and the other 10 consisting of the terrestrial forms commonly known as earthworms (cohort Terrimegadrili) (Jamieson, 1988). Species within the families Lumbricidae and Megascolecidae are ecologically the most important in North America, Europe, Australia, and Asia; some of these species have been introduced worldwide by human activities and now dominate the earthworm fauna in many temperate areas. Likewise, several tropical species in the families Glossoscolecidae, Eudrilidae, and Megascolecidae have become pantropical in distribution. Such "peregrine" or "anthropochorous" species are highly successful in many agricultural or otherwise disturbed areas, and often show significant effects on soil processes (Lee, 1985, Lavelle *et al.*, 1999). Different localities may be inhabited by all native species, all exotic species, a combination of native and exotic species, or by no earthworms at all. Relative abundance and species composition of local fauna depend greatly on soil, climate, vegetation, topography, land use history, and especially on past invasions by exotic species.

Whether introduced earthworms displace native species or occupy areas devoid of native species as a result of disturbance is a subject of debate (Kalisz and Wood, 1995; Hendrix and Bohlen, 2002). It is often suggested that the establishment of exotic earthworm populations proceeds through the stages of habitat disturbance, extirpation or reduction of native populations, introduction of exotic species, and colonization of vacant niche space by exotic species. Even in the absence of obvious habitat disturbance, some minimum habitat patch size may be required to maintain native earthworm assemblages; increased edges and potential vectors for invasion by exotic species into small ecosystem remnants may lead to displacement of native populations (Kalisz and Wood, 1995).

Earthworm Distribution and Abundance

As noted previously, earthworms occur worldwide in habitats where soil water and temperature are favorable for at least part of the year; they are most abundant in forests and grasslands of temperate and tropical regions, and least so in arid and frigid environments (e.g., desert, tundra, or polar conditions). Across this range of habitats, earthworms display a wide array of morphological, physiological, and behavioral adaptations to environmental conditions (Lee, 1985). Even in unsuitable regions, earthworms may inhabit local microsites where conditions are favorable (e.g., urban gardens, desert oases), especially if well-adapted species have been introduced (Gates, 1967). During unfavorable periods, many species are able to enter a temporary dormant state (aestivation or diapause) or produce resistant cocoons that hatch when conditions improve (Edwards and Bohlen, 1996). Within habitats, earthworms often show patchy spatial distributions corresponding with such factors as vegetation, soil texture, or soil organic matter; feeding preferences dictate vertical distributions of species within the soil profile.

Abundance and biomass of earthworms establish them as major factors in soil biology, leading Blakemore (2002) to remark: *"And while birdwatchers get excited about a few kilograms of birdlife, or the grazier is concerned about a couple of 100's kg per hectare of livestock in a pasture, almost totally ignored is an underground biomass of earthworms often far in excess of those above that may total 2 or 3 tonnes per hectare."* Earthworm densities in a variety of habitats worldwide range from less than 10 to more than 2000 individuals m^{-2}, the highest values occurring in fertilized pastures and the lowest in acid or arid soils (coniferous or sclerophyll forests) (Table 4.10). Typical densities from temperate deciduous or tropical forests and certain arable systems range from less than 100 to more than 400 individuals m^{-2}. Intensive land management (especially soil tillage and application of toxic chemicals) often reduces the density of earthworms or may completely eliminate them. Conversely, degraded soils converted to conservation management (e.g., no-tillage) often show increased earthworm densities and associated soil properties after a suitable period of time (Curry *et al.*, 1995; Edwards and Bohlen, 1996). Biomasses of lumbricid species in temperate regions of the world, where they have been spread by human activities, often exceed that of other animal groups. In the Piedmont region of Georgia in the United States, for example, Hendrix *et al.* (1987) reported an earthworm dry-matter biomass of 10-g carbon m^{-2} in no-tillage agricultural plots, a value larger than all other fauna combined.

Biology and Ecology

Earthworms are soft-bodied, segmented animals, ranging in length from a few millimeters (e.g., the American log worm, *Bimastos parvus*),

TABLE 4.10.   Typical Ranges of Earthworm Density and Biomass in Various Habitats

Habitat	Earthworms m^{-2}	Fresh wt. g m^{-2}
Temperate hardwood forest	100–200	20–100
Temperate coniferous forest	10–100	30–35
Temperate pastures	300–1000	50–100
Temperate grassland	50–200	10–50
Sclerophyll forest	<10–50	<10–30
Taiga	<10–25	≤10
Tropical rainforest	50–200	<10–50
Arable soil	<10–200	<10–50

Summarized from Lee, 1985, and Edwards and Bohlen, 1996.

to more than a meter (e.g., the giant Gippsland earthworm of Australia, *Megascolides australis*) (Fig. 4.61). Morphological details differ greatly among earthworm groups and many such details (e.g., position of reproductive organs) are used in taxonomic distinctions among species (Dindal, 1990). Nonetheless, a number of features are common to most earthworms (Fig. 4.62). In general, earthworms consist of a simple, tube-within-a-tube body plan, the outer tube constituting the body proper and the internal tube comprising the alimentary canal. Ingested material (e.g., mineral soil, particulate organic matter) is drawn through the mouth into a muscular buccal cavity and then through the pharynx into the esophagus. Many species have a muscular esophageal gizzard that grinds and mixes food material as it passes through. The esophagus in many species also contains a calciferous gland that functions in calcium metabolism and regulation of CO_2 levels in the blood. The remainder of the gut consists of the intestine that, in many endogeic species, has an infolding of the gut wall known as the typhlosole, which greatly increases the absorptive surface area of the intestine. The overall length of the gut and configuration of the typhlosole vary with species, probably as a function of diet.

Earthworms are hermaphroditic, each individual possessing male and female reproductive organs (testes, ovaries, and associated structures) (Edwards and Bohlen, 1996). During sexual reproduction, sperm is exchanged between two individuals and stored in sperm sacs or spermathecae. This sperm is later released, along with eggs, into cocoons secreted by the glandular clitellum, which is the characteristic thickening or saddle-shaped structure often seen around several anterior

FIGURE 4.61. The Australian giant Gippsland earthworm, *Megascolides australis*, measuring up to 3 m in length (from Blakemore, 2002). (Alan L. Yen, with permission.)

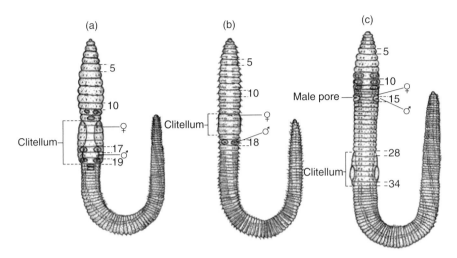

FIGURE 4.62.   General external characteristics of representatives from three earthworm families. (a) Acanthodrilidae. (b) Megascolecidae. (c) Lumbricidae (from Blakemore, 2002).

segments of sexually mature individuals (Fig. 4.63). One to several embryos may form within each cocoon, depending on earthworm species. Some earthworms reproduce parthenogenetically, whereby an ovum develops without fertilization by sperm. Parthenogenesis provides an effective means by which certain species can establish populations in new habitats; such species often are the successful peregrines and anthropochores discussed previously.

Earthworms are often grouped into functional categories based on their morphology, their behavior and feeding ecology, and their microhabitats within the soil (Lee, 1959, 1985; Bouché, 1977, 1983; Lavelle, 1983). These categories describe the ways by which different earthworm species utilize resources within a soil volume (Table 4.11) (Fig. 4.64). Epigeic and epi-endogeic species are often polyhumic (prefer organically enriched substrates) and utilize plant litter on the soil surface and carbon-rich upper layers of mineral soil; poly-, meso-, and oligohumic endogeic species inhabit mineral soil with high (e.g., the rhizosphere), moderate, and low organic matter content, respectively; and anecic species exploit both the surface litter as a source of food and the mineral soil as a refuge in which they make permanent burrows. The familiar *Lumbricus terrestris* is an example of an anecic species, constructing burrows and pulling leaf litter down into them. In contrast, *Bimastos parvus* (the American log worm) exploits leaf litter and decaying logs, with little involvement in the soil—an epigeic species. The ubiquitous

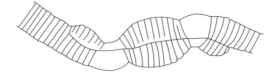

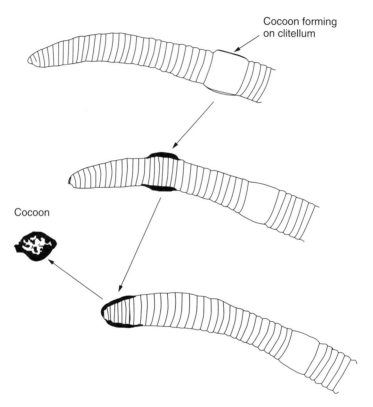

FIGURE 4.63.   Copulation and subsequent cocoon formation during sexual reproduction in lumbricid earthworms (from Edwards and Bohlen, 1996).

European lumbricid, *Aporrectodea caliginosa*, and several megascolecids (e.g., *Diplocardia* spp., native to eastern North America) are endogeic in life habits. Some earthworm species appear to be intermediate between these categories; for example the epi-endogeic *Lumbricus rubellus,* which can inhabit litter layers and form shallow horizontal burrows. Even though some species may not exactly fit, these categories

TABLE 4.11. Ecological Categories, Habitat, Feeding, and Morphological Characteristics of Earthworms

Category	Subcategory	Habitat	Food	Size and pigmentation
Epigeic	Epigeic	Litter	Leaf litter, microbes	<10 cm, highly pigmented
	Epi-endogeic /Epi-anecic	Surface soil	Leaf litter, microbes	10–15 cm, partially pigmented
Endogeic	Polyhumic	Surface soil or root zone	Soil with high organic content	<15 cm, filiform unpigmented
	Mesohumic	Upper 0–20 cm soil	Soil from 0–10 cm strata	10–20 cm, unpigmented
	Endo-anecic	0–50 cm soil, some make burrows	Soil from 0–10 cm strata	>20 cm, unpigmented
	Oligohumic	15–80 cm soil	Soil from 20–40 cm strata	>20 cm, unpigmented
Anecic	Anecic	Lives in burrows in soil	Litter and soil	>15 cm, anterodorsal pigmentation

Modified from Barois *et al.*, 1999.

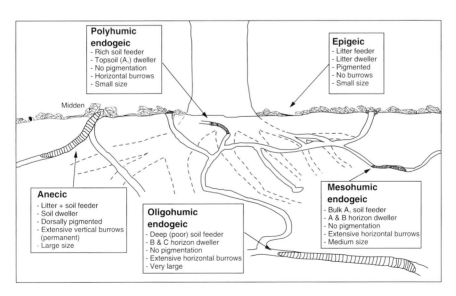

FIGURE 4.64. Pictorial representation of some of the characteristics of earthworm ecological strategies (categories) as proposed by Bouché (1977), Lavelle (1981), and Lavelle *et al.* (1989) (from Brown *et al.*, 1995).

have become a popular means for segregating earthworm communities into functional groups of species.

Within a particular soil, less than a half-dozen earthworm species are typically found, and the species within such an association often effectively partition the soil volume according to their functional categories. Further, the activities of earthworms influence soil processes in various ways according to these functional categories. For example, epigeic species promote the breakdown and mineralization of surface litter, whereas anecic species incorporate organic matter deeper into the soil profile and facilitate aeration and water infiltration through their formation of burrows.

Influence on Soil Processes

Earthworms, as ecosystem engineers (Lavelle et al., 1998), have pronounced effects on soil structure as a consequence of their burrowing activities as well as their ingestion of soil and production of castings (Lavelle and Spain, 2001; van Vliet and Hendrix, 2003). Casts are produced after earthworms ingest mineral soil and/or particulate organic matter, mix them together and enrich them with organic secretions in the gut, and then egest the material as a slurry or as discrete fecal pellets within or upon the soil, depending on earthworm species. Darwin (1881) observed that surface casting by earthworms buried chalk to considerable depths in soil over a 20-year period. Turnover rates of soil through earthworm casting range from $40–70 \, t \cdot ha^{-1} \cdot y^{-1}$ per hectare per year in temperate grasslands (Bouché, 1983) to $500–1000 \, t \cdot ha^{-1} \cdot y^{-1}$ per hectare per year in tropical savannas (Lavelle, 1978).

During formation in the earthworm gut, casts are colonized by microbes that begin to break down soil organic matter. As casts are deposited into the soil, microbial colonization and activity continue until readily decomposable compounds are depleted. Eventually, casts may harden into stable soil aggregates. Mechanisms of cast stabilization include organic bonding of particles by polymers secreted by earthworms and microbes, mechanical stabilization by plant fibers and fungal hyphae, and stabilization due to wetting and drying cycles and age-hardening effects (Tomlin et al., 1995). Earthworm casts are usually enriched with plant-available nutrients and thus may enhance soil fertility; plant-growth–promoting substances have also been suggested as constituents of earthworm casts. Castings from vermicomposting operations are sold commercially as soil amendments that purportedly enhance plant growth (Edwards, 1998).

Earthworm burrowing in soil creates macropores of various sizes, depths, and orientations, depending on species and soil type. Burrows tend to be similar in diameter to that of the earthworms that produced them, ranging from about 1 mm to larger than 10 mm in diameter and constituting among the largest of soil pores (Edwards and Shipitalo, 1998). Burrows of epigeic earthworms (e.g., *Dendrobaena octaedra*) are often small and limited to upper layers of soil; they may be horizontal to vertical in orientation. Endogeic species (e.g., *Diplocardia mississippiensis* or *Pontoscolex corethrurus*) may form networks of variously oriented burrows, as the earthworms ingest soil and cast behind them as they burrow. These networks may form continuous pores over some depth, but castings within the burrows may impede free water movement. Anecic earthworms (e.g., *Lumbricus terrestris*) may create deep vertical burrows that form continuous macropores to depths of 1 m or more (van Vliet and Hendrix, 2003). These burrows tend to be very stable because their walls are lined with organic matter drawn in or secreted by earthworms, and they often have higher bulk density than that of surrounding soil (Lee, 1985). Continuous macropores resulting from earthworm burrowing may enhance water infiltration by functioning as bypass flow pathways through saturated soils. These pores may or may not be important in solute transport, depending on soil water content, nature of the solute, and chemical exchange properties of the burrow linings (Edwards and Shipitalo, 1998).

The influence of earthworms on organic matter and nutrient cycling in soils is closely related to the density and feeding ecology of resident populations, as described previously (Lee, 1985; Barois *et al.*, 1999). Epigeic species typically inhabit the surface litter and the O and upper A horizons of soil, where they mix mineral soil and plant residues, fragment organic particles, inoculate them with microbes, and thus increase organic matter decomposition rates. Anecic earthworms pull surface litter into their burrows, thus transporting organic material deeper into the soil profile. They cast on the soil surface, mixing organic and mineral particles in the litter layer. The activities of both epigeic and anecic earthworms produce "mull" soil horizons, in which organic matter is intimately incorporated into the upper mineral soil of a well-developed A horizon overlain with a recently deposited litter layer. The extreme case is termed "vermimull," in which the Ah horizon is granular and characterized by organo-mineral complexes consisting of earthworm casts (Green *et al.*, 1994). Endogeic earthworms feed within the soil on organic matter and microbes associated with plant roots or mineral soil. As mentioned previously, they are termed oligo-, meso-, or polyhumic, depending on the level of organic enrichment of their substrate. Casts and burrows of endogeic earthworms are also sites of increased microbial activity and organic matter decomposition (Brown, 1995). Mineralization of organic matter in earthworm casts and burrow linings produces zones of nutrient enrichment that are different from those in bulk soil. These zones are referred to as the "drilosphere" and are often sites of enhanced activity of plant roots and other soil biota (Lavelle *et al.*, 1998).

Despite the many beneficial effects of earthworms on soil processes, some aspects of earthworm activities may be undesirable (Edwards and Bohlen, 1996; Lavelle, 1998; Parmelee *et al.*, 1998). Detrimental effects include removing and burying of surface residues that would otherwise protect soil surfaces from erosion; producing fresh casts that increase erosion and surface sealing; increasing compaction of surface soils; depositing castings on the surface of lawns and golf greens where they are a nuisance; dispersing weed seeds in gardens and agricultural fields; transmitting plant or animal pathogens; riddling irrigation ditches, making them less able to carry water; increasing losses of soil nitrogen through leaching and denitrification; and increasing soil carbon loss through enhanced microbial respiration. Furthermore, there have been reports of earthworms transmitting pathogens, either as passive carriers or as intermediate hosts, raising concerns that some earthworm species could provide a mechanism for the spread of certain plant and animal diseases.

Thus it is the net result of positive and negative effects of earthworms, or any other soil biota, that determines whether they have detrimental impacts on ecosystems (Lavelle *et al.*, 1998). An effect, such as mixing of O- and A-horizons, may be considered beneficial in one setting (e.g., urban gardens) and detrimental in another (e.g., native forests).

Earthworm Management

There is interest in managing earthworms to utilize their beneficial effects in organic waste reduction, in land reclamation, and in reduced-intensity agriculture (Lee, 1995; Edwards and Bohlen, 1996). Because of their effects on organic matter decay, earthworms are increasingly being used to accelerate decomposition of organic waste materials. Vermicomposting involves culturing of earthworms outdoors in beds or in confined chambers in the presence of waste materials, which are reduced in volume and carbon–nitrogen ratio as they are processed by earthworms and decomposed by enhanced microbial activity within the earthworms and their castings (Edwards, 1998). A variety of approaches and designs has been developed for vermicomposting systems, but the basic principle is the feeding of acceptable organic materials to earthworms in continuous or batch culture, and the collecting of processed wastes that ultimately consist of stabilized castings. Earthworm biomass is also harvested from vermicomposting systems for a variety of uses, including further composting operations, animal protein, and fishing bait. Organic wastes that have been used successfully in vermicomposting include animal manures, sewage sludge, food production wastes, and horticultural residues. Small-scale vermicomposting is becoming popular for reduction of household wastes such as kitchen scraps and yard trimmings. Earthworm species typically used for vermicomposting included the European lumbricids, *Eisenia fetida* and *Lumbricus rubellus*, often called "red worms"; the African "night-crawler," *Eudrilus eugeniae*; and the Asian "blue worm," *Perionyx excavatus*. The latter two species are tropical and best suited to composting under warm conditions (Edwards, 1998). A number of vermicomposting publications have appeared in the popular literature in the last decade, including the periodical *Worm Digest* and the ever-popular *Worms Eat Our Garbage* (Appelhoff *et al.*, 1993).

The potential for earthworms to ameliorate soils during land reclamation or in degraded agricultural sites is also of increasing interest (Lee, 1995; Baker, 1998). In many situations, it may be desirable to introduce earthworms. Techniques have been developed for large-scale inoculation of areas devoid of earthworms (e.g., reclaimed polders) and for introduction of species that may perform desired functions (e.g., epi-endogeic species for thatch removal from pastures). It is usually necessary that favorable soil conditions (e.g., adequate water and organic matter, appropriate temperatures) exist at the time of inoculation and/or that refugia (e.g., blocks of native sod or containers of native soil) are provided from which earthworms may disperse. Introductions of earthworms into unfavorable environments often fail.

Mixed-species assemblages of earthworms may influence a wider array of soil processes, such as organic matter turnover as well as soil structural properties, than a single species can (Lee, 1995). Introductions of such assemblages might include one or more anecic species that make deep vertical burrows and that cast on the surface and bury residues, and one or more endogeic species that feed belowground on dead roots and organic matter and that make horizontal burrows. Inclusion of epigeic species might accelerate decomposition of plant residues on the soil surface.

Earthworm Sampling and Identification

A variety of sampling methods have been used for collecting earthworms, both quantitatively and qualitatively (see Table 9.2). Hand digging and sorting of soil is the most commonly used method for quantitative sampling of earthworms. Pits of known dimensions (e.g., 25 by 25 by 25 cm) are dug with a shovel, often in layers of defined thickness, and the soil broken by hand. More elaborate modifications of this method, which may improve collection of juveniles and cocoons, include dry or wet sieving of soil through screens of known mesh size, and flotation of sieved material in high-density solutions to separate earthworms and other soil fauna. Earthworms may be collected alive or immediately preserved in 70% ethanol or 5% formalin for later counting and identification.

Earthworms are also collected by applying solutions of chemical irritants to the soil, which bring earthworms to the surface where they can be hand collected. A number of chemicals have been used, including $HgCl_2$, $KMnO_4$, formalin, and mustard powder slurry (Lee, 1985; Zaborski, 2003). The latter has received much attention recently because of its safety and availability. Chemical extraction techniques may be effective on anecic earthworms such as *Lumbricus terrestris* but may be less so on other species. Effectiveness varies with earthworm species and activity, soil water content, porosity, and temperature. Comparisons with hand sorting should be done before adopting extraction techniques for quantitative sampling.

Mechanical or electrical stimulation may also bring earthworms to the soil surface. A technique known as "grunting" employs a wooden stake driven into the soil, vibration of the stake with a bow or flat piece of metal, and collection of earthworms that emerge. Some megascolecid species have been sampled with this technique (Hendrix *et al.*, 1994) but it may not be effective on other earthworm groups.

Electrical extraction of earthworms uses metal rods connected to a source of electrical current and inserted into the soil; the current brings earthworms to the surface. Different voltages and amperages have been used with varying degrees of success; effectiveness of the technique is highly dependent on soil water content, electrolyte concentration, and temperature. As with mechanical vibration, the soil volume sampled with electrical current is not known and therefore these methods may be best suited for qualitative or comparative sampling (Lee, 1985). However, Schmidt (2001) used a commercially available "Octet" device to quantitatively sample earthworm communities in arable soils in Ireland; the electrical method gave estimates of species composition and population size comparable to those from hand digging, with the exception of one very small species. The electrical technique is potentially very dangerous and should only be used with extreme caution.

Earthworm populations also may be sampled with trapping techniques. Pitfall traps (described above) may give some idea of surface-active earthworm species present in an area. Baited traps consist of porous containers (e.g., clay flower pots) filled with bait such as animal manure and buried in the soil for appropriate periods of time. Trapping techniques are highly selective of certain earthworm species and thus are best suited for qualitative or comparative sampling (Lee, 1985).

For earthworm species that cast on the soil surface (e.g., *Lumbricus terrestris*), numbers and types of castings may give an indication of population activity. Because casting is dependent on soil water content and temperature, this technique is highly variable and not suitable for quantitative estimates of population density.

Many earthworm species in the family Lumbricidae can be identified from external body characteristics if the specimens are sexually mature. Several taxonomic keys are useful for the common lumbricids found worldwide (Reynolds, 1977; Sims and Gerard, 1985; Schwert, 1990). Most other earthworms require dissection for accurate taxonomic identification; position and characteristics of sexual organs, the gut and associated glands, and other structures are required. The procedures must be done carefully and require a degree of skill and practice. Additional general taxonomic references include Jamieson (1988), Fender and McKee-Fender (1990), James (1990), Edwards and Bohlen (1996), and Fragoso *et al.* (1999).

GENERAL ATTRIBUTES OF FAUNA IN SOIL SYSTEMS

In recent years, interest has been shown by soil scientists and ecologists in measuring "soil quality." This elusive concept has been the subject of entire symposia and volumes resulting from them (e.g., Doran *et al.*, 1994). As defined by soil scientists, soil quality can be considered as the degree or extent to which a soil can: (1) promote biological activity (plant, animal, and microbial); (2) mediate water flow through the environment, and (3) maintain environmental quality by acting as a buffer that assimilates organic wastes and ameliorates contaminants (Linden *et al.*, 1994). Many environmental scientists are attempting to use the concept of indicator organisms or indicator communities as a way to determine overall soil "health" (e.g., Bongers, 1990; Ettema and Bongers, 1993; Foissner, 1994; Linden *et al.*, 1994; Neher *et al.*, 1995; Ferris *et al.*, 2001). Because of their large size and public awareness of them, earthworms are often considered a sign of soil "health" (Linden *et al.*, 1994; Hendrix, 1995). All of the biota play important roles in affecting and influencing soil processes. As summarized in Table 4.12, each of the biotic groups has significant impacts. Among the fauna, microfauna have a principal role via interactions with the microflora. The mesofauna and macrofauna create fecal pellets, and produce biopores of various sizes, which affect water movement and storage as well as root growth and proliferation. Perhaps more important, over the longer term, they have marked effects on humification processes as well (Wolters, 1991). Based on biological characteristics, there are three general trophic systems: microtrophic (protozoa, nematodes, and some enchytraeids), mesotrophic (the mesofauna), and macrotrophic (the

TABLE 4.12.   Influences of Soil Biota on Soil Processes in Ecosystems

	Nutrient cycling	*Soil structure*
Microflora	Catabolize organic matter Mineralize and immobilize nutrients	Produce organic compounds that bind aggregates Hyphae entangle particles onto aggregates
Microfauna	Regulate bacterial and fungal populations Alter nutrient turnover	May affect aggregate structure through interactions with microflora
Mesofauna	Regulate fungal and micro- faunal populations Alter nutrient turnover Fragment plant residues	Produce fecal pellets Create biopores Promote humification
Macrofauna	Fragment plant residues Stimulate microbial activity	Mix organic and mineral particles Redistribute organic matter and microorganisms Create biopores Promote humification Produce fecal pellets

From Hendrix *et al.*, 1990.

large fauna capable of breaking through physical barriers of soil) (Heal and Dighton, 1985).

Further concerns about fauna as indicators of soil quality led Linden *et al.* (1994) to erect a hierarchical array of three categories in which fauna and soil quality interact, namely: (1) organisms and populations, relating to behavior, physiology, and numbers; (2) communities, with concerns about functional groups (i.e., guilds of burrowers and nonbur- rowers, trophic groups, and biodiversity); and (3) biological processes, relating to the several processes and properties listed in Table 4.13 (Lin- den *et al.*, 1994). These processes are considered in greater detail in later chapters on decomposition and nutrient cycling processes.

FAUNAL FEEDBACKS ON MICROBIAL COMMUNITY COMPOSITION AND DIVERSITY

Since the time of Darwin's epochal book on soil biology (1881), there has been considerable interest in the effects of fauna on microbial com- munities. Satchell (1983), in a centenary celebration of Darwin's book, stated that, by the culture methods existing up until then, there was lit- tle or no indication that earthworms have a qualitatively different flora

TABLE 4.13. Properties of Soil Fauna for Use as Indicators of Soil Quality

1. Organisms and populations
   Individuals
   Behavior, morphology, physiology
   Populations
   Numbers and biomass
   Rates of growth, mortality, and reproduction
   Age distribution
2. Communities
   Functional groups
   Guilds (e.g., burrowers vs. nonburrowers, litter vs. soil dwellers, etc.)
   Trophic groups
   Food chains and food webs (microbivores, predators, etc.)
   Biodiversity
   Species richness, dominance, evenness
   Keystone species
3. Biological processes
   Bioaccumulation
   Heavy metals and organic pollutants
   Decomposition
   Fragmentation of organic matter
   Mineralization of C and nutrients
   Soil structure modification
   Burrowing and biopore formation
   Fecal deposition and soil aggregation
   Mixing and redistribution of organic matter

From Curry and Good, 1992 and Linden *et al.*, 1994.

in their guts or in their castings. Yet there are numerous studies that have indicated an increase in microbial numbers or activity during or after passage through the gut and in the drilosphere (Barois, 1992; Daniel and Anderson, 1992; Kristufek *et al.*, 1992; Schoenholzer *et al.*, 1999). Recent studies using both molecular and culture-based analyses of agricultural soil and burrows and casts of the epigeic earthworm *Lumbricus rubellus* have revealed interesting differences between earthworm- and non-earthworm–influenced soils (Furlong *et al.*, 2002). Clone libraries of the 16S rRNA genes were prepared from DNA isolated directly from the soil and earthworm casts. Representatives of the *Pseudomonas* genus as well as the Actinobacteria and Firmicutes increased in number, and one group of unclassified organisms found in the soil library was absent in that of the cast. In fact, Singleton *et al.* (2002) isolated a new species of bacterium, *Solirubrobacter paulii*, from the intestinal wall of *Lumbricus rubellus*. This was not found in the soil library, and may represent the first known instance of a bacterium unique to the gut wall of an earthworm.

Studies of microbial community similarity have been conducted comparing termite mounds and nearby tropical soils. Harry *et al.* (2001) used RAPD (random amplified polymorphic DNA) molecular markers to estimate the similarity of microbial communities in the mounds of several termite species and surrounding soils. They studied four species of soil-feeding termites and one species of fungal-feeding termite in a tropical rain forest area of the Nyong River basin in Southern Cameroon. They found that microbial communities of the mounds of the soil-feeding termite species were clustered in the same clade, whereas those of the mounds of the fungus-growing species were distinct like those of the control soils. The microbial changes were dependent upon the species' behavior, with the soil feeding species including feces in their mound building and the fungal-feeding species using saliva as particle cement in its mounds.

SUMMARY

Animals in soils are a large, numerous, and diverse group of species, organized into complex food webs. In addition to a formal taxonomic classification, the soil fauna may be classified in several ways: persistence in the soil, distribution through the soil profile, body shape, and body size. The latter classification, body size, has the advantages of separating fauna into groups collected and quantified in similar manners. Methods for study of the microfauna including the protozoa are essentially the methods of microbiology. Among the mesofauna, the abundant and ubiquitous nematodes have significant impacts on microbial population and on roots. Another group, the microarthropods, contains mites and collembolans that feed on plant debris rich in fungus, nematodes, and other arthropods as well. The combination of microbes, nematodes, and microarthropods provides complex food webs, whose connections may vary opportunistically. The macrofauna contains a large group of arthropods, including the familiar isopods and millipedes as detritus feeders, and scorpions, spiders, and other predators. Pterygote (winged) insects are numerous in soils. The termites and ants are important soil movers (bioturbators) in many situations, as are earthworms. The earthworms may be the single most important groups of soil animals, in terms of their feeding upon detritus and their effects on soil structure. But the entire fauna is involved in maintenance of soil health. The microfauna and microfloral interactions, the feeding of the mesofauna on microbial-rich detritus, and the creation of biopores and the bioturbation effects of the larger mesofauna all interact in creating soil quality (see Chapter 8).

As noted in the summary to Chapter 3, the prospects of linking microbes and fauna by meaningful qualitative (structural) and quantitative (functional) techniques are growing rapidly, and the future is bright for synthesis in soil ecology studies.

5 Decomposition and Nutrient Cycling

INTRODUCTION

The bulk of terrestrial net primary production (NPP), along with the bodies and excretions of animals, is returned to the soil as dead organic matter. Some 90% of NPP eventually enters the soil system through dead plants in grasslands; through leaves, roots, and wood in forests; and through organic residue in agricultural fields. Indeed, ecosystems may be viewed as consisting of four functional subsystems: (1) the production subsystem, (2) the consumption subsystem, (3) the decomposition subsystem, and (4) the abiotic subsystem. The decomposition subsystem serves to reduce dead residues to carbon dioxide (CO_2) and soil organic matter, and to release nutrient elements for entry into soil food webs, and ultimately for reaccumulation by plants. The decomposition process drives complex belowground food webs, in which chemical forms of nutrient elements become modified. It is responsible for the creation of long- and short-lived organic compounds important in nutrient dynamics, and it fuels the formation of soil structure.

Terrestrial plant growth is highly dependent on the decomposition system, particularly in oligotrophic soils where nutrient stocks are held in litter and soil organic matter, rather than in mineral soil. Heterotrophic organisms in the soil are ultimately responsible for ensuring the availability of nutrients for primary production (Wardle, 2002). Thus the two subsystems, primary production and decomposition, are dependent upon each other. We emphasize, again, the necessity for evaluation of entire ecosystems when considering their respective parts. The soil subsystem performs crucial functions within terrestrial ecosystems, regardless of how modified the terrestrial ecosystems may be. Decomposition processes in highly modified agricultural systems still involve a significant variety of heterotrophic organisms with character-istic abilities (Wasylik, 1995).

Decomposition per se is the catabolism of organic compounds in plant litter or other organic detritus. As such, decomposition is mainly the result of microbial activities. Few soil animals have the enzymes that would allow them to digest plant litter. Animal nutrition depends upon the action of microbes, either free-living in the soil or specialized in the rhizosphere or in animal guts. However, the term "decomposition" is often used more generally to refer to the breakdown or disappearance of organic litter. In that context, the decomposition of organic residue involves the activities of a variety of soil biota, including both microbes and fauna, which interact together. The term "litter breakdown" has been applied to the interactive process, which results in the disappear-ance of organic litter.

Continuing interest in decomposition is apparent from the large number of studies of the process that have been published during the past 25 years. More than 1000 such publications have appeared in peer-reviewed journals, and the number would be much larger if symposia or reports on heterotrophs themselves were to be included (Heal *et al.*, 1997). Improved understanding of the decomposition process has accompanied the refinement of methods and conceptual models. The lit-terbag technique (Bocock and Gilbert, 1957; Shanks and Olson, 1961; Edwards and Heath, 1962; Crossley and Hoglund, 1962) has become a major tool in these studies, despite its limitations (Heal *et al.*, 1997). Radioactive tracers (Olson and Crossley, 1962) have been replaced by methods using stable isotopes of carbon and nitrogen (Nadelhoffer and Raich, 1992; Boutton and Yamasaki, 1996). Early models of mass loss (Jenny, 1941; Olson, 1963) defining a decomposition constant, k, are being supplanted by more sophisticated models that consider different constituents of litter (Jenkinson *et al.*, 1987; Parton *et al.*, 1994; Sinsabaugh and Moorhead, 1997).

INTEGRATING VARIABLES

In studies of soil systems, rates of litter breakdown have been used as integrating variables. That is, because litter breakdown rates are the result of the combined activities of the soil biota, breakdown rates may be used to evaluate effects of disturbance on the entire system. For example, conversion of agricultural systems into conservation tillage regimes will affect soil biology, notably by shifting the composition of microbial communities and increasing earthworm population densities, but with changes in other soil biota as well (Doran, 1980, Parmelee *et al.*, 1990, Beare *et al.*, 1992). We can evaluate the consequences of these

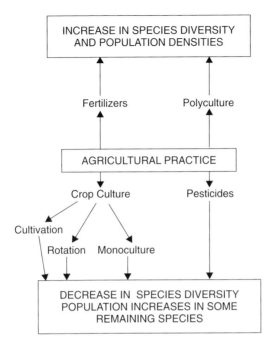

FIGURE 5.1. Changes in species diversity of soil microarthropods as a function of agricultural practice (from Crossley *et al.*, 1992).

changes in soil biota by measuring litter breakdown rates (Fig. 5.1) (Crossley *et al.*, 1992).

Other integrating variables include soil respiration, formation of soil structure, and nutrient dynamics. All of these variables are readily measured, and all are important for ecosystem function. Soil respiration estimates biological activity generally and is dominated by microbes, with an important contribution by roots (Cheng *et al.*, 1993; Kuzyakov, 2002). Soil structure is the result of combined actions of biota and climate on mineral substrates. Nutrient dynamics are the most valuable of the integrating variables for predicting primary productivity.

Although microbes are responsible for the biochemical degradation of organic litter, fauna are important in conditioning the litter and aiding in microbial actions. The soil scientist Hans Jenny characterized soil fauna as mechanical blenders: "They break up[1] plant material, expose organic surface areas to microbes, move fragments and bacteria-rich

---

[1]Although we are accustomed to the term litter break*down*, the term litter break*up* appears to be exactly equivalent.

excrement around, up, and down, and function as homogenizers of soil strata" (Jenny, 1980). Breakdown rates for organic litter integrate the effects of these various activities into a single set of variables. The combination of microbial and faunal activities results in a set of positive interactions of the type termed "facilitation" by Bruno *et al.* (2003). Results of the interactions are likely to be more significant than that by either component—animal or microbial—acting alone.

RESOURCE QUALITY, CLIMATE, AND LITTER BREAKDOWN

Litter breakdown rates vary between and among ecosystems on localized and broad geographic scales, as functions of soil biota, substrate quality, microclimate, and ecosystem condition. In general, we view breakdown and decomposition as the result of biota acting on substrates of varying quality within the constraints of climate and soil properties.

Resource quality is defined principally by the chemical composition of organic residues deposited on or in the soil. Sugars and starches (i.e., labile substrates) are easily digested by microbes and other soil biota, whereas tannins, lignins, and other compounds rich in polyphenols (i.e., recalcitrant substrates) can be utilized directly only by certain specialized organisms (e.g., white-rot fungi). Cellulose and hemicelluloses are intermediate in their degradability. Hence the relative proportions of these classes of compounds in organic materials greatly influence the overall rate of decomposition of those materials (Fig. 5.2) (Berg, 1986). Organic litter in most terrestrial ecosystems is a mixture of relatively labile and relatively recalcitrant substrates—thin, calcium-rich dogwood (*Cornus florida*) leaves versus thick, highly lignified, oak leaves (*Quercus* spp.) or conifer needles (*Pinus* spp.), for example. Even in agricultural systems, differences between leaves and stalks of corn (*Zea mays*), for example, represent different substrate qualities with different breakdown rates. Woody litter, high in tannins and lignins, may have breakdown rates measured in decades or even centuries for large logs in cool climates (Harmon and Chen, 1991). Fine root turnover may be measured in days, but coarse roots, with highly suberized tissues, turn over in years.

On a broad geographic basis, the change in breakdown rates as a function of latitude is generally predictable (Fig. 5.3) (Meentemeyer, 1978). However, the effect of latitude is not strictly a direct effect of climate; the abundance of the various soil biotas also changes with latitude (Fig. 5.4) (Swift *et al.*, 1979). For example, adaptations of the soil biota to desert conditions allow breakdown rates to proceed more rapidly than predicted by temperature–moisture considerations. Members of the desert soil biota are active nocturnally, when temperatures moderate and light dew may accumulate. Litter breakdown in tropical

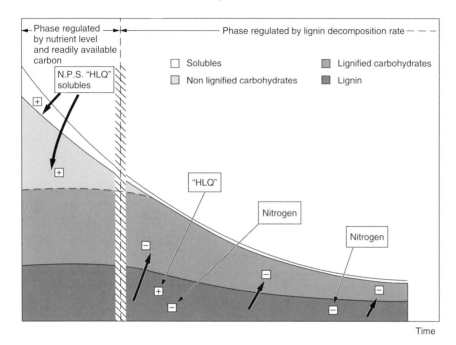

Time

FIGURE 5.2. Model for decomposition of some organic components in Scots pine (*Pinus silvestris*) needle litter. In the early phase of decomposition, high concentrations of nutrients such as nitrogen (N), phosphorus (P), and sulfer (S) exert a rate-enhancing influence on mass-loss of the nonlignified parts of the litter. Also, high concentrations of easily degraded solubles and celluloses influence a high mass-loss rate. In the late stage where mainly lignified material remains, lignin mass-loss is governing, which in its turn is negatively affected by high nitrogen concentrations and positively by high concentrations of celluloses in the lignified material. The negative effect of lignin on cellulose degradation is indicated by black arrows. A (+) indicates a rate-enhancing influence and (−) a negative one. HLQ designates the quotient between holocellulose and lignin plus holocellulose (from Berg, 1986).

systems may be strongly influenced by seasonality of litterfall as well as faunal abundance. González and Seastedt (2001) found that all three groups of factors (climate, substrate quality, and soil fauna) independently influenced the decomposition rate of leaf litter in tropical dry and subalpine forests. Soil fauna had a disproportionately larger effect on litter decomposition in a tropical wet forest than in tropical dry or subalpine forests.

Decomposition rates may vary along elevational gradients as well, but not as predictably. In a study conducted in Arizona in the United States, plant litter decomposition was measured along a gradient from desert to pinyon–juniper woodland and up into a ponderosa pine forest (Murphy *et al.*, 1998). Decomposition was more rapid at the upper, cooler elevation that was also moister. In these systems, moisture—not temperature—

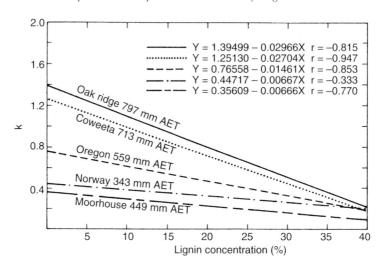

FIGURE 5.3. Simple correlation-regression between initial lignin concentration (%) and annual decomposition rate (k) for five locations ranging in climate from subpolar to warm temperate. AET, actual evapotranspiration (from Meentemeyer, 1978, with permission).

was the overriding variable. Similarly, in a hardwood forested ecosystem at Coweeta Hydrologic Laboratory, in North Carolina, decomposition did not vary predictably along an elevation gradient (Hoover and Crossley, 1995). Decomposition was slowest at a low-elevation, very mesic, cove hardwood site and was most rapid at intermediate elevation sites. Microclimate—temperature and moisture around decomposing substrates—regulates activity rates of the biota. Disturbed ecosystems and successional ones also may have litter breakdown rates that are slower than predicted from broad regional temperature–moisture conditions. Alteration of microclimates may reduce faunal activities, and substrate quality of foliage may change during plant succession. Furthermore, edaphic factors, particularly soil texture (i.e., relative proportions of sand, silt, and clay), greatly influence local microclimatic conditions by regulating the availability of surface water films for soil microbes and microfauna, and water holding capacity of the bulk soil for meso- and macrofauna. Thus soil–water relations exert indirect control on litter decomposition through their influence on soil biological activity.

DYNAMICS OF LITTER BREAKDOWN

The disappearance of litter on forest floors follows approximately a simple first-order equation:

$$dX/dt = -kX$$

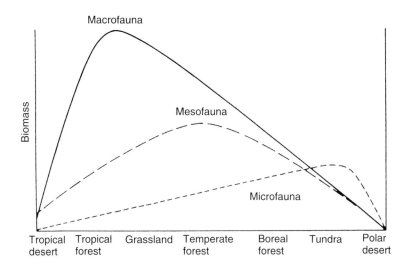

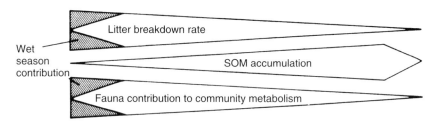

FIGURE 5.4. Hypothetical patterns of latitudinal variation in the contribution of the macro-, meso-, and microfauna to total soil fauna biomass. The effects on litter breakdown rates of changes in the relative importance of the three fauna size groups are represented as a gradient together with the faunal contribution to soil community metabolism. The favorability of the soil environment for microbial decomposition is represented by the cline of soil organic matter accumulation from the poles to the equator; soluble, or soil, organic matter (SOM) accumulation is promoted by low temperatures and waterlogging where microbial activity is impeded (from Swift *et al.*, 1979).

where X is the standing stock of litter and k is the annual fractional rate of disappearance. Olson (1963) proposed that it was a characteristic of mature forests that rates of litter production and disappearance were equal, so that annual production (L) would be balanced by breakdown $(-kX)$. Olson used the symbol X_{ss} to designate the standing stock of litter on the ground at steady state (i.e., when litter production and disappearance are equal). Then, the ratio of input (L) to standing stock (X_{ss}) provides an estimate of breakdown rate k:

$$k = L/X_{ss}$$

Olson (1963) estimated decomposition rates (k) for evergreen forests in various parts of the world (Fig. 5.5). Values for k ranged from 4 for rapid decomposition in tropical regions, through 0.25 for eastern United States pine forests, to 0.02 for higher latitude pine forests. Recent analyses using more sophisticated, mechanistic models have shown similar trends in litter decomposition across climatic gradients (e.g., Parton et al., 1989a; Moore and de Ruiter, 2000).

What is estimated here is the rate of leaf or needle litter breakdown. Olson's (1963), as most other models, did not consider inputs of organic litter belowground, although he did use the entire mass of carbon per square meters in estimates of X_{ss}. Current studies of root dynamics (see Chapter 2) are providing estimates of root breakdown rates, but these are more difficult to measure than leaf litter breakdown rates and consequently are less well known. Root death and decay may account for as much as one-half of the annual carbon addition to soils in forests or even more in grasslands; but as is often the case, dynamics within the soil are obscure.

The simple exponential model using a single constant, k, to represent decomposition rate continues to be widely used. It is not difficult to estimate k using litterbag techniques (see next section of this chapter). The simple model loses its attractiveness when patterns of litter breakdown are examined more closely. Leaf litter often is a combination of leaf species, each with different breakdown rates. Furthermore, each species contains both labile and recalcitrant fractions. Wieder and Lang (1982) examined several different models, and concluded that the single exponential model shown previously or double exponential models (including fast and slow components) best describe breakdown rates over time "with an element of biological realism." Jenkinson et al. (1987) and Paustian et al. (1997) considered single-pool, multiple litter pool, and continuous spectrum models of litter decomposition, and provided mathematical representations of decomposition rates. Their results show that litter quality is a key factor for accurately modeling decomposition dynamics.

DIRECT MEASUREMENT OF LITTER BREAKDOWN

In deciduous forests with annual pulses of leaf drop, it is possible to measure litter breakdown directly from litter samples taken through time. If combined with estimates of the mass of litterfall, these samples provide a good measure of the dynamics of litter breakdown (Fig. 5.6) (Witkamp and van der Drift, 1961). As the year progresses, the litter layer mass, L, becomes transformed into F layer (see Chapter 1). Sampling over several years reveals year-to-year variation in masses of

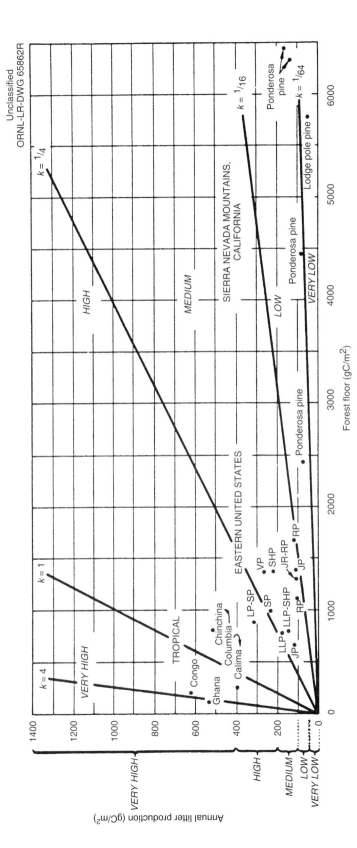

FIGURE 5.5. Estimates of decomposition rate factor k for carbon in evergreen forests, from the ratio of annual litter production (L) to (approximately) steady-state accumulation of forest floor (X_{ss}) (from Olson, 1963, with permission).

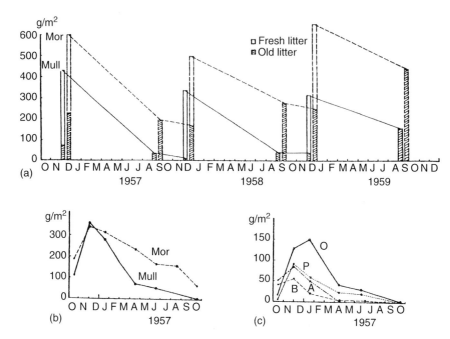

FIGURE 5.6.   (a) Rate of disappearance of litter in mull and mor. (b) Amount of litter on a cleaned surface in mull and mor. (c) Amount of fresh oak (O), birch (B), poplar (P), and alder (A) litter in mull (from Witkamp and van der Drift, 1961).

litter input and rates of breakdown (Table 5.1). It should be noted that two systems of organic layer horizonation are used in the soils literature: L, F, and H refer to litter, fermentation (or fragmented), and humus layers, respectively (see Green *et al.*, 1993, for a complete description of this biologically based system); these are equivalent to the O_i, O_e, and O_a layers often used in forest soils literature.

Rates of litter breakdown are measured more easily by using confined leaf litter. Mesh bags (litterbags) containing a known mass of leaf litter are placed on the forest floor at the time of leaf drop. Litterbags are then collected on a time schedule and the remaining mass is measured (Fig. 5.7). Litterbags have been a valuable tool for comparative studies of rates of litter breakdown (Fig. 5.8). Such studies include mass loss rates by different tree species, and have shown the importance of elemental contents, lignin, carbon–nitrogen ratios, and other resource quality factors (Table 5.2) (Blair, 1988a; Melillo *et al.*, 1982). Decomposition rates also vary between habitats and forest types, and litterbags have proved to be useful in delineating and analyzing differences (Table 5.3) (Cromack, 1973; Heneghan *et al.*, 1998, 1999).

TABLE 5.1. Summary of Litter Decomposition Experiments Conducted on Clear-cut (WS 7) and Control (WS 2) Watersheds, at the Coweeta Hydrologic Laboratory[a]

Species	74–75 WS 7, pre-cut		75–77 WS 2, pre-cut		77–78 WS 7, post-cut		WS 2, post-cut	
	Rate	% Remaining	Rate	% Remaining	Rate	% Remaining	Rate	% Remaining
Liriodendron tulipifera	-0.682	49.4	-0.656	49.8	-0.545	60.0	-0.814	47.3
Acer rubrum	-0.529	49.0	-0.477	57.9	-0.324	71.5	-0.368	67.9
Quercus prinus	-0.336	69.3	-0.285	72.2	-0.242	79.3	-0.3000	76.0
Cornus florida	-1.309	27.8	-0.711	47.8	-0.531	59.6	-0.825	43.4
Robinia pseudoacacia	-0.250	72.4	-0.530	49.7	-0.330	69.1	-0.330	70.7

[a] Data are summarized for a 1-year study on WS 2 and 7 (1974–75), a 2-year study on WS 2 (1975–77), and a 1-year study on WS 2 and 7 (1977–78). Values are shown for decay rate (per yr) and % remaining after 1 year. Only first-year decay results are presented here.
From D. A. Crossley, Jr., unpublished.

FIGURE 5.7.   View of leaf litterbag.

TABLE 5.2.   Initial Litter Quality Variables as Predictors of First-Year Decay Rates[a]

Initial litter quality variable	r^2	Slope	Y-intercept
% Nitrogen	0.271	−0.978	−1.29
C:N ratio	0.138	0.007	−0.12
% Lignin	0.987	0.029	−0.96
Lignin:N ratio	0.967	0.027	−1.05
% Water soluble	0.322	−0.015	−0.06
% Ethanol soluble	0.426	−0.040	−0.32

[a]Coefficients of determination (r^2), slopes, and Y-intercepts of regressions relating first-year decay rate constants (k) to initial litter quality variables for litter of the three species examined.
From Blair, 1988.

Use of litterbags does have its problems. Fine-mesh bags, with openings of 1–2 millimeters (mm), will exclude most macrofauna and thus underestimate decomposition rates. Larger meshes allow larger fragments to escape the bags, thus overestimating decomposition. The microclimate within litterbags tends to be moister than that of unbagged litter, and thus more favorable for microbial activity

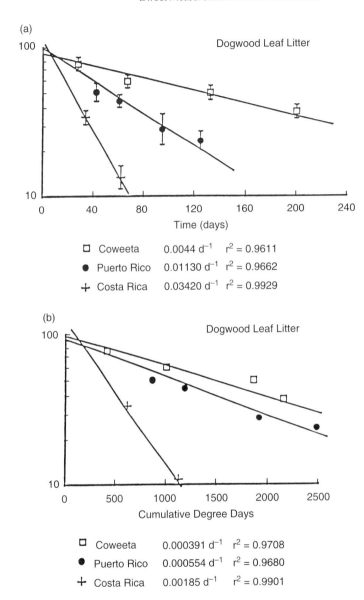

(a) Dogwood Leaf Litter

	Coweeta	0.0044 d^{-1}	r^2 = 0.9611
●	Puerto Rico	0.01130 d^{-1}	r^2 = 0.9662
+	Costa Rica	0.03420 d^{-1}	r^2 = 0.9929

(b) Dogwood Leaf Litter

	Coweeta	0.000391 d^{-1}	r^2 = 0.9708
●	Puerto Rico	0.000554 d^{-1}	r^2 = 0.9680
+	Costa Rica	0.00185 d^{-1}	r^2 = 0.9901

FIGURE 5.8. Litterbags and masses remaining showing days elapsed (a) and degree days (b) at one temperate (Coweeta) and two tropical (Puerto Rico and Costa Rica) sites (from Crossley and Haines, unpublished, 1990).

(Vossbrinck *et al.*, 1979). In most cases, litterbags probably underestimate actual breakdown rates and do not account for the fate of the fine particulate organic matter that falls from the bags and becomes part of the F, H, and mineral soil organic matter pools (Heal *et al.*, 1997). Their

TABLE 5.3.   First-Year Litter Breakdown Rates for Single Species[a]

Species	Year	% Weight Remaining	Exponential Loss Rate (k^a)	Correlation Coefficient (r)
White pine	1969–70	59.5	–0.52(12)	–0.888[b]
	1970–71	65.4	–0.42(19)	–0.480[c]
Chestnut oak	1969–70	57.8	–0.55(33)	–0.866[b]
	1970–71	51.4	–0.66(31)	–0.890[b]
White oak	1969–70	47.6	–0.74(36)	–0.887[b]
	1970–71	49.6	–0.70(32)	–0.906[b]
Red maple	1969–70	43.6	–0.83(36)	–0.934[b]
	1970–71	48.6	–0.72(30)	–0.839[b]
Dogwood	1969–70	30.8	–1.18(37)	–0.939[b]
	1970–71	26.0	–1.35(33)	–0.948[b]

[a]*Where* the annual exponential loss rate (k) is estimated from a semilogarithmic regression (base e) of monthly weight loss of litterbags.
[b]Denotes $p < 0.01$.
[c]Denotes $p < 0.05$.
From Cromack, 1973.

usefulness for comparative studies and for nutrient measurements makes them important tools, nevertheless.

Alternatives to litterbags and modifications of the standard approach have been reported. For example, individual leaves tied together by their petioles on a string ("trot-lines") were used by Crossley (unpublished data). Loss of weight (and area) by individual leaves measured through time yields estimates of litter breakdown rates. When biological activity increases in late spring and summer, rapid rates of loss are found. It is not clear whether these rapid losses are due to the separation of large fragments from the leaf, or if the unbagged rates allow for larger fauna to attack the decomposing leaf, or both. The simultaneous use of both techniques yields estimates of breakdown rates that doubtless bracket the true values. Blair and Crossley (1991) used "litter baskets," which confine an entire block of mineral soil, along with the L, F, and H layers, within wire mesh to study litter decomposition and nutrient transport down the soil profile. A further recent modification of the technique is that of "litter sandwiches" (Binkley, 2002), in which fiberglass mesh is placed each year on the annual accumulation of litterfall on a defined area of forest floor. Each subsequent year of decomposition can then be measured over an extended period of time. Binkley (2002) found that 80% of litter organic matter decomposed over 10 years in a loblolly pine forest, yielding $k = 0.1655$; the data predicted the steady state forest floor mass within 10% of the actual value.

PATTERNS OF MASS LOSS DURING DECOMPOSITION

A graph of mass retained in litterbags during decomposition of leaf litter in a temperate deciduous forest reveals a three-phase curve (Fig. 5.9). Initially, following autumnal leaf drop, there is a rapid decrease in weight, caused by the loss of rapidly metabolizable compounds or simply readily leachable substances. This initial phase is followed by a slow rate of loss during winter months. During late spring, rates again become accelerated as microclimates become more favorable

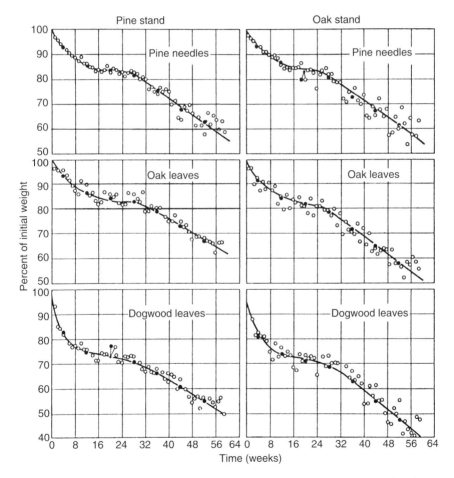

FIGURE 5.9. Three-phase mass loss curve. Weights of decaying leaves in litterbags, expressed as a percentage of initial weight of litter, through time. Three leaf species in two stands. Hollow circles are individual measurements; solid circles are averages for 8-week cycles. Lines fitted by eye (from Olson and Crossley, 1963).

for biological activity. During winter some microbial and faunal attack occurs, but the major abundance of fauna and microbes is found in litterbags during spring and summer.

Although rates do vary with season, the model using a single exponential constant (k) provides a good fit to these data (Fig. 5.10). The

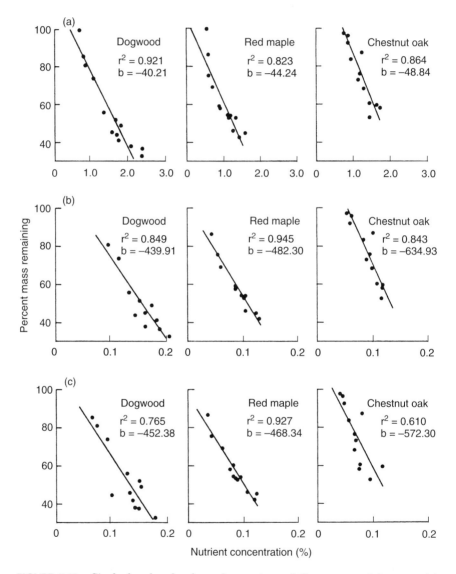

FIGURE 5.10. Single k value for decay [regressions of % mass remaining over (a) nitrogen, (b) sulfur, and (c) phosphorus concentrations in the residual litter for flowering dogwood, red maple, and chestnut oak]. (Reprinted from Blair, J. M. [1988a], with permission).

coefficient of determination (r^2) for these curves usually exceeds 85%. The constant k conceals seasonal dynamics but is a useful means for comparing leaf types or habitats, or geographical regions. More precision can be gained by calculating k from the spring–summer values alone. Some typical breakdown rates for forest litter are shown in Table 5.3.

Breakdown rates in agricultural systems are generally more rapid than in forested systems: crop residues, as a rule, tend to have fewer recalcitrant components. Figure 5.11 shows mass loss rates for rye litter from litterbags either placed on the soil surface (no-tillage) or buried

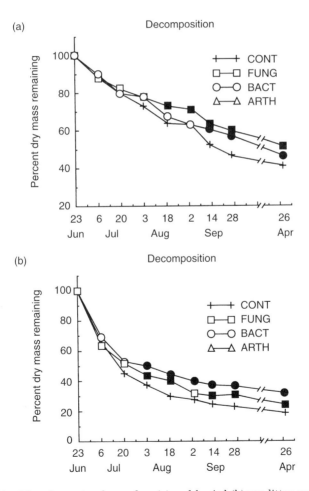

FIGURE 5.11. Mass loss rates for surface (a) and buried (b) rye litter over 320 days. CONT, control situation; FUNG, fungicide, with about one-half fungal population biomass; BACT, bactericide (oxycarboxin); ARTH, arthropod repellant (naphthalene). Filled symbols are significantly different from controls (p < 0.05) on a given date. (Modified from Beare *et al.*, 1992, with permission).

following plowing (conventional tillage) (Beare *et al.*, 1992). Loss rates were faster under conventional tillage ($k = 0.03$ per day) than in no-tillage soils ($k = 0.02$ per day). Usually, buried residues decompose more rapidly than surface residues because of more intimate contact with mineral soil and microbes and because of more moderated microclimate beneath the soil surface. In any case, rates for rye litter were much faster than for forest tree leaf litter.

To establish the extent of impact of faunal-moderated decomposition on plant residues, Tian *et al.* (1995) used gnotobiotic microcosms with residues of five plants from tropical agroecosystems (*Dactyladenia barteri*, *Gliricidia sepium* and *Leucaena leucocephala* prunings, maize stover, and rice straw) placed on the surface of large pots filled with an Oxic Paleustalf in Nigeria. The soil was defaunated by sun drying and the larger fauna removed by hand. Eighteen mature earthworms (*Eudrilus eugeniae*) and/or three millipedes (Spirostreptidae) were then added to a subset of pots. The experiment was run for 10 weeks, with an intermediate sampling of mass losses occurring after 4 weeks. Both earthworms and millipedes contributed more to the breakdown of the low-quality litter (*Dactyladenia*, maize stover, and rice straw) compared to the decomposition of the higher-quality leguminous tree prunings (Tian *et al.*, 1995). This effect has been observed in a number of forest ecosystems, as noted later in this chapter.

EFFECTS OF FAUNA ON LITTER BREAKDOWN RATES

The association of soil fauna with litter decomposition is an ancient one. Labandeira *et al.* (1997) reviewed the evidence concerning associations of soil fauna in the geologic record. The incidence of oribatid mite feeding in coal deposits from Illinois and Appalachian sedimentary basins occurred in all major plant taxa in Pennsylvanian coal swamps. Virtually every type of plant litter tissue was used by the mites. Evidence for termites and holometabolous wood-boring insects dates to the early Mesozoic. The illustrations published by Labandeira *et al.* (1997) provide striking evidence of the importance of detritivores in these primitive forests.

In more modern times, the Russian soil scientist Galina Kurcheva (1960, 1964) found that naphthalene (an insecticide) applied to oak leaf litter would drastically reduce the rate of breakdown. In the succeeding decades, various biocides and other techniques have been used to suppress various components of the soil biota, which is a measure of their importance in leaf litter breakdown (e.g., Parker *et al.*, 1984; Beare *et al.*, 1992). The upshot of these experimental manipulations has been to demonstrate that bacterial, fungal, and faunal members of the soil biota all have significant effects on litter breakdown (Fig. 5.11). Given that actual breakdown and decomposition rates are a function of the *interaction* among the various biota and with substrate quality and climate, the rate estimates derived from manipulations must be accepted with caution. The main effects, however, seem clear.

Seastedt (1984b) suggested that the equation describing litter breakdown might be partitioned into components, so that the constant k could be considered as the sum of several ks:

$$dX/dt = -kX = -(k_{bacteria} + k_{fungi} + k_{fauna})X$$

Seastedt reviewed studies in which microarthropods had been suppressed and found that a variable percentage of breakdown rates could be attributed to microarthropod activities. Table 5.4 shows the results of the Seastedt equation applied to forest tree litter in a floodplain forest in Athens, Georgia, in the United States. Litterbags with a 1-mm mesh size were used so that macrofauna were excluded from the bags. Naphthalene applications were used to reduce microarthropod populations in some of the litterbags. The results show that the importance of microarthropods varied with litter quality. Microarthropod activities were least significant for the more rapidly decomposing litter species (dogwood, tulip-poplar) and were most important for the slowest, most recalcitrant litter type (water oak). In a carefully controlled experiment, Couteaux *et al*. (1991) measured decomposition of litters of various qualities, namely with carbon–nitrogen ratios of 75 (low quality) versus 40 (higher quality). The soil fauna contributed more to the decomposition of the low-quality substrate, and the effect was significantly greater at later stages of incubation in the 24-week experiment, with greater faunal complexity accounting for greater amount of dry mass loss and total CO_2 evolution per unit time.

TABLE 5.4.   Percent of Leaf Litter Decomposition (Mass Loss) Attributable to Soil Fauna[a]

Leaf species	k_T	k_{NA}	k_F	Percent due to fauna
Dogwood	−0.00248	−0.00089	−0.00159	64.1
Sweetgum	−0.00248	−0.00089	−0.00159	71.4
Tulip-poplar	−0.00229	−0.00113	−0.00116	50.7
Red maple	−0.00125	−0.00069	−0.00056	44.8
Water oak	−0.00174	−0.00037	−0.00137	78.7
White oak	−0.00216	−0.00076	−0.00140	64.8

[a]Loss rate due to faunal activities calculates as: total rate (k_t) minus naphthalene rate (k_{NA}) equals rate due to fauna (k_F). Percent difference calculated as (k_F/k_t) × 100.
From D. A. Crossley, Jr., unpublished.

LIVERPOOL JOHN MOORES UNIVERSITY
LEARNING SERVICES

Experimental approaches such as these must be interpreted with caution. Usually more than one component of the system is modified by manipulations, be they chemical or physical ones (Crossley *et al.*, 1990). Other approaches, such as tracer methods and laboratory microcosms, need to be used in conjunction with manipulative experiments. Biocides such as naphthalene may alter other system components, sometimes to a large extent. Naphthalene, for example, may suppress microbes. However, González *et al.* (2001) reported higher microbial biomass when naphthalene was used to exclude soil fauna.

NUTRIENT MOVEMENT DURING DECOMPOSITION

Soils contain many of the same elements as found in their underlying substrate of rock, but the proportions differ greatly. Elements such as calcium (Ca), magnesium (Mg), potassium (K), and sodium (Na) are lost as soluble cations during weathering, depending on climatic conditions (especially precipitation). Some other elements, such as iron and aluminum, are resistant to leaching losses and their proportions may increase compared to rocks. Movements of cations are governed by the exchange properties of the soil, properties dependent upon the nature of the clays and amount and type of organic matter. Exchangeable cations in soils include Ca^{2+}, Mg^{2+}, K^+, NH_4^+ and Na^+, affinities for exchange sites (i.e., energy of adsorption) decreasing approximately in that order. Certain anions are not as tightly held in soils, again depending on the nature of clay colloids and on soil pH. Phosphate ions, multiply charged, are more tightly fixed by anion exchange properties than are singly charged ions such as nitrate (Bowen, 1979; Foth, 1990).

During the decomposition process, elements are converted from organic to inorganic forms (mineralized) and may enter the exchangeable pools, from which they are available for plant uptake or microbial use. Cellulose and hemicellulose account for more than 50% of carbon in plant debris and help to fuel microbial processes such as transformations of nitrogen (Fig. 5.12) and sulfur (Fig. 5.13), which gradually reduce the carbon–nitrogen and carbon–sulfur ratios in decomposing materials.

As plant litter decomposes, the elemental mix changes because of differential mobility and biological fixation. Carbon is lost through microbial respiration, as cellulose and other labile organic compounds are hydrolyzed and utilized in growth and maintenance. Potassium is highly mobile until it encounters exchange sites, where it can become fixed. Sodium ions, which are more mobile in soils, are not accumulated in plants but are essential for animals. The "herbivore exclusion hypothesis" (McNaughton, 1976; McNaughton *et al.*, 1998) proposes that plants discriminate against sodium and thereby limit herbivory. Sodium does

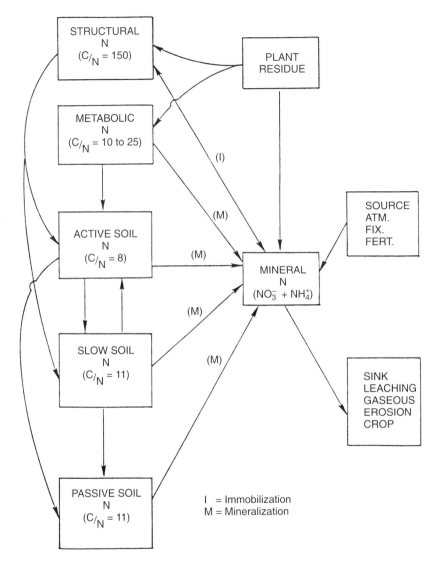

FIGURE 5.12. Soil nitrogen (N) cycle showing active (1–1.5 years), slow (10–100 years), and passive (100–1000 years) fractions. Flow diagram for the nitrogen submodel of the Century model (from Parton *et al.*, 1987, with permission).

accumulate in food chains, often increasing by a factor of 2–3 between trophic transfers.

The nitrogen content of decomposing litter increases during the initial stages of decomposition and then declines (Fig. 5.14) (Berg and

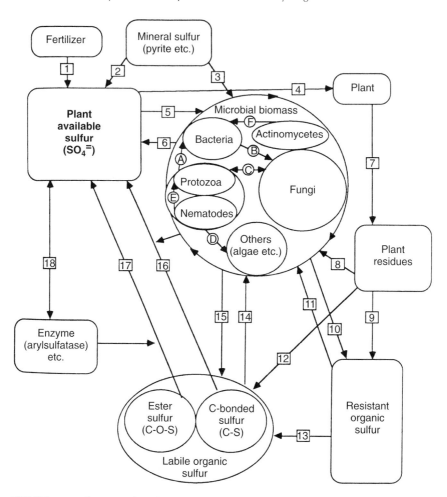

FIGURE 5.13.   Conceptual model for microbial and faunal components of the sulfur cycle in soil (from Gupta, 1989).

Staaf, 1981). Nitrogen is mineralized during decomposition and is simultaneously immobilized by microbes, resulting in an increase in the concentration of nitrogen in the litter, and in the absolute amount of nitrogen if it is transported into the litter from soil or by atmospheric nitrogen-fixation. As decomposition proceeds, the carbon–nitrogen ratio declines until the substrate becomes more suitable for microbial action. In some forests, the period of nitrogen increase may extend for 2 years or more (Fig. 5.15) (Blair and Crossley, 1988). Phosphorus and sulfur also show increases in absolute amounts during decomposition of some species of tree leaf litter (Fig. 5.16) (Blair, 1988a), even though mass is

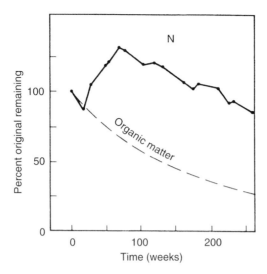

FIGURE 5.14. Rates of nutrient immobilization and mineralization from decomposing Scots pine (*Pinus sylvestris*) litter over a 5-year period (from Staaff and Berg, 1982). Note initial influx of nitrogen (N) into litter.

being lost. Calcium and magnesium concentrations in decomposing litter change only slightly through time. There may be an initial decrease in concentration followed by a slight increase (Blair, 1988b). Thus the absolute amounts of these elements during decomposition approximately track the loss of mass (Cromack, 1973). Potassium is not a structural element, and is lost via solubilization more rapidly than mass is lost from decomposing leaf litter. Decomposing woody litter, in contrast, accumulates calcium and phosphorus, evidently as a result of fungal invasion and translocation from soil.

The nitrogen pool in decomposing litter is a dynamic one. Although nitrogen is accumulating, there is evidently a large amount of turnover taking place. When tracer amounts of ^{15}N [as $(NH_4)_2SO_4$] were added to leaf litter, significant losses of tracer took place even as total nitrogen accumulated (Fig. 5.17) (Blair *et al.*, 1992). Nitrogen evidently became incorporated from exogenous sources, in amounts greater than those lost through biotic factors. Inputs of nitrogen via rainfall or canopy throughfall are a potential source of added nitrogen. However, these would appear to be inadequate to account for the amount of nitrogen immobilized in litter. Fungal translocation from lower layers (F, H, or mineral soil) is another possibility. Finally, lateral transport to and from "hot spots" in the forest floor may contribute to the dilution of tracers.

As noted in Chapter 2, one of the main sources of particulate organic matter to soils is that from decomposing roots. Researchers have often

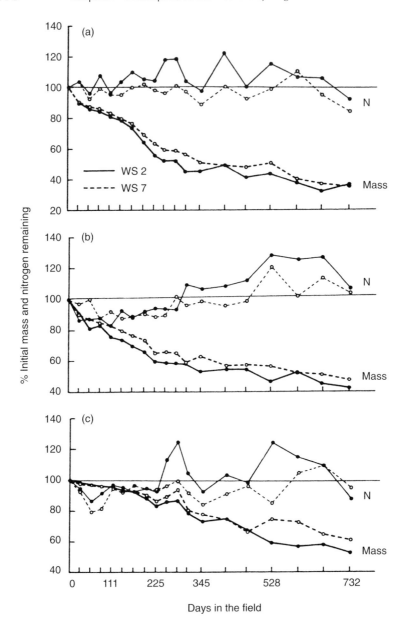

FIGURE 5.15. Mean percentage of initial mass and nitrogen remaining over time in (a) *Cornus florida*, (b) *Acer rubrum*, and (c) *Quercus prinus* litter on uncut WS 2 (solid line) and clearcut WS 7 (dashed line) at Coweeta Hydrologic Laboratory from January 1975 to January 1977 (from Blair and Crossley, 1988).

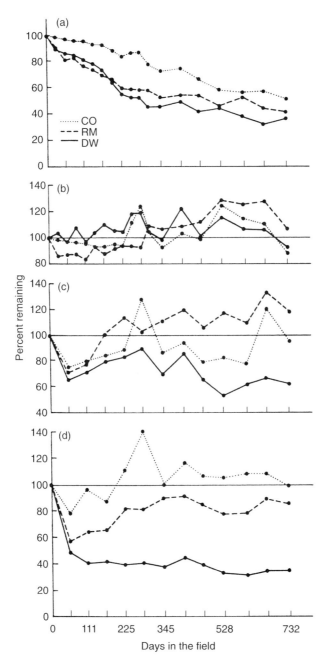

FIGURE 5.16. Mass, nitrogen, phosphorus, and sulfur changes in (a) mass and absolute amounts of (b) nitrogen, (c) phosphorus, and (d) sulfur in flowering dogwood (DW), red maple (RM), and chestnut oak (CO) litter decomposing over a 2-year period. (Reprinted from Blair, J. M. [1988a], with permission.)

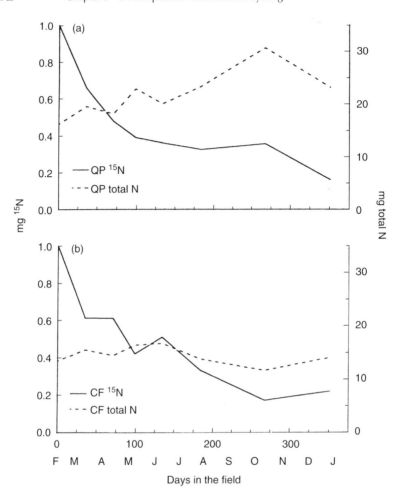

FIGURE 5.17. ^{15}N and Total N. Changes in the amount of added ^{15}N recovered in the litter (solid line) versus changes in the total amount of nitrogen (N) (dashed line) over time in (a) *Quercus prinus* (QP) and (b) *Cornus florida* (CF) litter from control litter baskets. (Reprinted, with permission, from Blair, J. M., Crossley, D. A., Jr., and Callaham, L. C. (1992).]

used root-litter bags, and followed dry matter and nutrient loss over several months to a few years. Unfortunately, preparation of the root tissues for decomposition studies represents a significant departure from *in situ* conditions. Dornbush *et al.* (2002) developed an intact-core technique that retains natural rhizosphere associations, maintaining *in situ* decay conditions. Cores (15 cm long by 5.3 cm in diameter) were taken under monospecific stands of silver maple, maize, and winter wheat, and covered at the top and bottom with 160-micrometer mesh polyethylene caps. The same mesh was used to make litterbags to hold an amount of roots similar to that in the soil cores. After reinstallation in the field sites, cores and bags were retrieved seasonally at time intervals up to 1 year and the decay rates compared. After 1 year, mass loss was 10–23% greater and nitrogen release was 21–29% higher within intact cores than in litter bags (Fig. 5.18). Dornbush *et al.* (2002) attributed a majority of the differences to alterations in litterbag-induced dynamics of decomposer organisms and unavoidable changes to fine-root size-class composition (less than or equal to 1 mm in diameter) in the bags.

Nutrient movement through the L, F, and H layers and through mineral soil have been studied intensively by both biologists and organic geochemists. Tracking nutrient movements in soil is difficult because of nonlinearities of flow due to the existence of preferential flow paths compared to the bulk soil. For example, in a carefully instrumented study in a mixed beech and spruce forest in Switzerland, Bundt *et al.* (2001) measured organic carbon concentrations from 10–70% higher in the

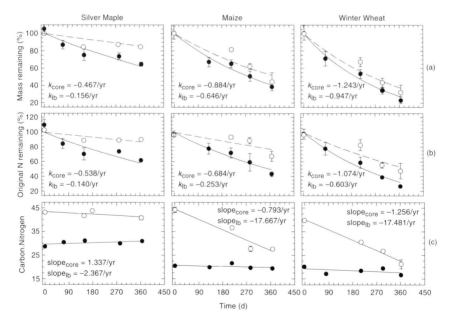

FIGURE 5.18. (a) Mass loss, (b) nitrogen loss, and (c) changes in carbon–nitrogen ratio for fine roots of silver maple, maize, and winter wheat decomposed in a riparian meadow in central Iowa in the United States. Solid circles represent intact cores, and open circles represent litterbags (lb); *k* values are based on exponential decay models. In the upper six panels, the intact-core data are normalized such that 100% equals the *t* = 0 intercept of the exponential decay regression; error bars are +/–1 SE (from Dornbush *et al.*, 2002).

preferential flow paths than in the soil matrix. In addition, organic nitrogen concentrations, effective cation exchange capacity, and the base saturation were all increased in the preferential flow paths. DNA concentrations and direct cell counts showed similar patterns, but there were no changes in domain-specific organisms such as Eukarya or Archaea. However, *Pseudomonas* showed increased abundances in the preferential flow paths, indicating that it responds to increased organic matter status in this "hot spot" similarly to that in other locations such as the rhizosphere (Bundt *et al.*, 2001).

Nutrients and organic matter also move through soils in soluble form, for example as dissolved organic matter (DOM). In general, sorptive interactions between DOM and mineral phases contribute to the preservation of soluble, or soil, organic matter (SOM). However, the situation is considerably more complex than this, due to the existence of several structural events in the soil profile. In a comparison of movement in the profiles of seven forest soils, Guggenberger and Kaiser (2003) considered the location of organic matter (OM) within the soil profile. They contrasted the amount and concentrations of OM in soil water with that on "fresh," or exposed, mineral surfaces versus that on "natural" soil surfaces that have a considerable amount of microbially derived biofilms. Based on their study, they produced a conceptual model of entities in the soil profile that differ in their biological activity. In the soil solution, soil microorganism density is low, OM concentration is low, and there is little biodegradation of DOM. On the "fresh" mineral surfaces, there is OM sorption to minerals, including complexation of functional groups, changed conformation, incalation in small pores, and sorptive stabilization. On the "natural" soil surfaces with high microorganism density, there is OM sorption into biofilms; the sorption concentrates OM, which appears to be a prerequisite for decomposition (Fig. 5.19) (Guggenberger and Kaiser, 2003). Research is currently underway to determine the dynamics of exchange phenomena between DOM and biofilms. The OM input enhances the heterotrophic activity in the biofilm, converting the DOM into either organic compounds by resynthesis or inorganic mineralization products. Iron hydrous oxides embedded within the biofilms may serve both as a sorbent and a shuttle for dissolved organic compounds from the surrounding aqueous media.

McDowell (2003) notes that our knowledge of the ecological significance of dissolved organic nitrogen relative to that of dissolved organic carbon is yet in its infancy. We also need more information on how fluxes of these dissolved substances are altered in human-dominated environments. New analytical techniques are being developed to better quantify dissolved organic compounds in soils and their effects on microorganisms, particularly saprophytic fungi and mycorrhiza. These

Soil water   "Fresh" mineral   "Natural" soil
                              surfaces   surfaces

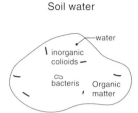      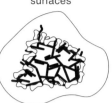

- Microorganism density   Sorption to minerals: Sorption into biofilms:
  low   • Complexation of   • Microorganism
- Organic matter   functional groups   density high
  concentration low   • Changed conformation   • Sorption concentrates
→ Little biodegradation   • Incalation in small   organic matter
  of dissolved organic   pores   → Sorption as prerequisite
  matter   → Sorptive stabilization   for decomposition

FIGURE 5.19. Conceptual model of the fate of organic matter in the soil solution and sorbed to soil surfaces differing in their biological activity (from Guggenberger and Kaiser, 2003).

new approaches are giving us better insights into soil ecosystem function, as noted in Chapters 3 and 8.

NUTRIENT CYCLING LINKS IN SOIL SYSTEMS

In addition to the translocation abilities of saprophytic fungi mentioned previously, there is a rapidly growing literature on the roles of ectomycorrhizal and other symbiotic fungi in mobilizing nitrogen and phosphorus from organic pools (Chalot and Brun, 1998; Bending and Read, 1996; Northup *et al.*, 1995). This work has extended to the clearly demonstrated roles of ectomycorrhizal fungi in mineral weathering, that is, mobilizing inorganic nitrogen as ammonium from the interstices of feldspar minerals and solubilizing P from volcanic rocks (Landeweert *et al.*, 2001). As noted in Chapter 4 about the nutrition of oribatid mites, the fungal matlike structures formed by ectomycorrhizal fungi (e.g., *Hysterangium*, *Hydnellum,* and *Gautieria* spp.) at the interface of the surface humus layer and upper A horizon may cover several square meters of forest floor. The mineral soil within this concentrated mass of mycorrhizal hyphae is more strongly weathered than the surrounding soil as a result of the excretion of oxalic acid by the fungus. Within the mat, calcium oxalate crystals are abundant and decomposition rates and nutrient availability are increased relative to the nearby soil (Entry *et al.*, 1992). The calcium oxalate crystals are a readily available source of calcium ions for the mites, as was demonstrated elegantly by

Cromack *et al*. (1988). The fact that this inorganic "hot spot" serves as a possible source of both inorganic and organic nutrients for the microbivorous fauna is further proof of the impressive nutrient feedback loops operating in soils. For more extensive coverage of nutrient dynamics in soil profiles over centuries and millennia, see the extensive synthesis of Richter and Markewitz (2001).

ROLE OF SOIL FAUNA IN ORGANIC MATTER DYNAMICS AND NUTRIENT TURNOVER

For the last several decades, there has been interest in the role of soil fauna in litter and organic matter turnover in ecosystems. The pioneering studies of Darwin (1881) and P. E. Müller (1887) emphasized the prominent signs left in many temperate forest and grassland communities by earthworm, mesofauna, and biotic activities in general. A very prescient account of the "biotic" structure of soils was given by Jacot (1936). Signs of faunal activity include coating of mineral grains, which has a significant effect on promoting the formation of aggregates (Kubiëna, 1938). Termites in semitropical and tropical regions have similar functions as well (e.g., Lee and Wood, 1971; Wood *et al*., 1983). Only a few ecologists are aware, however, that often the soil meso- and microfauna are vastly more numerous—and usually more active in terms of respiratory activity—than the large soil fauna (Wolters, 1991; Coleman, 1994, 2001).

Our concerns as ecosystem researchers should include both an understanding of which organisms are present and the major processes that they carry out in a wide range of terrestrial ecosystems. Following the flow of energy and nutrients in the system (as noted by Volobuev, 1964) will enable us to concentrate on key processes that occur, avoiding the pitfall of what is obvious to the naked eye being singled out for study. We must get to the appropriate level of resolution to ascertain the roles of participants in soil processes (Macfadyen, 1969; Coleman, 1985). This requires exploring the myriad of surfaces and volumes that occur in a few cubic millimeters of soil and organic matter (Elliott, 1986; Elliott and Coleman, 1988).

Fauna are members of the "organism" category in Jenny's (1941) factors of soil formation (recall from Chapter 1 [Fig. 1.5]: soil = f (cl, o, r, p, t), where cl = climate, o = organisms, r = relief, p = parent material, and t = time). As noted by Crocker (1952), only a few of these factors are independent variables, so we are dealing with a multicause, interdependent subset of a terrestrial ecosystem. To simplify matters, let us consider organisms alone, that is, vegetation, organic matter inputs therefrom, and the array of heterotrophic organisms feeding upon and decomposing organic detritus. The factors plus ecosystem processes acting over time lead to ecosystem properties (Coleman *et al.*, 1983; Elliott, 1994).

The immediate result of faunal feeding activity is the production of fecal pellets, some of which can be identified as species- or group-specific (Kühnelt, 1958; Jongerius, 1964; Zachariae, 1965; Rusek, 1975; FitzPatrick, 1984; Pawluk, 1987). For example, the fecal pellets of collembola and oribatid mites are surrounded by a chitin-rich layer called the "peritrophic membrane" (Krantz, 1978), which acts to retard the rate at which a fecal pellet disappears (Fig. 5.20). A comprehensive review (Bal, 1982) of soil fauna activities in soil refers to "zoological ripening" as faunal movement of organic matter and mineral materials in previously uncolonized soil. This soil maturation and development process has been of great significance in Dutch polder regions, and has been demonstrated in Canadian (Nielson and Hole, 1964) and New Zealand (Stockdill, 1966) soils as well. These processes are reviewed extensively in Brussaard and Kooistra (1993).

In addition to physical signs, there are chemical indicators of faunal presence and activity. For example, in certain cool, moist New Zealand tussock grassland soils, nearly 10% of the organic phosphorus is comprised of phosphonates (carbon–phosphorus [C–P] bonded) (Newman

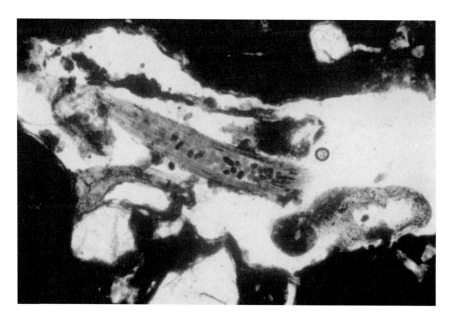

FIGURE 5.20.   Fecal pellets in the soil profile from the Horseshoe Bend agroecosystem site, Athens, Georgia, United States, at a depth of 5–10 cm (from Larry T. West, personal communication).

and Tate, 1980; Tate and Newman, 1982), as contrasted with the more prevalent phosphate esters. Phosphonates are produced by ciliates, and their subsequent rates of input and flow through the soil phosphorus cycle remain unknown (Stewart and McKercher, 1982).

Several authors have reviewed work on experimental pedogenesis (soil formation), examining roles of primary colonizing plants, including dissolution of rock minerals by lichens and fungi, as well as faunal impacts on mineral or soil movement, and organic matter transformation (Hallsworth and Crawford, 1965; Bal, 1982; Landeweert *et al.*, 2001). Webb (1977) studied the effects of particle size and decomposability of macrofaunal and microfaunal fecal pellets. There are differing effects of comminution (breaking up) of leaf litter by large and small fauna and they play different roles in facilitating further leaf litter decomposition. Webb (1977) noted that fecal pellets of *Narceus annularis* (Diplopoda: Spirobolidae) had a lower surface-to-mass ratio, whereas those of microarthropods such as oribatids had a greater surface-to-volume ratio than the original leaf litter. This should lead to greater decomposition per unit time (Fig. 5.21) (see previous comments about the peritrophic membrane).

Physical interpretation of organic matter decomposition should be tempered with careful observation of life-history details, such as likelihood of localized aggregation of mite or collembolan fecal pellets that may decompose locally at a much slower rate than hypothesized from *in vitro* laboratory studies. Substrate quality plays an important role here. Dunger (1983) noted that macroarthropods ingest mineral soil along with litter material. Kilbertus and Vannier (1981) and Touchot *et al.* (1983) demonstrated ingestion of argillic (clay) material by *Tomocerus* and *Folsomia* sp. (collembola), a trait that was particularly evident when they ingested polyphenol-rich *Quercus* leaves. This detoxification process presumably led to greater decomposition of the leaf material, with enhanced bacterial growth in the pellets with clay particles versus those without the clay adsorbent material. The impact of Collembola is greatest in mor soils, which may have entire layers in the F or H horizon filled with collembolan fecal pellets (Pawluk, 1987).

Research over the last 8 to 10 years has shown a significant impact of root-associated organisms on nutrient dynamics of phosphorus and nitrogen in experimental microcosms. These studies are reviewed in Coleman *et al.* (1983) and Anderson *et al.* (1981a, b). The use of both microcosm and mesocosm (i.e., field enclosures larger than a square meter) in soil ecological studies has proliferated in recent years and greatly increased our understanding of biological effects on nutrient cycling in soils (Ingham, 1985; Ingham, 1986a, b; Ingham *et al.*, 1989; Parmelee *et al.*, 1990; Beare *et al.*, 1992; Moore *et al.*, 1996).

In the laboratory, groups of rhizosphere bacteria, fungi, and microbivorous nematodes were grown singly or in combination, all with

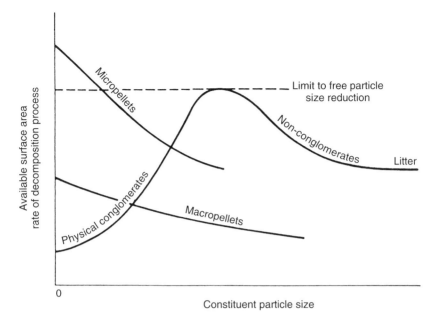

FIGURE 5.21. Graphical representation of physical conglomerate feces differentiation theory (from Webb, 1977). As particle size of litter (right to left) is reduced, surface area and decomposition increase until constituent particles are small enough to aggregate into more stable conglomerates (limit to free particle size reduction). Physical conglomerates increase in size as constituent particle size decreases, but arthropod pellets decrease in size because of the direct relationship of body size to degree of pulverization and pellet (conglomerate) size. Micropellets are therefore able to maintain a much smaller conglomerate size and break the limit to free particle size reduction (from Webb, 1977).

growing seedlings of the shortgrass prairie grass, *Bouteloua gracilis* (blue grama). In all treatments that had the root, microbe and microbial grazer (*Pelodera* sp. as bacterial-feeder), and *Aphelenchus avenae* as fungal-feeder, there was an enhanced shoot growth and dry-matter yield, when compared to the plant-alone control (Ingham *et al.*, 1985).

Other work using mesofauna (nematodes) (Ingham *et al.*, 1986a, b) and macrofauna (isopods) (Anderson *et al.*, 1985) has shown significant enhancement of nutrient cycling (nitrogenous compounds) in field experimental situations. Thus an enhanced (20–50%) nutrient return (mineralization) occurs in the presence of the fauna, compared with experiments in which they are present in very low numbers, or completely absent (Anderson *et al.*, 1983). This work was further amplified by simulation models of detrital food webs, which showed a significant (about 35%) contribution to mineralization of nitrogen by microfauna (amoebae and flagellates) and bacterial-feeding nematodes (Hunt *et al.*, 1987; De Ruiter *et al.*, 1993; Moore *et al.*, 1996). More detailed studies using ^{13}C and ^{15}N tracers in microcosms with varying degrees of organic matter accumulation ("hot spots") and microbes alone, microbes and protozoa, or nematodes and microbes with both faunal groups in combination (Bonkowski et al., 2000) have revealed similar patterns to earlier microcosm studies (e.g., Ingham et al., 1985). Rye grass seedlings significantly increased in dry matter and nitrogen content, with protozoa and nematodes and protozoa present (Bonkowski et al., 2000). Interestingly, the pattern of decomposition of labeled litter closely followed the nitrogen dynamics, with protozoa, and nematodes and protozoa, showing significantly more ^{13}CO$_2$-C respiration between weeks 2 and 4 of the 6-week experiment. This study was conducted concomitantly with a detailed analysis of microbial community composition. Griffiths et al. (1999) found significant selectivity of protozoa for species of soil bacteria, with a definite preference shown for several Gram-positive species.

FAUNAL IMPACTS IN APPLIED ECOLOGY—AGROECOSYSTEMS

There are several areas in the interface between theoretical and applied ecology where our knowledge of soil physics, chemistry, and biology can, and should, be put to good use. One of these is in the area of agroecosystem studies. The essentials of decomposition and nutrient dynamics in temperate agroecosystems were reviewed by Andren et al. (1990), Hendrix et al. (1992), Coleman et al. (1993), and in Africa by Vanlauwe et al. (2002).

It is generally acknowledged that zero, or reduced, tillage has several effects on abiotic and biotic regimes in agroecosystems. Retention of litter keeps the surface of the soil cooler and moister than in a conventionally tilled plot (Fenster and Peterson, 1979; Phillips and Phillips, 1984), and also leaves more substrate available in the 0–7.5 cm depths for nitrifiers and denitrifiers (Doran, 1980a, b). This abiotic buffering seems to promote a slower nitrogen cycle, one that continues over a longer time span but at a lower rate per unit time (House et al., 1984; Elliott et al., 1984). Soil invertebrate populations, particularly microarthropods (Stinner and Crossley, 1980; House et al., 1984), and earthworms (Parmelee et al., 1990) are enhanced as well (Table 5.5) (Coleman and Hendrix, 1986). The microarthropods are undoubtedly responding to increased populations of litter-decomposing fungi, which tend to concentrate nitrogen by hyphal translocation (Holland and Coleman, 1987). In fact, dominant families of fungivorous mites responded by markedly decreasing in numbers in field mesocosm plots treated with captan, which brought fungal populations down to about 40% of normal levels (Mueller et al., 1990).

A total of 22 agroecosystem components and processes were compared in no-till and conventional tillage in Georgia. In many instances, there was greater resilience in the no-till system, as shown by greater invertebrate species richness, greater soil organic matter, and ecosystem nitrogen turnover time (Table 5.6) (House *et al.*, 1984).

These findings were confirmed and extended by Elliott *et al.* (1984), who examined dynamics in long-term stubble-mulch and no-till plots on a silty-loam soil in eastern Colorado that underwent alternate crop and fallow regimes. These plots had been under cultivation for more than 75 years, and no-till had been an experimental treatment for nearly 20 years. Nitrate accumulated to a greater extent in the fallow than in the cropped rotation (Table 5.7). Ammonium-N was usually at very low levels (about $1.0 \mu g \, NH_4\text{-N per } g^{-1}$ soil), but on one date the concentration reached $4.6 \mu g \, NH_4\text{-N per } g^{-1}$ soil in the top $2.5 \, cm$ of the no-till plots just

TABLE 5.5. Numbers and Estimated Biomass of Soil Fauna in Conventional Tillage (CT) and No Tillage (NT) Agroecosystems at Horseshoe Bend

	$Numbers \cdot m^{-2}$		$mg \, dry \, wt \cdot m^{-2}$	
	CT	NT	CT	NT
Nematodes[a]				
Bacterivores	1836[e]	909	237	117
Fungivores	227[e]	500	14	31
Herbivores	945[e]	1064	93	104
	3008	2473	344	252
Microarthropods[b]				
Mites	41,081[e]	78,256	118	303
Collembola	6244[e]	14,684	17	40
Insects	2105	2548	—	—
	49,430	95,489	135	343
Macroarthropods[c]				
Ground beetles	7[e]	33	6	30
Spiders	1[e]	17	1	14
Others	6[e]	28	—	—
	14	78	7	44
Annelids[d]				
Earthworms	149[e]	967	3129	20,307
Enchytraeids	1867	520	592	17
	2016	1487	3721	20,324
Total	**54,468**	**99,526**	**4207**	**20,980**

[a]Means of samples from June–October 1983; numbers are $\times 10^{-3}$.
[b]Means of samples from May–December 1983.
[c]Means of samples from April–June 1983.
[d]Means of samples from April 1983.
[e]For numbers of organisms, tillage treatments differ significantly at $P = 0.05$.
From Hendrix *et al.*, 1986.

TABLE 5.6.   Comparison of Agroecosystem Components and Associated Agroecosystem Processes from Conventional Tillage (CT) and No-Tillage (NT) Systems

Component or process	CT versus NT
Crop yields	NT = CT (except during drought)
Crop biomass	Decreasing in both CT and NT
Weed biomass	NT > CT
Plant nitrogen dynamics	CT > NT (nitrogen flux)
Shoot to root ratios	CT > NT
Nitrogen fixation	NT > CT (?)
Surface crop and weed residues	NT >> CT
Litter decomposition rates	CT > NT
Surface litter (%N)	NT > CT
Soil total N	NT > CT in upper soil layer
Nitrification activity	NT > CT in upper soil layer
	CT > NT in middle soil layer (?)
Soil organic matter	NT > CT
Soil moisture	NT > CT
Ground water leaching (nitrate-N)	CT > NT (?)
Foliage arthropods	CT = NT
Crop herbivory by insects	CT > NT
Nitrogen content of crop foliage	CT > NT
Arthropods species diversity	NT > CT
Soil arthropods (no. of individuals)	BT >> CT
Nitrogen contained in arthropods:	
soil	NT > CT
foliage	CT > NT
Ecosystem N turnover time	NT > CT
Ecosystem N efficiency	NT > CT (?)

From House *et al.*, 1984.

prior to the highest rate of NO_3-N accumulation in the no-till than in the stubble-mulch treatments. However, it is possible that there was more mineralization in the stubble-mulch plots earlier in the year before the first sample date, and this mineralized nitrogen was moved below the sampling depth (20 cm) as NO_3-N during a rainfall event. Interactions between modification of system structure and major nutrient processes need more study. Certainly soil fauna are sensitive to increased nutrient inputs from fertilizers and manures, and this needs to be considered in experimental work (Marshall, 1977; Hendrix *et al.*, 1992).

APPLIED ECOLOGY IN FORESTED ECOSYSTEMS

There are some interesting comparisons and analogies to be drawn between no-till agriculture and forested ecosystems of the "mor" type, which have a distinct stratification of L, F, and H layers (O_i, O_e, and O_a

TABLE 5.7. Faunal Carbon, Microbial Biomass, Mineralized Carbon and Nitrogen, Nitrogen and Phosphorus Under Stubble Mulch, and No-Till Treatments of the Fallow Phase of Dryland Wheat Plots

		Date				
Variable	*Treatment*	*8 June*	*6 Jul*	*2 Aug*	*23 Aug*	*13 Sept*
Faunal carbon[a]						
Collembola	Stubble mulch	6.69	1.03	2.18	0.61	1.49
×100	No till	5.58	8.29	4.01	1.60	1.49
Acari	Stubble mulch	3.77	0.47	0.91	0.40	0.92
×10	No till	3.60	3.31	1.24	1.32	0.54
Holophagous	Stubble mulch	0.88	0.46	1.12	0.51	0.56
nematodes						
×10	No till	0.48	1.69	0.96	0.57	0.32
Protozoa	Stubble mulch	1.76	0.67	0.74	1.96	1.92
×1	No till	2.20	0.96	0.93	2.29	1.92
Microbial biomass[b]						
Carbon	Stubble mulch	245	204	271	186	255
	No till	329	273	299	194	256
Nitrogen	Stubble mulch	81	57	53	60	56
	No till	92	72	49	62	48
Phosphorus	Stubble mulch	5.7	7.5	10.1	4.6	7.2
	No till	5.5	6.8	10.1	4.9	7.1
Mineralized C and N[c]						
Respired C	Stubble mulch	52	77	57	42	68
(0–10 days)	No till	97	110	98	41	67
Respired C	Stubble mulch	49	42	56	53	24
(10–20 days)	No till	80	69	84	57	24
Mineralizable N	Stubble mulch	10.63	6.39	5.79	2.62	1.95
(0–20 days)	No till	12.68	5.79	8.90	−3.3	−1.38
N and P[d]						
NH$_4$-N	Stubble mulch	3.1	1.0	2.0	1.8	0.8
	No till	4.0	1.8	4.6	1.2	1.4
NO$_3$-N	Stubble mulch	7.2	10.6	16.3	17.8	16.1
	No till	7.8	23.8	21.6	45.1	46.2

[a]Soil fauna biomass C (kg C ha^{-1} to 10 cm) for four categories. (Note differences in the multiplier for each category.)
[b]Microbial biomass C, N, and P (kg element ha^{-1} to 10 cm).
[c]Mineralizable C (CO_2-C) and NO_3-N as kg ha^{-1} to 10 cm in unchloroformed 0–10 or 10–20 day incubations of soils sampled from the field.
[d]NH$_4$-N, NO$_3$-N and extractable inorganic and organic P amounts (kg ha^{-1}) in the top 10 cm of soil.
From Elliott *et al.*, 1984.

in the U.S. terminology). It is generally recognized that abundance of fungi and fungivorous arthropods is greater in these soils than in soils with a less-pronounced litter layer (Kühnelt, 1976; Wallwork, 1976; Pawluk, 1987; Blair et al., 1992). However, it is important to determine the amount of activity occurring in these surface layers as well. Ingham et al. (1989) and Coleman et al. (1990) have shown significantly greater fungal biomass and microarthropod biomass in L, F, and H layers under Pinus contorta (lodgepole pine), compared with mountain meadow. There was also a greater amount of fungal activity, as demonstrated by FDA-positive fungal hyphae (Söderström, 1977; Ingham and Klein, 1982). This forest experience is corroborated by Verhoef and De Goede (1985), who noted greater activity of Collembola in pine forests in Holland, contrasted with habitats which had a thin or nonexistent litter layer.

Because energy flow is fundamental to the function of decomposer organisms and ecosystems, energetics could provide some fundamental constraints on soil carbon dynamics (Currie, 2003). Often, carbon is considered as a surrogate for energy in studies of detrital decay and carbon turnover in soils. By testing relationships between carbon and energy across samples of forest detritus above- and belowground, across decay stages, and between a deciduous and coniferous forest at the Harvard Forest in the United States, Currie (2003) found that energy and carbon concentrations were closely related (within 10%), as were ratios of heterotrophic energy dissipation to carbon mineralization across types of detritus (within 16%). These relationships should be borne in mind when we explore the energetics of detrital food webs, in Chapter 6.

Other areas of interest in applied soil ecology include revegetation of mine spoils. Extensive studies in the United Kingdom, Germany, and elsewhere have been made of decomposition ecology and of microbial parameters in strip-mined coal lands (Bentham et al., 1992). Intentional manipulations (especially introductions) of earthworm populations have been used to enhance productivity of crop and pasture lands (Lee, 1995) and to speed organic waste decomposition via vermicomposting (Edwards, 2004), as discussed in Chapter 4.

Several researchers (Tisdall and Oades, 1982; Rothwell, 1984; Jastrow and Miller, 1991; Jastrow et al., 1998) have investigated the roles of saprophytic and VAM fungi in stabilizing macroaggregates. Rothwell (1984) suggests that there is a biochemical coupling reaction between glucosamines in the hyphal walls of the fungus with phenolic compounds released during lignin degradation from leaf and root tissues. An additional possibility, little investigated yet, is the apparently widespread occurrence of interspecific physical linkages that enable transfers of nutrients via mycorrhizae of various annual and perennial plants (Chiariello *et al.*, 1982; Read *et al.*, 1985; Read, 1991). Physical, chemical, and biological contacts may be operating simultaneously in mycorrhizal-mediated interactions. Recall the comments about the role of glomalin in promoting aggregate stability in soils as noted by Wright *et al.* (1999) in Chapter 2.

Some of the latter examples may seem a bit removed from the general theme: the role of soil fauna in soil processes. However, it is apparent from studies by Warnock *et al.* (1982), Moore *et al.* (1985), and Curl and Truelove (1986) that soil mesofauna, for example, Collembola, show considerable preference for, and have an impact on AM fungal growth, just as they do for saprophytic fungi (Newell, 1984a, b) and plant pathogenic fungi (Lartey *et al.*, 1994). This impact undoubtedly extends to nematodes (Ingham *et al.*, 1985) and soil amoebae as well (Chakraborty and Warcup, 1983; Chakraborty *et al.*, 1983; Gupta and Germida, 1988; Gupta and Yeates, 1997).

SUMMARY

The major lesson to be learned for soil ecologists is one of paying attention to details yet considering them in a holistic perspective. Certainly we are past the time when measurement of the "soil biomass" (referring to the microbial biomass) alone, by whatever method, is considered adequate (Coleman, 1994a). Small groups of organisms, perhaps highly aggregated within the ecosystem, may be facilitating (or retarding) turnover of other organisms, or of major nutrients such as nitrogen, phosphorus, and sulfur. In fact, as Darwin observed more than a century ago, these seemingly small biological processes operating over long time periods and large spatial scales can make profound changes in the world around us, including the formation of soil.

Decomposition rates, along with nutrient dynamics, soil respiration, and formation of soil structure, are integrating variables. They are generalized measurements of the functional properties of ecosystems, and they summarize the combined actions of soil microflora, fauna, abiotic variables, and resource quality factors. Litter breakdown rates can be compared using simple first-order models, so that rate variations between ecosystems or between different substrates may be compared. Litter breakdown rates are easily measured using bagged leaf litter ("litterbags"). Decomposition per se is due to microbial activities, but experiments show that fauna have a strong influence on litter breakdown rates, especially for more resistant substrates. The interaction between microflora and fauna is especially important for nutrient cycling mechanisms. Organic matter dynamics are strongly influenced by soil fauna. Termites and earthworms are well known for their influences on nutrient dynamics, soil organic matter, and soil structure. But the entire soil fauna is involved in these processes and, through their interactions with soil microbes, must be considered in a holistic perspective.

6

Soil Food Webs: Detritivory and Microbivory in Soils

INTRODUCTION

The traditional studies of food webs and food chains began with pioneering efforts of Summerhayes and Elton (1923), in Spitsbergen, Norway. This early study explicitly linked detrital biotic interactions with other parts of the terrestrial and aquatic food web (Fig. 6.1). Work on detrital food webs progressed slowly for the next 20 years, although Bornebusch (1930) carried out some pioneering studies of detrital food webs and their energetics. Further insights were gained from the studies of Lindeman (1942), who developed the concept of trophic levels.

Building on the soil ecology studies funded by the U.S. Atomic Energy Commission in the late 1950s (Auerbach, 1958) and into the early 1960s, soil ecologists recognized a clear need for a more holistic study of energetics and interactions of organisms in ecosystems. This led to the ambitious effort known as the International Biological Program (IBP). The overall intent was to bring working groups together, addressing how carbon and energy flow in a wide range of terrestrial and aquatic ecosystems, with the ultimate goal being a better understanding of how ecosystems work and could be manipulated for the benefit of mankind (Blair, 1977). The main findings of the IBP were that for a wide range of grassland, desert, and forested ecosystems, the net flow into the aboveground grazing (consumer) component is only 5% or less, with the remainder entering the detrital-decomposer food web (Coleman *et al.*, 1976). This research led to several post-IBP studies in North America and Europe to follow up on the initial results.

In the late 1970s and 1980s, a series of investigations of detrital food webs were carried out in the semiarid and arid grasslands and desert lands of Colorado and New Mexico (Coleman *et al.*, 1977; Coleman *et al.*,

(a)

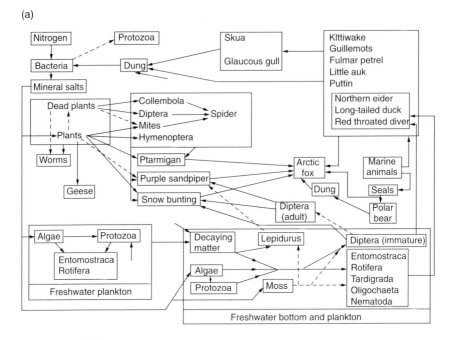

(b)

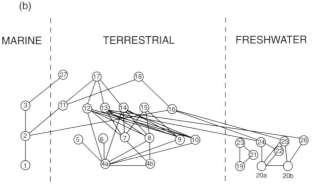

FIGURE 6.1. Arctic food web (a) as described by Summerhayes and Elton (1923) and as diagrammed (b) by Pimm (1982). (b): (1) plankton, (2) marine animals, (3) seals, (4a) plants, (4b) dead plants, (5) worms, (6) geese, (7) Collembola, (8) Diptera, terrestrial, (9) mites, (10) Hymenoptera, (11) seabirds, (12) snow bunting, (13) purple sandpiper, (14) ptarmigan, (15) spiders, (16) ducks and divers, (17) arctic fox, (18) skua and Glaucous gull, (19) planktonic algae, (20a) benthic algae, (20b) decaying matter, (21) protozoa, (22) invertebrates, (23) Diptera, freshwater, (24) other invertebrates, (25) Lepidurus, and (27) polar bear. [From Pimm (1982) and Pimm and Lawton (1980).]

1983; Parker *et al.*, 1984; Whitford *et al.*, 1983; Hunt *et al.*, 1987; Moore *et al.*, 1988) (Fig. 6.2). These studies and several in the Netherlands (Brussaard *et al.*, 1990; De Ruiter *et al.*, 1993; Moore and De Ruiter, 2000), Sweden (Persson, 1980; Bååth *et al.*, 1981, Andrén *et al.*, 1990), and the United Kingdom (Anderson *et al.*, 1985) found that microbial–faunal interactions have significant impacts on nutrient cycles of the major nutrients, namely nitrogen, phosphorus, and sulfur (Gupta and Germida, 1989). Some of these studies used assemblages of a few species in microcosms but were beginning to delineate the mechanisms that are important in soil systems in general. Among the fauna, the protozoa were often overlooked, despite the findings by Cutler *et al.* (1923) that there are important predator–prey interactions between protozoa and bacteria in soils. Clarholm (1985) noted that soil protozoa are avid microbivores and turn over an average of 10–12 times in a growing season, in contrast to many other members of the soil biota, which may turn over only once or twice in an approximately 120- to 140-day growing, or activity, season. These findings were further extended (Kuikman *et al.*, 1990) with the observation that nitrogen uptake by plants may increase from 9 to 17% when large inocula of protozoa are present. The demographics and microbial–faunal interactions provide much of the driving force in the models of nitrogen turnover in semiarid grasslands (Hunt *et al.*, 1987) and arable lands (Moore and de Ruiter, 1991, 2000).

Recent studies have noted the more complex nature of food webs when detrital components are included (Polis, 1991; Hall and Raffaelli, 1993; Scheu and Setälä, 2002). DeAngelis (1992), in his treatise on nutrient cycling, devoted an entire chapter to nutrient interactions of detritus and decomposers. His ideas have provided insights into decomposition–nutrient cycling processes. This chapter addresses several aspects of soil biota and nutrient cycling in soils, namely demography and "hot spots" of activity, which are often overlooked in energetics studies of soil systems. These factors are crucial to understanding how organisms and soils interact, and contribute to ecosystem function.

PHYSIOLOGICAL ECOLOGY OF SOIL ORGANISMS

Given the physiological ecology of the microbes and fauna involved, are long food chains energetically possible? There are several theoretical reasons why long food chains could be expected. Let us take, as an example, the energetically most dominant interactions between microbes and fauna, which occur in many terrestrial ecosystems, summarized by Hunt *et al.* (1987) (Fig. 6.2). The flow of organic carbon or nitrogen moves from initial organic substrates (labile or resistant) to the primary decomposer, either bacteria or fungi, and then on into micro-

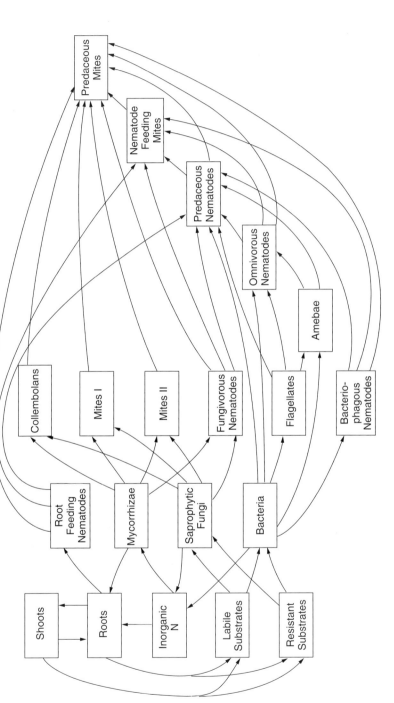

FIGURE 6.2. Representation of detrital food web in shortgrass prairie. Fungal-feeding mites are separated into two groups (I and II) to distinguish the slow-growing Oribatids from faster-growing taxa. Flows omitted from the figure for the sake of clarity include transfers from every organism to the substrate pools (death) and transfers from every animal to the substrate pools (defecation) and to inorganic nitrogen (N) (ammonification) (from Hunt *et al.*, 1987, reprinted with permission).

bivorous microfauna (flagellates and amoebae) or microbivorous meso-fauna (feeding on fungi) and, in turn, to omnivorous or predaceous nematodes, and on to nematode feeding mites and predaceous mites. Further predation upon the mites by ants (E. O. Wilson, personal communication) or lithobiomorph Chilopods (centipedes) is possible, although not explicitly represented by Hunt *et al.* (1987). There are at least eight links in the bacterial-based detrital food chain, with considerable evidence of omnivory. For example, many fungivorous mites require a nematode "supplement" to complete their life cycles (Walter *et al.*, 1991). Note that Figure 6.2 is a rather ecosystem-specific diagram. One could draw another for decomposition in a coniferous or oak/beech forest, with a significant proportion of the total decomposition being mediated by ectotrophic mycorrhizae, operating perhaps in competition with the saprophytic fungi (Gadgil and Gadgil, 1975).

For desert and estuarine food webs, reviewed by Hall and Raffaelli (1993), the detrital food chain length noted previously is comparable to the average length of five to seven links (Polis, 1991), with maximal recorded of eight. In contrast, Hairston and Hairston (1993) assert that the usual food chain length in detrital systems seldom exceeds three. As noted later in this chapter, these long chain-lengths of five to seven links are not only feasible, but also thermodynamically possible at several times and in several locations in the soil matrix, particularly the rhizosphere and other "hot spots" of activity.

What levels of taxonomic resolution are both most useful and appropriate for detrital food web studies? Our inability to sort out the details of microbial taxonomy *in situ* (see Furlong *et al.*, 2002, and other references in Chapter 3 for insight into molecular probing techniques in agroecosystems) and limited knowledge of many of the soil invertebrates, particularly the immature stages (Behan-Pelletier and Bissett, 1993), requires use of rather coarse functional groups for taxonomy of the soil biota. Interestingly, this sort of separation enabled Wardle and Yeates (1993) to identify competition and predation forces operative in an assemblage of detritus–microbial–nematode trophic groups in an agricultural field. Using a correlation analysis, they noted that predatory nematodes reflected most closely the changes in primary production, and the microbivorous nematodes seemed to be more dependent on substrate quality in the microbial (bacterial and fungal) community.

ENERGY AVAILABLE FOR DETRITAL FOOD CHAINS AND WEBS

If one considers the variance (i.e., range around the mean values,), of the assimilation and production efficiencies of the biota (Table 6.1), the amount of energy that will move from primary decomposers all the way

TABLE 6.1.   Physiological Data on Major Biotic Groups in Soil

Trophic Group	*Fraction of food assimilated*			*Production– assimilation ratio*		
	max	**$\overline{X}$**	**min**	**max**	**$\overline{X}$**	**min**
Bacteria	?	1.0	<0.01?	0.7	0.4	<0.01?
Saprophytic Fungi		1.0		0.7	0.4	
Arbuscular Mycorrhiza (AM)		?		0.8?	0.4	
Amoebae		0.95		0.8	0.4	
Flagellates		0.95		0.8	0.4	
Nematodes Phytophagous		0.25		0.5?	0.37	
Nematodes Fungivorous		0.38		0.5?	0.37	
Nematodes Bacterivorous		0.6		0.5?	0.37	
Nematodes Omnivorous/ Predaceous		0.55		0.5?	0.37	
Mites Fungivorous (*r*)		0.5		0.5?	0.35	
Mites Fungivorous (*k*)		0.5		0.5?	0.35	
Mites Nematophagous		0.9		0.5?	0.35	
Mites Predaceous		0.6		0.5?	0.35	
Collembola		0.5		0.5?	0.35	
Enchytraeids		0.28			0.4	
Earthworms		0.2			0.45	
Termites		0.4?			0.15?	

r = rapid growth strategy
k = slow growth strategy
Modified from Humphreys, 1979; Hunt *et al.*, 1987; Payne, 1970; de Ruiter *et al.*, 1993.

up the food chain can be calculated. There is indeed energy to spare for such elaborate food chains. Using the maximal values for production efficiency, such as that for bacteria of 70% (Payne, 1970) and 80% for soil amoebae and flagellates (Humphreys, 1979), and moving on to protozoan-consuming nematodes (Fig. 6.3) (which doubtless occurs in

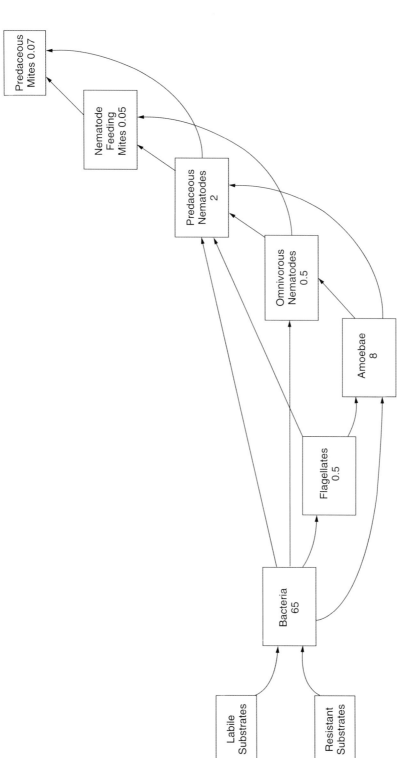

FIGURE 6.3. Calculations of annual carbon flows along a bacteriophagic food chain assemblage, exhibiting considerable omnivory. Average standing crops are indicated. Flows via protozoan feeding are estimated as probably two to three times greater than via nematodes in that ecosystem. Proportions may be reversed in lower-pH forested systems, and more flows via fungi (not shown) (from Hunt *et al.*, 1987, reprinted with permission).

certain "hot spots," e.g., at the zone of elongation of a growing root), then there is an adequate amount of carbon available for passage through the four- and five-membered detrital food chains of interest. Considerable omnivory is prevalent in these soil systems (DeAngelis, 1992). The protozoa and nematode feeding pathway highlighted in Figure 6.3 (Hunt *et al.*, 1987) accounted for 37% of the total nitrogen mineralization and some 82% of the total mineralization resulting from soil fauna. Similar percentages were obtained for a wide range of agroecosystems in the United States and Europe (de Ruiter *et al.*, 1993; Moore and de Ruiter, 2000).

The relative contributions of the soil fauna to microbial turnover and nutrient mineralization are directly related to the demographics of the soil biota (Coleman *et al.*, 1983, 1993), as noted for average standing crops and energetic parameters and turnover times per year for microorganisms, micro-, meso-, and macro-fauna in a grassland and a no-tillage agroecosystem (Coleman *et al.*, 1993) (Table 6.2). Thus the protozoa, and naked amoebae in particular, turn over 10 or more times per season, and consume several times their mass of living microbial tissues. The microbes and several other faunal groups have much lower turnover rates, on average. Although the amoebae are considered to be primarily bacterial feeders, there are important instances when other amoebal species will feed on protoplasm in fungal hyphae, or even on the fungal spores themselves (Chakraborty and Warcup, 1983; Chakraborty *et al.*, 1983). When considered in combination with the information in Table 6.1 on the range of assimilation and production efficiencies, the impacts of these small organisms are very marked. It should be noted that extensive studies in Sweden on arable lands (Andrén *et al.*, 1990) have reached similar conclusions. The increasing miniaturization of sensors, so that one can carry out microcalorimetry (Battley, 1987) at localized microsites, will enable us to measure direct energetic transformations more readily *in situ*.

The practical implications of soil food webs in agroecosystems have been of interest to researchers in several countries, notably the Netherlands, Sweden, and the United States. In a major synthesis of several research papers (some of which are cited earlier in this chapter), Bloem *et al.* (1997) calculated the impact of microbivorous invertebrate fauna in agroecosystems. Using a combination of experimental results and simulation modeling runs, they calculated that in fields that had greater additions of organic matter, including manure, average nitrogen mineralization was 30% higher than in fields that did not have such organic matter additions. This reflected the activities of protozoa and nematodes, which were 64% and 22% higher numbers, respectively, in the fields with organic additions. Nitrogen mineralization was performed mainly by the bacteria, which dominated in these fields, but

TABLE 6.2. Average Standing Crop and Energetic Parameters for Microorganisms, Mesofauna, and Earthworms in a Lucerne Ley and Georgia No-Tillage Agroecosystem[a]

	Naked amoebae	Flagellates	Ciliates	Bacteria	Fungi	Microbivorous nematodes	Collembola	Mites	Enchytraeids	Earthworms
Typical size in soil	30 μm	10 μm	80 μm	0.5–1× 1–2 μm	Ø 2.5 μm 1.0–5.5 μm	Ø ~40 μm	Ø 5000 μm	Ø 1000 μm	Ø 1000 μm	Ø 5000 μm
Mode of living	In water films on surfaces	Free–swimming in water films	Free–swimming in water films	On surfaces	Free and on surfaces	In water films, free, and on surfaces	Free	Free	Free	Free in soil
Biomass (kg dw ha⁻¹)	95%	5% (50[b])	<1%	500–750[c]	700–2700[d]	1.5–4[e]	0.2–0.5[e]	2–8[e]	1–8[e]	25–50[e]
% active	0–100	10		15–30	2–10	0–100	80–100	80–100	?	0–100
Estimated turnover times, season⁻¹	10			2–3	0.75	2–4	2–3	2–3	?	3
No. of bacteria division⁻¹ × 10⁻³	3–8	0.6–1	20–2000							
Minimum generation time in soil (hours)	2–4			0.5	4–8	120	720	720	170	720

[a] Modified from Clarholm (1985), Hendrix et al. (1987), and Beare et al. (1992). Reprinted with permission from Coleman et al. (1993). Copyright Lewis Publishers, an imprint of CRC Press, Boca Raton, Florida.
[b] MPN technique.
[c] Direct counts plus size class estimations.
[d] Direct estimation of total hyphal length and diameter.
[e] Extractions and sorting.

the nitrogen mineralization was increased by protozoa by 30%, on a growing season average. Interestingly, the protozoa did not enhance carbon mineralization, because their impacts, as were those of the nematodes, were by direct grazing upon and lysing the bacterial cells (Bloem *et al.*, 1997).

ARENAS OF INTEREST

Soils are best considered as the extremely heterogeneous entities they are. This requires that we "let the soil work for us" (Elliott and Coleman, 1988), and stratify, in a statistical sense, the regions of the soil that are "hot spots" of activity. These zones include the rhizosphere, aggregates, litter and organic detritus, and the "drilosphere," which is that portion of the soil volume influenced by secretions of earthworms (Bouché, 1975) (Fig. 6.4). Each region is a relatively small subset of the total soil volume, but may contain a preponderance of numbers, and more importantly, activity of the soil biota (Beare *et al.*, 1995). Examples include: The 5–7% of the total soil that was root-influenced or rhizosphere in extensive pot trials of Ingham *et al.* (1985) contained a majority (greater than 70%) of the bacterial- and fungal-feeding nematodes. Ingham *et al.* (1985) also measured higher biomasses of rhizosphere bacteria in microcosms with large numbers of microbivorous nematodes (greater than 4000 per gram of rhizosphere soil) than in microcosms without these nematodes. Yet the extent of mineralization of nitrogen in the microcosms with nematodes reflected that they were ingesting large quantities of microbes as well. Thus there was a net enhancement of microbial production, in a fashion similar to that measured by Porter (1975), who found a net stimulation of phytoplankton growth after the cells had undergone transit through the guts of *Daphnia* sp. in freshwater incubations. As an example of the dynamic nature of shifting "hot spots," Griffiths and Caul (1993) found that more nematodes were active in the rhizosphere, and they moved readily to new concentrations of fresh organic matter (leaf litter) in short-term trials. Other examples of "hot spots" that have shown enhanced microbial activity include the drilosphere and worm castings, which show enhanced carbon and nitrogen (Syers *et al.*, 1979a; Daniel and Anderson, 1992) and phosphorus mineralization (Syers *et al.*, 1979b; Lavelle *et al.*, 1992). *Lumbricus terrestris* "middens" (small patches of plant litter and casts gathered around the burrow entrance) in experimental field sites in Ohio were found to be functionally different, with enhanced acetate incorporation and microbial cell synthesis, compared with surrounding non-earthworm–influenced soil (Bohlen *et al.*, 2002). Another center of activity is the aggregatusphere (Fig. 6.4), or region of micro- and macroaggregates

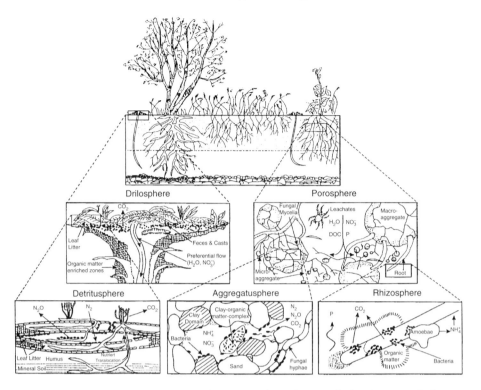

FIGURE 6.4. Arenas of activity in soil systems. These "hot spots" of activity may be less than 10% of the total soil volume, but represent more than 90% of the total biological activity in most soils worldwide (from Beare *et al.*, 1995, reprinted with permission).

(Elliott, 1986; Elliott and Coleman, 1988; Beare *et al.*, 1995; Six *et al.*, 1999). This zone of influence is less well studied and is a major source of some of the dynamic yet highly patchy behavior found in soils. Foster (1985) and Foster and Dormaar (1991) have demonstrated, using electron microscopy (Fig. 3.3), amoebal pseudopodia extending into very small pore-necks and pores (only a few tenths of micrometers in diameter) in well-structured soil, attacking bacterial colonies which seemed to be inaccessible to the smallest nematodes and amoebae or other protozoa. A study using a combined approach to rhizosphere and soil cracks for locations of "hot spots" of labile organic matter was used by van Noordwijk *et al.* (1993) to good effect.

A HIERARCHICAL APPROACH TO ORGANISMS IN SOILS

Because of the need to deal with soil heterogeneity in space and time, arenas of interest, noted in the previous section, are represented in

Figure 6.4 (Beare *et al.*, 1995) showing the volumes and biotic groups of concern. The aggregatusphere shows bacteria, amoebae, and some nematodes, having varying degrees of success in gaining access to the prey biota of interest (Vargas and Hattori, 1986). Moving up to a coarser level of resolution, to the rhizosphere, a few millimeters or less in scale one sees the microbes and fauna associated with them, and the considerable feeding and activity which has been documented numerous times. The activities are strongly influenced by abiotic, i.e., wetting and drying events, and the intrusion of new organic substances from growing root tips (Cheng *et al.*, 1993; Kuzyakov, 2002), or deposited feces from microarthropods, enchytraeids, or other mesofauna. The next level of resolution expands from many centimeters to several meters across the landscape, when any of the macrofauna such as earthworms or burrowing beetles come into play. There is then a qualitative shift, brought about by the ingestion of soil, which includes considerable amounts of micro- and mesobiota, that is, protozoa and nematodes (Yeates, 1981; Piearce and Phillips, 1980) as food. Interestingly, even with earthworms, the drilosphere *sensu stricto* is only 2–3 millimeters in thickness (Bouché, 1975) but the burrow extends laterally for many centimeters or meters through the soil. As a consequence of this activity, there can be major short-term decreases in viability of the existing biota, but possibly longer-term stimulation by enhanced microbial activity, as noted previously, and also from the considerable input of mucopolysaccharide-containing mucus (Marinissen and Dexter, 1990).

An additional aspect of altered species makeup of bacteria in earthworm-influenced soils has been explored using molecular probing techniques. Using 16S rRNA probes and "libraries" of soil bacteria at the Horseshoe Bend site in Athens, GA, Furlong *et al.* (2002) and Singleton *et al.* (2002) found enhanced percentage occurrences of Actinobacteria, Firmicutes, and gamma-Proteobacteria in castings of *Lumbricus rubellus*, an epigeic earthworm.

In addition, considerable amounts of ammonia and urea, as nitrogenous end-products of metabolism, may be voided either externally through nephridiopores, or internally into the gut cavities of earthworm genera that have that mode of nitrogen excretion (Lavelle *et al.*, 1992). In tropical regions, certain endogeic earthworms will process and assimilate end-products of the breakdown (from 2 to 9%) of soil organic matter in a wide range of ecosystems (Lavelle and Martin, 1992).

Similar sorts of activities may be catalyzed by certain termites, particularly those in the advanced family Termitidae, which are truly geophagous. These geophages utilize soil organic matter, deriving significant amounts of nutrition from this low-quality substrate by processing the organic matter in a high-pH chemical milieu in the region between the midgut and the first proctodaeal segment (see the Isoptera section in Chapter 4) (Bignell, 1984; Bignell *et al.*, 2000). The additional influence of microbial enzymes on insect digestive processes and, indeed, enhancement of nitrogen fixation in downed branches and logs (Martin, 1984) are well known. Finally, the impacts of ant and termite nests are significant, and certainly have an influence at the landscape scale. The impacts of the macrofauna, sometimes termed "ecosystem engineers" (Jones *et al.*, 1994), can extend for many meters beyond the immediate zones that they occupy. It has been contrasted with the impacts of smaller fauna, with smaller fauna more influential in energy flow and immediate nutrient recycling, noted previously, versus the longer-term effect of the "engineering" by the macrofauna (Scheu and Setälä, 2002) (Fig. 6.5). Scheu and Setälä (2002) and Wardle (2002) note that "trophic cascades," the term denoting the effects of predation on the biomass of organisms at least two trophic levels removed, occur in soil systems. Although developed principally for systems with living net primary production as the energy base, there are numerous examples in soil systems, particularly ones dominated by fungi. Scheu and Setälä (2002) comment on the limited number of studies of trophic cascades in soil systems to date, and that the fungal-based energy channel may be

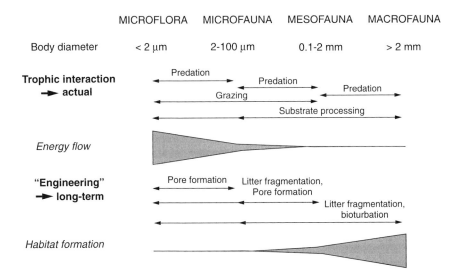

FIGURE 6.5. Size dependent interactions among soil organisms. Trophic interactions and interactions caused by "engineering" are separated; both are indicated by arrows. Note that trophic interactions and interactions caused by engineering are strongly size dependent but complement each other (tapering and widening triangles). Both function at different scales: trophic interactions drive the current energy flow, engineering sets the conditions for the existence of the soil biota community in the long term (from Scheu and Setälä, 2002).

much more prone to trophic cascades than the bacterial-based channel. This assertion is certainly a candidate for further experiments in the future.

Conceptualizations of detrital food webs are undergoing a considerable shift early in the third millennium. Following up on earlier ideas of Wardle (1995) and Lavelle *et al.* (1999), Pokarzhevskii *et al.* (2003) note that a definite nested element exists such that different compartments feed into others. For example, the bacteria–algae–protozoa compartment is nested inside a fungi–microarthropod compartment, and this in turn is contained within an earthworm–rhizosphere compartment. Animals at higher levels consume communities of the lower levels as a whole (Pokarzhevskii *et al.*, 2003) (Fig. 6.6). This arises from the dependence of all animals on microorganisms for their supply of proteins and

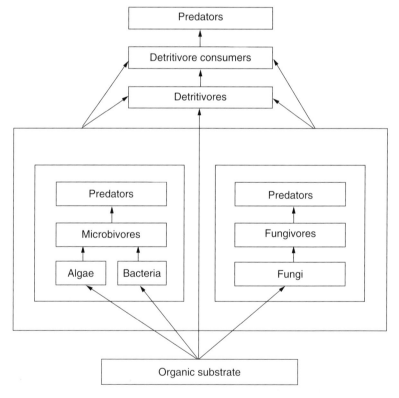

FIGURE 6.6. A conceptual scheme illustrating the nested structure of detrital food webs. A distinction is made between bacteria–algae–protozoa communities (left), fungi–microarthropod communities (right), and earthworm–plant communities (top). Communities of the higher levels consume communities of the lower levels as a whole (indicated by arrows) (from Pokarzhevskii *et al.*, 2003).

scarce minerals. The concept of "ecological stoichiometry," which concerns the roles of interactions between several major nutrients such as nitrogen, phosphorus, and/or sulfur, has been discussed at length by Sterner and Elser (2002). Much of Pokarzhevskii *et al.*'s (2003) paper discusses the need to consider the effects of limiting nutrients, which may be in shorter supply than the carbon or energy that characterize the outlook of many of the previously developed detrital food webs.

FUTURE RESEARCH PROSPECTS

It is becoming more and more imperative to bring small working groups, or teams of investigators, together to make further progress in food web studies. The real breakthroughs are certain to come from efforts that include the more transitional fauna between above- and belowground such as ants, dipteran larvae, and ground beetles, or cryptozoans such as the isopods, centipedes, and millipedes, linking them to the truly belowground fauna and microbes.

Various techniques noted in several papers in this volume should be extended as well. Stable isotopes, introduced in an initially enriched substrate such as labeled glucose or acetate, will be useful in delineating food webs. The effective use of carbon-13 and nitrogen-15 (^{15}N) was reviewed extensively by Scheu (2002). An innovative use of ^{15}N tagging in a microcosm study detected significant predation on springtails by an ectomycorrhizal fungus, *Laccaria bicolor* (Klironomos and Hart, 2001). The ectomycorrhizal fungus immobilized the animals before infecting them. Springtails (*Folsomia candida*), alive or already dead and labeled with ^{15}N, were added to the microcosms containing mycorrhizal or nonmycorrhizal *Pinus strobus* plants. Only the fungus and not the roots made contact with the animals. Amounts of nitrogen were determined in plant tissues and extraradical fungal hyphae over a 2-month period. Up to 25% of plant nitrogen was derived from springtails when they were in the presence of *L. bicolor*. At the end of the experiment, less than 10% of the number of animals were present compared to at the start. Using the same system, growing *Pinus strobus* seedlings with a different ectomycorrhizal fungus, Klironomos and Hart (2001) measured less than 5% of plant nitrogen acquired from the springtails. This experiment demonstrates a much greater range of possible interactions between mycorrhiza and fungal-grazing animals, and is yet another example of the tight linkages existing in forest nutrient cycling.

Opportunities for use of radiotracer carbon-14 (^{14}C) must also be kept in mind. For example, Kisselle *et al.* (2001), Garrett *et al.* (2001), Fu *et al.* (2001), and Coleman *et al.* (2002) described the detrital food web and its dynamics in an agroecosystem as a function of the impacts of above-ground experimentally induced herbivory. They measured increased microbial biomass production in no-tillage treatments that experienced moderate levels of aboveground herbivory (grasshoppers grazing on corn leaves). This was transmitted up the food chain to bacterial-feeding nematodes, with significantly more ^{14}C activity being taken up in the low grazing-intensity treatments, similar to the findings of Holland and Cheng (1996) (Fig. 6.7). Another notable finding was the higher ^{14}C activity in microarthropods extracted from rhizospheres of weed plants, compared to that of corn (Fig. 6.8). Garrett *et al.* (2001) suggest that weed rhizospheres may be more important than crop rhizospheres in supporting soil food webs. This might be expected, because crop plants are selected to maximize their aboveground net primary production (NPP), unlike weeds. If this pattern is general, weeds may be a signifi-cant factor for the protection of soil biodiversity, especially in conven-tionally tilled agroecosystems. The linkages between above- and belowground food webs is an exciting new topic for the first decade of the 21st century (Hooper *et al.*, 2000; Wolters *et al.*, 2000).

Hall and Raffaelli (1993) suggest two major areas of food web research that would be most beneficial to follow: (1) focusing on commu-nity assembly and (2) documenting the strength of trophic interactions between elements in webs. Examining the latter objective, Neutel *et al.* (2002) studied interaction strengths organized in trophic loops (defined as the product of interaction strengths in a food web). Using seven docu-mented soil food webs, Neutel *et al.* (2002) introduced the term "loop weight," which is the geometric mean of the absolute values of the inter-action strengths in the loop (Fig. 6.9). This enables one to compare loops of different lengths and to use the maximum of all loop weights as an indicator for matrix stability. They used the conservative figure of 0.1 for

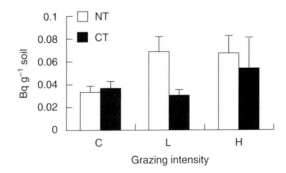

FIGURE 6.7.   The ^{14}C specific activity (Bq g^{-1} soil) of soil nematodes under different levels of aboveground herbivory. C: control, no grazing; L: low-grazing level (four grasshoppers per plant); H: high grazing level (eight grasshoppers per plant; NT: no-tillage; CT: con-ventional tillage (from Fu *et al.*, 2001).

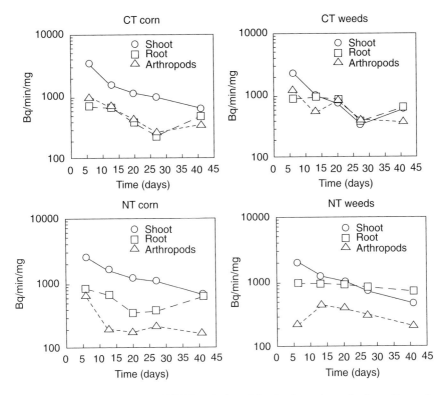

FIGURE 6.8. Concentrations of ^{14}C (Bq min^{-1} mg^{-1}) in shoots, roots, and microarthropods from rhizospheres of corn and weed plants grown under no-tillage (NT) and conventional tillage (CT) regimes (from Garrett *et al.*, 2001).

trophic transfer efficiencies, from one trophic level to another. They innovatively compared the community matrix, including the patterned interaction strengths ("real matrices") with several randomizations of this matrix ("randomized matrices"). This was done by randomly exchanging predator–prey pairs of interaction strengths, keeping these pairs intact and preserving the sign structure of the matrix. Stability was measured as the minimum degree of relative intraspecific interaction needed for matrix stability (s). Matrices with a smaller s value were considered "more stable."

Loop weights of the longer loops were low in the real matrix and tended to be heavier in the randomized matrices than the shorter loops [Fig. 6.9(a), 6.9(b)]. Interestingly—although absolute values of effects of predators on their prey are generally two orders of magnitude larger than effects of prey on their predators, as shown in the randomized matrices—in the real matrices the long loops with many top-down

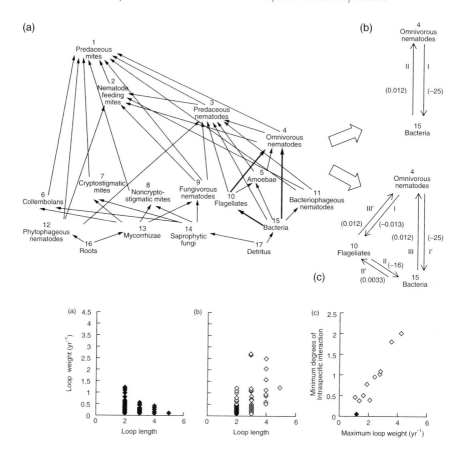

FIGURE 6.9. Loop length, loop weight, and stability in the Central Plains Experimental Range (CPER) food web and randomizations of this matrix. (a) Loop weight versus loop length in the real matrix. (b) Loop weight versus loop length in a randomized matrix (a typical example). Long loops with a relatively small weight—those with many bottom-up effects—are not shown because they are not relevant for maximum loop weight. (c) Maximum loop weight and stability of the real matrix (solid diamond) and of 10 randomized matrices (open diamonds). Stability was measured as the value s that leads to a minimum level of intraspecific interaction strength needed for matrix stability. In a sensitivity analysis, variation in the parameter values within intervals between half and twice the observed value led to only a small variation in stability (from Neutel *et al.*, 2002).

effects had a relatively low weight. This revealed that not all top-down effects were equal. With the maximum loop weight in the real matrix being markedly lower, the real matrix was much more stable [Fig. 6.9(c)]. Neutel *et al.* (2002) explored the ramifications of omnivory. For a three-species omnivorous interaction (Fig. 6.10), the omnivore feeds on two prey types, which are at different trophic levels. Assuming that it feeds according to prey abundance, and that the biomass of the prey on

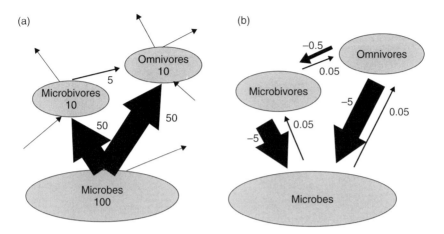

FIGURE 6.10. Interaction strengths and loop weights in an omnivorous food web. (a) Equilibrium feeding rates and population sizes. Feeding rates were assumed to be proportional to the population sizes of the prey. (b) Interaction strengths. In the example, efficiencies were assumed to be 0.1 for all species. The loop weights of the two loops of length 3 are $([-0.5] \times [-5] \times 0.05)^{1/3} = 0.5$ (anticlockwise loop, starting with the omnivores) and $([-5] \times 0.05 \times 0.05)^{1/3} = 0.23$ (clockwise loop). The relatively small top-down effect (-0.5) keeps the weight of the loop with two top-down effects relatively low (from Neutel *et al.*, 2002).

the lower trophic level is significantly larger than that of the prey on the higher trophic level, then the omnivore feeds largely on the lowest trophic level. Consequently, it exerts a relatively large top-down effect on its lowest prey and relatively small top-down effect on its higher prey, because the top-down effect is the feeding rate per unit of predator biomass. This approach was extended further to a wide range of published food webs, and their findings held true even for aboveground-oriented food webs.

In a seminal review paper, Moore *et al.* (2003) noted that predators within the rhizosphere alter the interactions between microbes and plants in two contrasting but probably equally important ways. Predators regulate their prey in a traditional "top-down" fashion but in doing so, they alter the release of nutrients that may limit plant productivity and thereby affect plant growth in a "bottom-up" fashion as well. They note that the interdependence between the aboveground and belowground realms can be explained in terms of the patterning of trophic interactions within the rhizosphere and the influence of these interactions on the supply of nutrients and rates of nutrient uptake by plants.

We suggest that a useful approach will include a melding of the two objectives named earlier, in terms of documenting the extent of soil food webs, the relative impacts of the trophic interactions at the various hierarchical levels of organization, and the location in the landscape (Coleman and Schoute, 1993; Hooper *et al.*, 2000). For example, does the soil system in the absence of earthworm or termite activity operate at more or less of a background or maintenance level? When the macrofauna move through the soil matrix, literally consuming and chaotically reassembling it, does this represent a more intensive level of activity? Certainly it is at a different level of resolution, but one dependent upon the myriad interactions of the microbes and the micro- and mesofauna.

SUMMARY

There is an interesting convergence occurring in aboveground and belowground portions of detrital food webs. In both locations, particularly in arid habitats (i.e., in deserts), the food webs are long (7–8 membered) and show extensive amounts of omnivory. By including members of the microfauna (protozoa) and mesofauna (microbivorous nematodes) that have been overlooked often in the past, there are ample amounts of food, because secondary production passes up the food chains. Production efficiencies may reach or exceed 70%, and trophic transfer efficiencies may exceed 20% in various "hot spots" such as rhizospheres, drilospheres, or in any other concentrations of reduced, labile, organic matter. The diversity of nutrient retention and recycling strategies in soil systems continues to increase as more innovative experiments are carried out using a variety of isotopic tracer techniques.

7 Soil Biodiversity and Linkages to Soil Processes

INTRODUCTION

Interest in soils as reservoirs of biodiversity has increased in the last several years. It is important to define biodiversity, which is an inclusive concept. Biodiversity encompasses a wide range of functional attributes in ecosystems in addition to being concerned with the numbers of species present in a given ecosystem. Within terrestrial ecosystems, soils may contain some of the last great "unknowns" of many of the biota. This includes such relatively well-studied fauna as ants (Hölldobler and Wilson, 1990), as well as the more numerous and less studied meso-fauna, such as microarthropods (Behan-Pelletier and Newton, 1999) and nematodes (Ettema and Yeates, 2003), that interact with elements of the microbiota, such as mycorrhiza, in several ways, including mutu-alistic ones (Wall and Moore, 1999). Much has been learned in the last decade about prokaryotic genetic diversity in soils; see the review by Hugenholtz *et al.* (1998).

BIODIVERSITY IN SOILS AND ITS IMPACTS ON TERRESTRIAL ECOSYSTEM FUNCTION

There is increasing concern among biologists in the fates of the very diverse array of organisms in all ecosystems of the world. What do we know of the full species richness, particularly in soils, to make even edu-cated guesses about the total extent of the organisms, or how many of them may be in an endangered status (Hawksworth, 1991a, 2001; Coleman *et al.*, 1994b; Coleman, 2001)? Soil biodiversity is best consid-ered by focusing on the groups of soil organisms that play major roles in ecosystem functioning. Spheres of influence of soil biota are recognized; these include the root biota, the shredders of organic matter, and the soil bioturbators. These organisms influence or control ecosystem processes and have further influence via their interactions with key soil biota (e.g., plants) (Coleman, 2001; Wardle, 2002). Some organisms, such as the fungus and litter-consuming microarthropods, are very speciose. For example, there are up to 170 species in one Order of mites, the Oribatida, in the forest floor of one watershed in western North Carolina. Hansen (2000) measured increased species richness of Oribatids as she experimentally increased litter species richness in experimental enclosures from one to two, four, and finally seven species of deciduous tree litter. This was attributed to the greater physical and chemical diversity of available microhabitats, which is in accord with the mechanisms suggested earlier by Anderson (1975).

Only 30–35% of the Oribatids in North America have been adequately described (Behan-Pelletier and Bissett, 1993), despite many studies carried out over the last 20–30 years. The studies suggest that there may be more than 100,000 undescribed species of oribatid mites yet to be discovered. Particularly in many tropical regions, Oribatids and other small arthropods are very little known in both soil and tree canopy environments (Behan-Pelletier and Newton, 1999; Nadkarni *et al.*, 2002). This difficulty is compounded by our very poor knowledge of identities of the immature stages of soil fauna, particularly the Acari and Diptera. Solution of this problem may require considerable application of molecular techniques to more effectively work with all life stages of the soil fauna (Behan-Pelletier and Newton, 1999; Coleman, 1994a; Freckman, 1994). We concur with Behan-Pelletier and Bissett (1993): "Advances in systematics and ecology must progress in tandem: systematics providing both the basis and predictions for ecological studies, and ecology providing information on community structure and explanations for recent evolution and adaptation." Chapin *et al.* (2000) note that 12% of birds and nearly 20% of mammals are considered threatened with extinction, and that from 5 to 10% of fish and plants are similarly threatened. With many of the soil invertebrates yet undescribed, it is impossible to affix a numerical value to losses of these members of the biota.

There are currently 70,000 species of fungi described (Table 7.1). By assuming that a constant ratio of species of fungi exists to those plant species already known, Hawksworth (1991b, 2001) calculated that there may be a total of 1.5 million species of fungi described when this mammoth classification task is completed.

Indeed, it may be possible to gain insights into biotic functions belowground by considering a "universal" set of functions for soil and sediment biota that include the following: degradation of organic matter,

TABLE 7.1. Comparison of the Numbers of Known and Estimated Total Species Globally of Selected Groups or Organisms

Group	Known species	Estimated total species	Percentage known
Vascular plants	220,000	270,000	81
Bryophyes	17,000	25,000	68
Algae	40,000	60,000	67
Fungi	69,000	1,500,000	5
Bacteria	3,000	30,000	10
Viruses	5,000	130,000	4

From Hawksworth, 1991.

cycling of nutrients, sequestration of carbon, production and consumption of trace gases, and degradation of water, air, and soil pollutants (Groffman and Bohlen, 1999).

What are the consequences of biodiversity? Does the massive array of hundreds of thousands of fungi and probably millions of bacterial species make sense in any ecological or evolutionary context? As was noted in Chapter 3 on microbes, the numbers of bacterial species are greatly underestimated because most investigations have relied on culturing isolates and examining them microscopically. There have been two key developments in studies of microbial diversity. First, the use of signature DNA sequences has greatly increased the numbers of identified taxa, with hundreds of novel DNA sequences being identified yearly. Two bacterial divisions, which appear to be abundant and ubiquitous in soils but have very few cultured representatives, are Acidobacterium and Verrucomicrobium (Hugenholtz et al., 1998). Second, we have only recently come to an appreciation of the incredibly wide distribution of prokaryotes (both Archaea—methanogens, extreme halophiles living in hypersaline environments, and hyperthermophiles living in volcanic hot springs and mid-sea oceanic hot-water vents—and Bacteria) worldwide. Prokaryotes constitute two of the three principal domains, or collections of all organisms, with Eucarya consisting of protists, fungi, plants, and animals (Fig. 7.1) (Pace, 1999; Coleman, 2001). The total numbers of bacteria on earth in all habitats is truly mind-boggling: $4–6 \cdot 10^{30}$ cells, or 350–550 petagrams (10^{15} g) of carbon (Whitman et al., 1998). The amount of the total bacteria calculated to exist in soils is approximately $2.6 \cdot 10^{29}$ cells, or about 5% of the total on earth. A majority of bacteria exist in oceanic and terrestrial subsurfaces, especially in the deep mantle regions, extending several kilometers below the earth's surface. Some of these organisms, which are the most substrate-starved on earth, may have turnover times of centuries to millennia (Whitman et al., 1998).

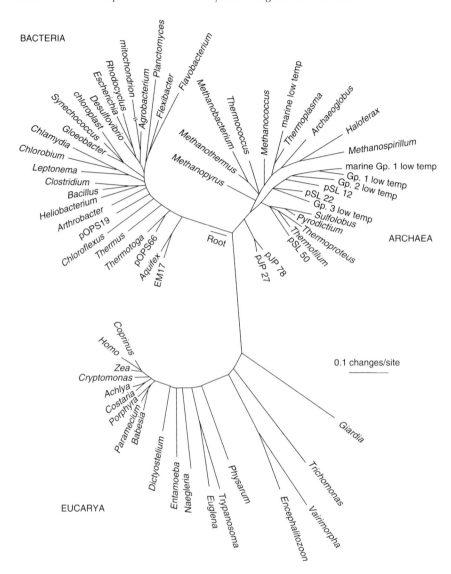

FIGURE 7.1.   The universal tree of life. The three domains of life—Bacteria, Archaea, and Eucarya—are shown here represented by small-subunit rRNA sequences of various organisms within each domain. The domain Bacteria (Eubacteria) and the domain Archaea (previously the archaebacteria) are entirely microbial, and the domain of Eucarya is predominantly microbial (from Pace, 1999).

What is the implication of the apparent "excess" of species diversity of soil microflora, where many species exist at a very low frequency and in an inactive state? If considerable species richness and accompanying large genetic pools are maintained in soils, what are the impacts on the evolution of new taxa? What are the implications for ecosystem function if this degree of redundancy exists; does it imply that some of the organisms are somehow vestigial remnants or relics of bygone conditions (Coleman et al., 1994b)? What are the functional roles of such hidden or apparently cryptic organisms? Are they performing some essential but unknown functions, perhaps at microsites that we don't observe or work with? One approach that may show promise is the use of reporter genes linked to gene promoters; this technique is used to measure in situ the activity of specific enzymes related to defined processes (Wilson et al., 1994). We need to link specific methods such as those noted here with soil thin-section studies, such as those of Tippkoetter et al. (1986), Postma and Altemueller (1990), and Foster (1994). Such means will enable the inclusion of spatial dimensions to soil ecological studies; addition of a temporal one provides the much-needed aspect of time as well. Soils are rife with historical signs and legacies, as has been made evident by studies using radiotracers, and stable isotope studies (Stout et al., 1981; Nadelhoffer et al., 1985; Gaudinski et al., 2000).

What are the linkages between biodiversity and ecosystem function? It should be possible to look for natural "experiments" such as regions with low species richness, e.g., on an island, versus sites at similar latitudes that are on continents, where one can measure key ecosystem processes such as rates of decomposition or nutrient cycling. Under such conditions, all of the major abiotic factors are held reasonably similar, allowing study of the impacts of species richness of key indicator microflora or fauna on ecosystem processes of interest. Such experiments are certainly performable, and might yield some surprising results.

Studies of the interactions of climate change and biological diversity have been reviewed by Vitousek (1994) and, in a modeling context, by Smith et al. (1998). By using species–area curves from island biogeography it is possible to estimate the fraction of species whose loss is entrained by loss of habitat (land use change), even without knowing how many species exist (Wilson, 1992, cited in Vitousek, 1994). Unfortunately, there is little information available on soil organisms to conduct such a comparative study.

More than 20 studies of the empirical evidence of relationships between ecosystem processes and different components of plant diversity (species richness, functional richness, and functional composition) were followed in natural and synthetically assembled groups of grassland species worldwide (Diaz and Cabido, 2001). The linkage was found to be neither simple nor universal, but some significant trends were noted. The range and more particularly the values of functional traits carried by plants (e.g., whether they are nitrogen-fixing, warm-season grasses or rosette forbs) are generally strong drivers of ecosystem processes. These studies combined simplified microcosms and natural field sites, so extrapolation from them is limited (Table 7.2). However, it is noteworthy that most of the studies showed that species richness and functional composition had positive effects on aboveground biomass.

Numbers of species aboveground and belowground may be correlated when taxa in both habitats respond similarly to the same or correlated environmental driving variables, in particular across large gradients of disturbance, climate, soil conditions, or geographic area. Differentiating between simple correlation and causation may be problematic, however. High diversity in plant species can result in high diversity of litter quality or types of litter entering the belowground system. This resource heterogeneity can lead to a greater diversity of decomposers and detritivores (Hooper *et al.*, 2000). In contrast, a high diversity of resources and species in soil could feed back to a high diversity aboveground, where certain species or functional groups are closely linked to groups belowground. A useful example of this was noted by van der Heijden *et al.* (1998), who found a positive correlation between the diversity of endomycorrhizal species and plant diversity, perhaps because different species of fungi infect different species of plants to different degrees, although alternative explanations have been offered for these patterns (Wardle *et al.*, 1999). Interestingly, Hartnett and Wilson (1999) and Smith *et al.* (1999), working in a Kansas tallgrass prairie, showed that mycorrhiza promoted obligately mycorrhizal C_4 grasses, resulting in competitive exclusion of facultatively mycorrhizal C_3 species, reducing overall plant species diversity. A similar mechanism seems to operate in tropical rainforests, in which ectomycorrhizal (ECM) tree species competitively exclude arbuscular mycorrhizal (AM) species (Connell and Lowman, 1989, cited in Wardle, 2002). It should be noted that at the level of functional types of mycorrhiza, this pattern does not hold: low-diversity AM can be associated with high diversity of plants, and high-diversity ECM communities can be associated with low diversity of plants (Allen *et al.*, 1995).

In an extensive experiment carried out under field conditions, Porazinska *et al.* (2003) tested aboveground–belowground diversity relationships in a naturally developed tallgrass prairie ecosystem by comparing soil biota and soil processes occurring in homogeneous and heterogeneous plant combinations of C_3 and C_4 photosynthetic pathways. Some bacterial and nematode groups were affected by plant characteristics specific to a given plant species, but no uniform patterns emerged. Interestingly, invasive and native plants were quite similar with respect to the measured soil variables (e.g., phospholipid fatty acids, protozoa, and nematode functional groups). Contrast these results with those of Belnap and Evans (2001) given toward the end of Chapter 8.

Text continued on page 259

TABLE 7.2. Empirical Evidence of Relations between Ecosystem Processes and Different Components of Plant Diversity[a]

Ecosystem	Experimental setup	Ecosystem processes[b]	Positive effects reported[c]			Functional types (sensu lato)
			Species richness	Functional richness	Functional composition	
Synthetic assemblages						
Serpentine grassland, United States	Plant mixtures planted in the field	N retention in ecosystem Aboveground biomass Inorganic N pools in soil	NA NA NA	No No Yes	Yes Yes Yes	Bunchgrasses N-fixers, early and late-season annual forbs
Savannah grassland, United States	Plant mixtures planted in the field	Aboveground biomass, light penetration, and plant % and total N	No	Yes	Yes	C_3 grasses, C_4 grasses, legumes, forbs, and woody plants
Mesic grassland, United Kingdom	Plant mixtures planted in the field	No. of invading species and total biomass of invasives	No	NA	Yes	Perennial grasses and forbs
Grasslands, Germany, Portugal, Switzerland, Greece, Ireland, Sweden, UK	Plant mixtures planted in the field	Total aboveground biomass	Yes	Yes	Yes	Grasses, legumes, herbs
Annual grassland, France	Plant mixtures planted in the field	No. of invasives from soil seed bank and survival of seedlings of the exotic and annual forbs *Coniza bonariensis* and *C. canadensis*	No	No	Yes	Annual grasses, annual legumes, and annual Asteraceae
Acid grassland, United Kingdom	Plant mixtures planted in the field	Decomposition of standard material Decomposition of litter mixtures	Yes No	No No	No Yes	Grasses, legumes, and herbs

LIVERPOOL
JOHN MOORES UNIVERSITY
AVRIL ROBARTS LRC
TEL. 0151 231 4022

TABLE 7.2. *Continued.*

Ecosystem	Experimental setup	Ecosystem processes[b]	Positive effects reported[c]			Functional types (sensu lato)
			Species richness	Functional richness	Functional composition	
Grassland on old fields, Switzerland and Sweden	Plant mixtures planted in the field	No. of leafhoppers (Cicadellidae)	No	No	Yes	Grasses, legumes, land forbs
		No. of wingless aphids (Aphididae)	No	Yes	Yes	
		No. of hymenopteran parasitoids	No	No	No	
		No. of grasshoppers (acrididae) and slugs (Gastropoda)	No	No	No	
		No. of carabid beetles (Carabidae) and spiders (Araneae)	No	No	Yes	
Calcareous grassland on old field, Switzerland	Plant mixtures planted in the field	Preference by voles[d]	Yes	No	Yes	Grasses, legumes, and forbs
		Earthworm biomass[d]	Yes	Yes	No	
		Plant aboveground biomass, soil microbial biomass, LAI, plant light absorbance per unit ground area	Yes	Yes	Yes	
		Mesofauna feeding activity	No	No	No	
		Decomposition of standard material	No	No	Yes	
		Soil moisture	No	No	Yes	
Grassland, Greece	Plant mixtures planted in the field	Total aboveground biomass	Yes	NA	Yes[e]	Annuals and perennial grasses, geophytes, and legumes

Location/system	Method	Response variable				Species composition
Serpentine grassland, United States	Plant mixtures planted in the field	Aboveground biomass of invasive forb *Centaurea solstitialis*	No	Yes	Yes	Annual grasses, perennial grasses
		Impact of invader on aboveground biomass of resident species and whole-system evapotranspiration	Yes	Yes	Yes	Bunchgrasses, early-season and late-season annual forbs
Grasslands on old fields, Czech Republic, the Netherlands, United Kingdom, Sweden, and Spain	Plant mixtures planted in the field	Total aboveground biomass	Yes	NA	Yes	Grasses, forbs, and legumes
		Suppression of natural colonizers	Yes	NA	Yes	
Grassland, United States	Plant mixtures planted in greenhouse microcosms	Aboveground biomass	Yes	NA	Yes	C_3 grasses, C_4 grasses, legumes, and forbs
		N retention	No	NA	Yes	
Annual grassland, France	Plant mixtures in greenhouse microcosms	Invasibility (establishment of the forb *Echium plantagineum*)	No	No	Yes	Grasses, legumes, and rosette dicots
Prairie grassland, United States	Plant mixtures in greenhouse microcosms	Above- and belowground biomass, light transmission, and water retention in soil	Yes	Yes	Yes	Grasses, legumes, and forbs
		Decomposition of standard material	No	No	Yes	
Prairie grassland, United States	Plant mixtures planted in the field and in greenhouse microcosms	Resistance to invasion (total biomass of invasive)	Yes	NA	Yes	C_3 grasses, C_4 grasses, legumes, and forbs

TABLE 7.2. Continued.

Ecosystem	Experimental setup	Ecosystem processes[b]	Positive effects reported[c]			Functional types (sensu lato)
			Species richness	Functional richness	Functional composition	
Grassland-crop site, New Zealand	Litterbags placed in the field	Decomposition rate of, rate of N release from, and active microbial biomass on litter	No	NA	Yes	Grasses, weedy forbs, forbs from grasslands, and trees
Grasslands, United Kingdom	Litterbags placed in indoor soil microcosms	Soil microbial biomass	No[f]	NA	Yes	Dominant species in intensively managed fertile grasslands, or traditionally managed unfertilized grasslands
Manipulation of natural communities						
Grassland, Argentina	Mostly perennial grassland in neighboring paddocks under different grazing regimes	Aboveground net primary production	No	No	Yes	Cool-season graminoids, warm-season grasses, cool-season and warm-season forbs
Boreal forest, Sweden	Vegetation on islands of different area, subjected to different frequencies of wildfires	Aboveground biomass, litter decomposition, N mineralization, and humus accumulation	No	NA	Yes	Early versus late successional species

System	Study description	Response variable				Dominant community
Savannah grasslands, India	Vegetation along a productivity, diversity, and disturbance gradient, with different burning and grazing experimental treatments	Resistance to compositional change across communities	No[g]	NA[g]	Yes[g]	Not explicit, communities dominated by the grasses *Cymbopogon flexuosus* or *Aristida setacea*
		Resistance to species turnover across communities	Yes[g]	NA	No	
		Resistance to compositional change and to species turnover within communities	No[g]	NA	No	
Calcareous grasslands, United Kingdom	Contrasting grassland subjected to temperature and precipitation manipulations in the field	Resistance of total aboveground biomass and species compositions	No	NA	Yes	Communities dominated by fast-growing early successional species or by slow-growing, stress-tolerant perennial grasses and sedges
Mediterranean shrublands, Greece	Sites naturally differing in species diversity and growth-form composition	Aboveground biomass	Yes	NA	Yes	*Cistus* sp., other shrubs, and herbs
Sand-prairie grassland, United States	Experimental removal from natural communities on old fields	No. of individuals and cover of invaders	NA	Yes	Yes	C$_3$ graminoids, C$_4$ graminoids, and forbs
		Light transmittance through canopy	NA	Yes	Yes	
		Soil moisture, soil extractable N, and aboveground biomass	NA	No	Yes	

TABLE 7.2. *Continued.*

Ecosystem	Experimental setup	Ecosystem processes[b]	Positive effects reported[c]			Functional types (sensu lato)
			Species richness	Functional richness	Functional composition	
Dairy grasslands, New Zealand	Grasslands differing in climate and seasonal vegetation, subjected to experimental extreme temperature and rainfall events	Stability of biomass production after extreme events	No	NA	Yes	C_3 or C_4 species
Sand-prairie grassland, United States	Old-field communities subjected to removal of different functional types	Total aboveground biomass	NA	No	Yes	C_3 graminoids, C_4 graminoids, and forbs
		Community drought resistance	NA	No	Yes	

From Diaz and Cabido, 2001.

[a] Only studies assessing the impact of at least two components of plant diversity on ecosystem processes, and published in 1995 or later, were considered. Comparisons are qualitative and should be taken with caution, because unless a study explicitly has a test for species richness, functional richness, and functional composition in its design, it might lead to underestimation or misrepresentation of different components of diversity. Field studies differ markedly among themselves and with synthetic assemblages studies in approach, design, and intervening factors and thus strict comparison is not possible.

[b] Abbreviations: LAI, leaf area index; N, nitrogen.

[c] In the case of species and functional richness, only positive effects were considered: *No*, either no effect or a negative effect. In the case of functional composition: *Yes*, any significant (positive or negative) effect; *NA*, not assessed.

[d] Species: vole, *Arvicola terrestris*; earthworms, *Octolasion synaeum, Nicodrilus longus, Allolobophora rosea, A. chlorotica, Lumbricus terrestris*; and *L. castaneum*.

[e] Species richness effect obvious only when annuals were included in analysis.

[f] Effect of increasing litter diversity on soil microbial biomass was not unidirectional: two- and four-species litter treatments decreased it, whereas five- and six-species treatments increased it.

[g] Shannon Diversity Index.

HETEROGENEITY OF CARBON SUBSTRATES AND EFFECTS ON SOIL BIODIVERSITY

A step-by-step process for the ways in which increased heterogeneity of carbon (C) substrates from aboveground will positively influence belowground diversity is as follows (Fig. 7.2) (Hooper *et al.*, 2000): (1) diversity of primary producers leads to diversity of C inputs belowground, (2) C resource heterogeneity leads to diversity of herbivores and detritivores, and (3) diversity of detritivores or belowground herbivores leads to diversity of organisms at higher trophic levels in belowground food webs. The critical point is the nature and extent of trophic interactions (Hooper *et al.*, 2000). There are three general categories of interactions by which organisms in one compartment can affect biodiversity in another one: (1) obligate, selective interactions (one-to-one linkage), through mutualism for example; (2) one-to-many species linkages, via keystones and dominants; and (3) causal richness, or many-to-many linkages. The nature and extent of these interactions varies a great deal depending on the systems studied and the spatial scales at which the mechanisms are being considered.

There is a strong interaction between ecosystem function, organismal abundance and diversity, and the nature of humus forms in soil. Ponge (2003) compared more than 20 ecosystem attributes, and the nature of the processes and organisms occurring in mull, moder, and mor soils (Table 7.3) (Ponge, 2003). The table is a useful means of comparing many soil attributes across a broad range of physical, chemical, and biological traits. It shows a marked gradient from high (mull) to low (mor) biodiversity and rapid to slow and very slow rates of humification. Not surprisingly, a key determinant of litter decomposability, phenolic content, varied inversely across the same sequence of three humus types. Of course, we have yet to see how well these generalizations hold up when including a detailed analysis of the microbial communities in all three humus types.

Studies of biodiversity should include assessments of the nature and extent of anthropogenic disturbance. In a recent multistate and provincewide study of snail distributions and diversity in 443 sites, anthropogenic disturbance was found to be a major factor in decreases in species richness in forested ("duff") versus grassland ("turf") sites. This indicates that the conservation of faunas in the former will require protection of the soil surface architecture (Nekola, 2003).

IMPACTS OF SPECIES RICHNESS ON ECOSYSTEM FUNCTION

Recent studies of Wall and colleagues in the McMurdo Dry Valleys of Antarctica may offer some insights into the impacts of species richness. The dry valley ecosystems contain only three species of nematode: one

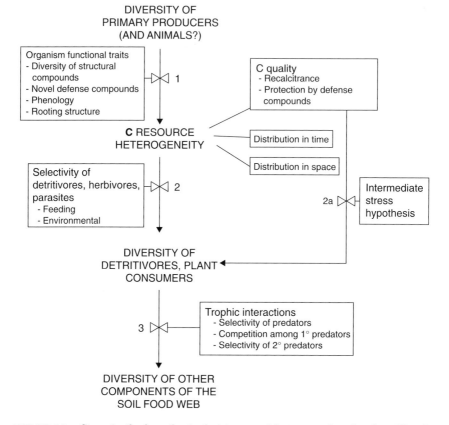

FIGURE 7.2. Steps in the hypothesis that increased heterogeneity of carbon (C) substrates from aboveground organisms will positively influence belowground diversity. This mechanism postulates strong bottom-up control of diversity in belowground communities; it should be tested in the context of other potential (e.g., top-down) controls (Hunter and Price, 1992). Step 1. Diversity of primary producers leads to diversity of C inputs belowground. Step 2. Carbon resource heterogeneity leads to diversity of herbivores and detritivores. (Alternative Step 2. Carbon resource quality, rather than heterogeneity, leads to diversity of detritivores.) Step 3. Diversity of detritivores or belowground herbivores leads to diversity of organisms at higher trophic levels in belowground food webs. (From Hooper *et al.*, 2000; see paper for more details of the complex interactions involved in aboveground and belowground diversity.)

bacterial feeder, one microbial feeder, and one omnivore-predator that are present in very low numbers ($2–5$ per kg^{-1} soil) (Wall and Virginia, 1999). These systems have very low precipitation (the equivalent of about 10 centimeters of rainfall per year), and make the usually harsh climate of the Chihuahuan desert of New Mexico seem like an oasis, with 7 plant parasites, 10 genera of microbivores, 2 omnivore genera,

TABLE 7.3. Main Biological Features of the Three Main Humus Forms

Ecosystems	Mull Grasslands, deciduous woodlands with rich herb layer, Mediterranean scrublands	Moder Deciduous and coniferous woodlands with poor herb-layer	Mor Heathlands, coniferous woodlands, sphagnum bogs, alpine meadows
Biodiversity	High	Medium	Low
Productivity	High	Medium	Low
Litter horizons	OL, OF	OL, OF, OH	OL, OM
Soil type	Brown soils	Grey-brown podzolic soils	Podzols
Phenolic content of litter	Poor	medium	High
Humification	Rapid	Slow	Very slow
Humified organic matter	Organo-mineral aggregates with clay-humus complexes	Holorganic faecal pellets	Slow oxidation of plant debris
Exchange sites	Mineral	Organic (rich)	Organic (Poor)
Mineral weathering	High	Medium	Poor
Mineral buffer type	Carbonate range	Silicate range	Iron/aluminum range
Impact of fire	Low (except in Mediterranean ecosystems)	Medium	High
Regeneration of trees	Easy (Permanent)	Poor (cyclic processes)	None (fire needed)
Dominant mycorrhizal types	VA-mycorrhizae	Ectomycorrhizae	Ericoid and arbutoid mycorrhizae
Mycorrhizal partners	Zygomycetes	Basidiomycetes	Ascomycetes
Nitrogen forms	Protein, ammonium, nitrate	Protein, ammonium	Protein
Nutrient availability to plants	Direct (through absorbing hairs)	Indirect (through extramatrical mycelium)	Poor
Nutrient use efficiency	Low	Medium	High
Fauna	Megafauna, macrofauna, mesofauna, microfauna	Macrofauna (poor), mesofauna (rich), microfauna	Mesofauna (poor), microfauna (poor)
Faunal group dominant in biomass	Earthworms	Enchytraeids	None
Microbial group dominant in biomass	Bacteria	Fungi	None
Affinites with polluted condition	Low	Medium	High

From Ponge, 2003.
Abbreviations: OF, fermentation layer; *OH,* humifaction layer; *OL,* litter layer; *OM,* matted organic matter just above the mineral soil, *VA,* vesicular-arbuscular (endo)mycorrhizae.

and 3 genera of predators (Fig. 7.3) (Wall and Virginia, 1999). The latter system contains numerous vascular plants, with considerable organic inputs both above- and belowground. In the McMurdo Dry Valleys, the sources of organic matter are restricted to allochthonous inputs from algae in nearby lakes or streams, or small amounts of indigenous soil algae and cyanobacteria. Although depauperate in species, their distributions spatially are markedly different, and highly correlated with differences in tolerances to desiccation and salinity, with the omnivore-predator and bacterivore being more water-requiring, concentrating in stream beds, and the microbivorous (bacteria and yeast spp.) endemic species *Scottnema lindsayae* restricted to the drier uplands (Treonis *et al.*, 1999). Although complicated in terms of life-history details, the fact that the number of species is so small makes it seem likely that a

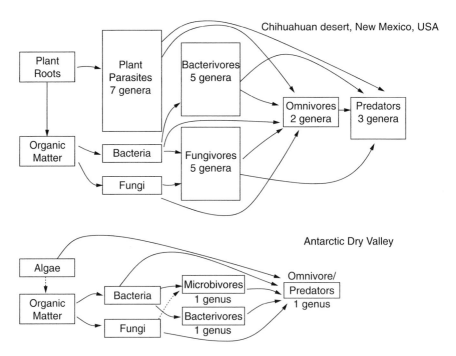

FIGURE 7.3. Complexity of soil nematode food webs in a hot desert (Chihuahuan, Jornada Long-Term Ecological Research [LTER], New Mexico) with 22 nematode genera, and a cold desert (Taylor Valley, McMurdo LTER in Antarctica) with three genera. For the nematodes, the height of the boxes illustrates the number of genera. The Antarctic Dry Valley has one species of a microbivore, *Scottnema lindsayae*, that feeds on bacteria and yeast; one bacterivore, *Plectus antarcticus*, that feeds on bacteria; and an omnivore-predator, *Eudorylaimus antarcticus*, that probably feeds on algal cells, bacteria, yeast, fungi, nematodes, and other small fauna (from Wall and Virginia, 1999).

fuller understanding of microbial and faunal interactions related to diversities is possible.

The role of redundant species and the functional roles played by them are crucial to understanding the interplays between biodiversity and ecosystem function. Without detailed knowledge of the biology of species involved, it can be difficult to decide how many functional types are present in a system or determine the functional roles of individual species (Bolger, 2001). Pathogen protection benefits of arbuscular mycorrhizas may be as significant as the nutritional benefits to many plants growing in temperate ecosystems (Newsham *et al.*, 1995, cited in Bolger, 2001).

MODELS, MICROCOSMS, AND SOIL BIODIVERSITY

Hunt and Wall (2002) modeled the effects of loss of soil biodiversity, viewed from a functional group perspective, on ecosystem function. They constructed a model for carbon and nitrogen transfers among plants, functional groups of microbes, and fauna. They used 15 functional groups of microbes and soil fauna: bacteria; saprophytic and mycorrhizal fungi; root-feeding, bacteria-feeding, fungal-feeding, omnivorous, and predaceous nematodes; flagellates and amoebae; collembola; r- and k-selected fungal-feeding mites; and nematophagous and predaceous mites (see Fig. 6.2) (Hunt *et al.*, 1987). The 15 functional groups were deleted one at a time and the model was run to steady state. Only 6 of the 15 deletions led to as much as a 15% change in abundance of a remaining group, and only deletions of bacteria and saprophytic fungi led to extinctions of other groups. By this analysis, no single faunal group had a significant effect on subsequent ecosystem behavior. However, the authors caution that, despite numerous compensatory mechanisms that occurred, it is premature to assume that the system is inherently stable even with the loss of several faunal groups. In fact, earlier analyses of similar food webs by Moore *et al.* (1993) and Moore and De Ruiter (2000) showed that loss of top predators had much greater impacts on lower trophic levels than their low biomasses might indicate.

Another approach to biodiversity and its linkages to soil processes is by use of experimental microcosms. Building on results of earlier studies of Setälä *et al.* (1997), Liiri *et al.* (2002) established microcosms with litter, humus, and mineral layers, and controlled access from the outside soil allowed by using either 45-micrometer or 1-millimeter mesh screens on the side of the microcosms. The microcosms were then half-buried to the top of the mesh in the side of the funnel, and then the upper portion left open to provide light for the pine seedling in the microcosm (Fig. 7.4) (Liiri *et al.*, 2002). Microcosms were watered at regular

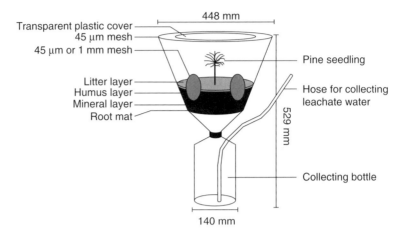

FIGURE 7.4.   A scheme of the lysimeter used with forest soils (from Liiri *et al.*, 2002).

intervals, or alternatively run through drought cycles, and leachates drawn off from the collecting bottle underneath to analyze for inorganic nitrogen and organic carbon in them. The experiment was run for 152 weeks, or nearly 3 years. The authors followed microbial community composition using phospholipid fatty acid analysis (PLFA) and BIOLOG to differentiate between bacteria and fungi, and sampled periodically for nematodes, enchytraeids, and microarthropods. They varied pH regimes by applying wood ash to some microcosms and not to others. They observed significant decreases in microarthropod numbers in the first year, followed by gradual increases in numbers of organisms with small body sizes. Enchytraeid numbers followed similar patterns. Nematodes had ready access to all microcosms, and were quite numerous, ranging from 67 to 191 g soil^{-1} in the controls, and from 98 to 545 g soil^{-1} soil in the ash-treated microcosms. The ash had significant effects on the microbial community makeup inside, but not in the soil outside the microcosms. The main effects on pine seedling growth and nutrient dynamics were governed by the abiotic factors of pH and availability of water. These seemed to govern the overall dynamics, in spite of the functional complexity of the soil biota. It would be most instructive to see this experimental design repeated in other habitats and biomes to ascertain the generality of the findings. In addition, it would be useful to compare the fauna at least to the family or genus level to see if finer-grained responses to the experimental manipulations occurred during this nearly 3-year-long experiment.

   In an extensive comparison of seven food webs of native and agricultural soils, de Ruiter *et al.* (1998) modeled energetics and stability, evaluating the roles of various groups of organisms and their interactions in energy flow and community stability. They measured feeding rates, interaction strengths, and impacts of the interactions on food web stability arranged according to trophic position in seven belowground food webs: one from Central Plains, Colorado, in the United States; two tillage manipulations at Lovinkhoeve in the Netherlands; two tillage manipulations at Horseshoe Bend, Athens, Georgia; and no fertilizer and fertilizer additions at Kjettslinge in southern Sweden. De Ruiter *et al.* (1998) found that only a fraction of the species manipulations had a strong effect on food web structure. Also there was an absence of correlation between the impacts on stability and feeding rates, meaning that interactions representing a relatively low rate of flow of materials can have a large impact on stability, and interactions having a high rate of material flow can have a small impact. Thus the higher-level predatory mites and nematodes had an impact far out of proportion to their biomass, and the contrary was true of the high biomass organisms, namely bacteria and fungi. De Ruiter *et al.* (1998) urge that future research be focused on the energetic properties of the organisms forming the basis of the patterning of interaction strengths. This is a big order, and one that will require innovative experiments under both laboratory and field conditions. The stakes are high, however, because these studies should help to provide further insights into the nature of biodiversity and ecosystem function.

EXPERIMENTAL ADDITIONS AND DELETIONS IN SOIL BIODIVERSITY STUDIES

Additional studies of biotic roles of soil fauna and bacteria and fungi have been approached in two ways. One approach is by gamma irradiating sieved soils and inoculating them with suspensions of full-strength, 10^2, 10^4, and 10^6 dilutions of soil organisms (Griffiths *et al.*, 2001). The other approach subjects unsterile soils to chloroform fumigation and incubation (Griffiths *et al.*, 2000), and tracks the subsequent changes in functional variables such as ammonium, nitrate, soil respiration, etc., in relation to microbial biomass and diversity, as measured by DNA patterns on denaturing gel electrophoresis (DGGE). Results were divergent in the two studies, with the chloroform fumigation simplification of community biomass and species diversity having a direct impact on the functional stability, as measured by the physiological response variables. In contrast, although there were progressive declines in biodiversity of the soil microbial and protozoan populations, there were no consistent changes in functional parameters. Some functions showed no trend (thymidine and leucine incorporation, nitrate accumulation, respiratory growth response), some a gradual increase with increasing dilution (substrate induced respiration), some declined only at the highest dilution treatment (short-term respiration from added grass, potential nitrification rate, and community level physiological profile), while others varied even more idiosyncratically. At no stage were any of the physiological functions eliminated completely. The final commentary on this by Griffiths *et al.* (2001) is that within any realistic sort of range of changes in biodiversity to be experienced by soils, there will be no direct effect on any soil functional parameters measured. Other authors, for example, Wardle *et al.* (1999) suggested that it is possible to overcome selective species effects by (1) measuring the effects of all species in monoculture and (2) by species removal experiments. Neither of these approaches is feasible with current technology, so this problem awaits the attention of a future generation of soil ecologists.

Another approach to microcosm studies was taken by the large group working in the Ecotron controlled-environment facility at Silwood Park in the United Kingdom. Constructing analogs of a temperate, acid, sheep-grazed grassland in northern Britain, Bradford *et al.* (2002) established terrestrial microcosms of graded complexity, with soil, plant, and microorganisms, and then assemblages of microfauna, micro- and mesofauna, and then micro-, meso-, and macrofauna. This functional group approach provided a range of metabolic rates, generation times, population densities, and food size. The microcosms were maintained in the Ecotron for a period of 8.5 months. Bradford *et al.* (2002) found significant increases in decomposition rate in the most complex faunal treatment, but both mycorrhizal colonization and root biomass were less abundant in the macrofauna treatments. Interestingly, plant growth was not enhanced in these treatments, despite higher nutrient (nitrogen and phosphorus) availability. Contrary to initial hypotheses, neither aboveground net primary production (NPP) (plant biomass) nor net ecosystem production (net CO_2 uptake) were enhanced in the most complex microcosms. Bradford *et al.* suggested that respiration was most likely buffered by the combined stimulatory effect of both meso- fauna and macrofauna on microbes (see Chapters 4 and 5), which served to maintain microbial activity at a level equivalent to that in the micro- fauna and mesofauna communities. This study has served as a bench- mark in large-scale microcosm studies, but as Bradford *et al.* (2002) note, it is not a substitute of longer-term *in situ* field studies, as difficult to conduct and interpret as they may be.

PROBLEMS OF CONCERN IN SOIL BIODIVERSITY STUDIES

An alternative to the functional approaches just discussed is taken by André *et al.* (2002), who note that most investigators use inadequate sampling designs or sample too shallowly in the soil profile to get a complete sample of microarthropods to provide the information used in the models noted previously. In an extensive survey of the worldwide literature on microarthropods, they claim that, on average, at most 10% of the soil microarthropod populations have been explored and 10% of the species described, due to the use of inefficient extraction procedures. This is supported by Walter and Proctor (2000), who suggest that perhaps only 5% of the species of mites worldwide are described so far. André et al. (2002) make the very valid point that ecologists need to be aware of the numerous pitfalls and possible flaws inherent in many extraction procedures; that is, none of them are 100% efficient. In the section on field studies and laboratory analyses, we explore some of these concerns more extensively.

There is an understandable concern that some quantifiable relationship be given to the relationship between ecosystem function and diversity. This is portrayed in Figure 7.5 (Bengtsson, 1998), which contrasts two curves of ecosystem function as a function of increasing numbers of species. Type 1, a continually ascending curve, represents the hypothesis that all species are important for ecosystem function. Type 2, initially convex and then flat, represents the species redundancy hypothesis. Bengtsson (1998) argues that it is more informative to consider specific functions in ecosystems, namely decomposition, nutrient mineralization, or primary production, thus focusing on phenomena that are more amenable to scientific inquiry. Bengtsson (1998) argues strongly that diversity does *not* play a role in ecosystem function. He goes so far as to assert that: "correlations between diversity and ecosystem functions—which may very well exist—will be mainly non-causal correlations only." As we are trying to show in this chapter, the truth may well lie in some midpoint between these extremes. The fact that certain functions may be linked to just a few genera or species, such as autotrophic and heterotrophic nitrifiers, for example, means that this might well be a "pressure point" for concern about long-term ecosystem function. The "natural insurance capital" concept of Folke et al. (1996), also discussed in detail by Bolger (2001), suggests that it is essential to retain as much species richness as possible to ensure that complete ecosystem services exist as human needs or environmental changes occur.

As in all areas of ecology, there is a spatial dimension to the biodiversity of soil organisms. It is essential to know not only which species are present, but also where the counted species occur in relation to one another. Do species occur together at every microsite, or do they occur mostly individually in separate sites? This has an important bearing on competition and other interactions, with functional consequences for the ecosystem. Ettema and Yeates (2003) measured patterns of small (centimeter) and intermediate (meter [m]) scales in nematode communities in a forest compared to a pasture system on a similar soil type in New Zealand. Using geostatistical techniques and mathematical calcu-

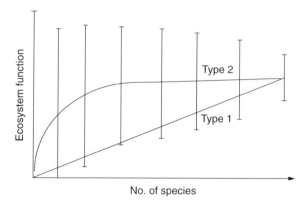

No. of species

FIGURE 7.5.   A hypothetical example of an attempt to quantify the form of the relationship between ecosystem function and diversity. The Type 1 curve represents the hypothesis that all species are important for ecosystem function, while Type 2 is the species redundancy hypothesis. The bars indicate the range of responses as different numbers of species are randomly drawn from a source pool of species, given that species' effects on ecosystem function are mainly species-specific (idiosyncratic) and not related to diversity. Note that although an average response may be observed, it neither allows a distinction between the two different hypotheses (Type 1 and Type 2), nor does this average response allow any useful prediction of what will happen in individual cases of species deletions (from Bengtsson, 1998).

lations of species turnover, they compared nematode genera in forestland, then in pasture. The forestland was assumed to have greater variation in vegetation and hence belowground inputs, on small and intermediate scales, than in the pasture. Thus they hypothesized that nematode genera are more strongly aggregated (occurring in "hot spots") in the mixed forest than in the ryegrass/white clover pasture. Applying an optimization method for sampling in geostatistical studies called spatial simulated annealing (SSA) developed by Van Groenigen and Stein (1998), Ettema and Yeates (2003) sampled along 40-m-long transects for the meter scale, with distance classes of 3 m, reflecting the scale of tree spacing. The centimeter scale transect was one-tenth of the large scale, or 4 m. The total number of nematodes per soil core volume was more than five times higher in the pasture (2800 ± 1234) than in the forest (430 ± 252), but the average number of genera in the forest (23.7 ± 3.3) was higher in the forest than in the pasture (19.1 ± 2.5). Also, many more nematode genera occurred in the forest (53) than in the pasture (37). Dissimilarity analysis showed that generic turnover was significantly greater in the forest than in the pasture, both at the small and intermediate scales (Fig. 7.6) (Ettema and Yeates, 2003). Because increasing distance in the forest led to increasing dissimilarity between communities, and no plateau was reached, it is possible that there is additional species turnover on scales larger than those explicitly sam-

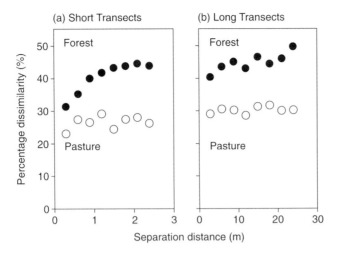

FIGURE 7.6. Mean dissimilarity (%) as a function of separation distance along the short (a) and long (b) transects. Each point is an average of p = 35–42 observation pairs. Open symbols represent pasture, closed symbols forest (from Ettema and Yeates, 2003).

pled. The amount of work required for larger scale studies would be much greater, and should be kept in mind when considering work on spatial scales even with soil fauna of relatively small size.

WHY IS SOIL DIVERSITY SO HIGH?

We arrive at the end of this chapter with an understanding of the phenomenon that high species equals functional diversity (e.g., Anderson, 1975, 2000), but its root causes are yet unknown. As noted by Wardle (2002), the belowground environment provides numerous niche axes in the Hutchinsonian (1957) hyperspace, concerning numerous microhabitats, microclimatic properties, soil chemical properties, and phenologies of the organisms themselves. When one adds in the fact that many of the organisms may exist in quiescent or dormant stages (Coleman, 2001, and noted in Chapter 4), there is considerable niche space for the impressive belowground species diversity.

BIOGEOGRAPHICAL TRENDS IN DIVERSITY OF SOIL ORGANISMS

Interestingly, with the exception of termites, whose diversity declines significantly over a large geographical gradient, numerous taxa of soil organisms, ranging from ciliate protozoa (Foissner 1987a, b) to earthworms (Hendrix, 1995) do not decrease from 0 to 60° latitudinal range. Wardle (2002) suggests two principal reasons: (1) With increasing latitude there is a general trend for greater amounts of organic matter accumulation, and higher amounts of carbon and nutrients are stored in the soil relative to the amount of plant biomass present. Greater humus depth may provide greater habitat heterogeneity and greater amounts of nutrients present in the soil. (2) Diversity of soil organisms may be governed by local factors rather than by regional pool size. If one adds in the fact that numerous smaller soil organisms (soil microfauna and microflora including both fungi and bacteria) can be transported by wind currents and macrobiota over intercontinental distances, one would expect to see pandemic distributions, and this is what is observed (Wardle, 2002).

8 Future Developments in Soil Ecology

INTRODUCTION

There are several areas of rapid change that are of interest to soil ecologists in the 21st century. The effects of soil processes and soil biota on global change, particularly with relation to global greenhouse gases, are of concern to resource managers and globally oriented ecologists (Coleman *et al.*, 1992). More recently, as noted in Chapter 7, there has been a rising current of interest in soils and biodiversity (Coleman, 2001). Within terrestrial ecosystems, soils may contain some of the last great "unknowns" of many of the biota. This includes such relatively well-studied fauna as ants (Hölldobler and Wilson, 1990), as well as the more numerous and less studied mesofauna, such as microarthropods and nematodes. The role of soils in the ecology of invasive species is an area of rapidly increasing findings. We also consider the roles of soils and the "Gaia" mechanism, and finally, ways to evaluate soil quality.

ROLES OF SOILS IN CARBON SEQUESTRATION

Soils are probably the last great frontier in the quest for knowledge about the major sources and sinks of carbon (C) in the biosphere. The direct effects of deforestation on global patterns of carbon cycles are relatively minor; the effects of changed sink strengths, with deforestation decreasing rates of carbon dioxide (CO_2) uptake, may be much larger. Another source of carbon input to the atmosphere has come from the oxidation of soil organic matter during cultivation of native lands such as the Great Plains region of North America and the Eurasian steppes of eastern Russia (Wilson, 1978; Houghton *et al.*, 1983). The standing stocks of soil carbon are twice as large as all of the standing crop biomass of all of the terrestrial biomes combined (Fig. 8.1) (Post *et al.*, 1990; Anderson, 1992). However, the plant and soil systems are strongly coupled, and the rates of inflows and outflows are significantly controlled by rates of above- and belowground herbivory in forests (Pastor and Post, 1988) and in grasslands (Schimel, 1993). The feedback effects of the principal greenhouse gases, namely CO_2, methane, and nitrous oxide, are very large (Mosier *et al.*, 1991; Rogers and Whitman, 1991), with the effects of CO_2 being some 56% of the total impact (Anderson, 1992). However, the rate of increase of methane is almost twice that of CO_2 (Houghton *et al.*, 1987, 1990) and is being closely observed by atmospheric scientists. One of the major concerns of scientists interested in global change is the extent of involvement by soils and soil processes in the evolution of greenhouse gases, and roles of soil biota and organic matter in the global carbon cycle.

We examine next the ways in which soils operate over ecological and geological time spans, and how they may be influenced by, or have an effect on, global change processes. Soil development and change may be viewed as the result of the basic processes of additions, removals, transformations, and translocations (Anderson, 1988). A given landscape will experience runon, runoff, transformations, and transfers up and down in the profile, and additions and losses either aerially or pedologically (Fig. 8.2). These processes may be very dynamic for processes such as movement of soluble salts, which vary within seasons, or be measured in thousands of years, for example, for clay weathering processes. The microbial portion of the organic matter cycle will have mean net turnover times of 1–1.5 years, whereas humification processes such as the interactions of clay–humic compounds may be considered intermediate (centuries) in time scale (Stewart *et al.*, 1990) (Table 8.1). These processes can be envisioned readily via the carbon, nitrogen, and phosphorus submodels of the Century model (Fig. 8.3). This model was developed to simulate the additions and losses in agricultural lands and grasslands worldwide (Parton *et al.*, 1987, 1989a), but has now been extended to a wide range of ecosystems including tundra and taiga (Smith *et al.*, 1992) and tropical ones as well (Parton *et al.*, 1989b; Schimel *et al.*, 1994; Smith *et al.*, 1998).

FIGURE 8.1. Pools and fluxes of carbon in major terrestrial ecosystem types: (a) distribution of net primary production, (b) biomass, and (c) soil carbon pools. The total area occupied by each ecosystem type is represented by the horizontal axis with flux or density of the vertical axis; the area is therefore proportional to the global production or storage in each ecosystem type (from Anderson, 1992). *Note:* The numbers inside the boxed areas are measured in petagrams C (Pg C).

(a)

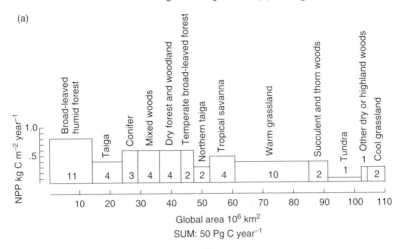

(b)

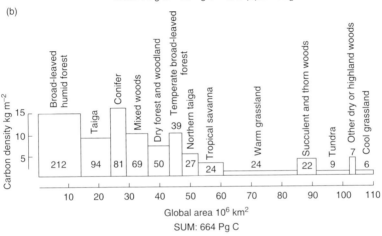

(c)

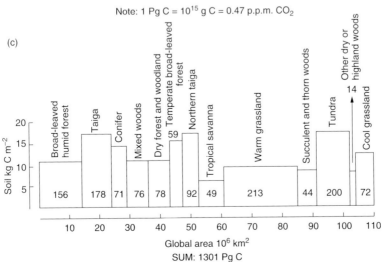

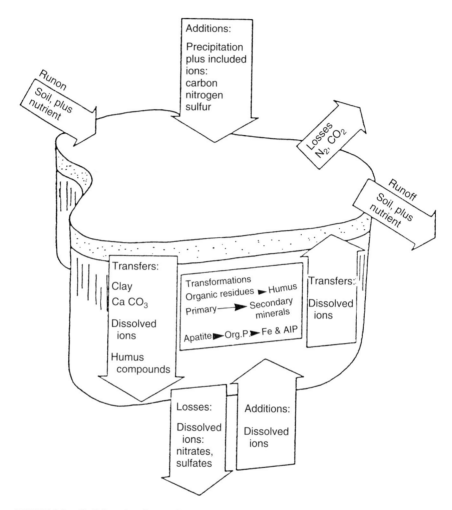

FIGURE 8.2. Soil-forming factors based on the concepts of Simonson (1959) as described by Anderson (1988) (from Stewart *et al.*, 1990).

TABLE 8.1. Grouping of Soil Related Processes and Components Based on Time

Highly dynamic	Dynamic	More static, slow
Soluble nutrients	Adsorbed nutrients	Nutrient reserves in minerals
Active or soluble organic matter	Labile organic matter adsorbed to clay	Chemically stabilized organic matter
Solution and movement of soluble components	Weathering of carbonate minerals	Weathering of silicates and clay minerals
Microbial growth	Microfauna and mesofauna plant growth	Vegetation, i.e., forest

From Stewart *et al.*, 1990.

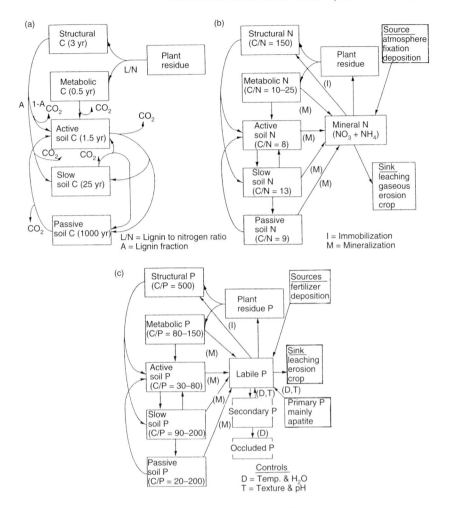

FIGURE 8.3. Flow diagram for the (a) carbon submodel, (b) nitrogen submodel, and (c) phosphorus submodel of Century (adapted from Parton *et al.*, 1987, 1989b) (from Stewart *et al.*, 1990).

ROLES OF SOILS IN THE GLOBAL CARBON CYCLE

What patterns and processes of global change are most likely to affect the global carbon cycle in soils? What are the effects of climate change on vegetation? Are there possible changes in sink strengths (pools of organic matter, active roots, etc.) in various parts of the globe? Do we know enough about the dynamics of carbon in the 13 or more major biomes that comprise the terrestrial biosphere? For example, consider the size of the live biomass in broad-leaved humid forest, which amounts to 212 petagrams ($Pg = 10^{15}$ grams) versus warm grasslands that have only 24 Pg live biomass. When comparing the amounts of soil carbon stored with carbon in live biomass, there is relatively less storage in the humid broad-leaved forest (156 Pg), giving a ratio of 212 : 156, or 1.36, of live biomass to soil organic matter (SOM) (Fig. 8.1) (Anderson, 1992). Warm grasslands, with 213 Pg in soil organic matter, have a ratio of 24 : 213, or 0.11, in biomass versus that in the SOM. Tundra, with only 9 Pg in live biomass versus 200 in the SOM, has a ratio of only 0.05 in living biomass versus SOM (Anderson, 1992). What are the climatological versus plant physiological and microbiological implications of such differences in these widely different biomes? Research in this area requires considerable effort in soil science and also microbial ecology, because we are faced with problems of measuring substrate quality, covered earlier in Chapter 5, and its feedback effects on future primary production and nutrient dynamics. Of course the modes of growth of grasses versus trees are also influential, because more of the total growth effort is invested belowground in both grassland and tundra soils.

Recent reviews have addressed key aspects of the terrestrial carbon cycle: carbon fixation by primary production, and then mechanisms for either sequestering the carbon during organic matter decomposition and transformation processes, or mechanisms for mineralization via human-induced or natural processes (e.g., Lal, 2002; Houghton, 2003). A central concern for ecologists and soil scientists is that soil organic carbon is the second largest pool in the terrestrial organic carbon cycle, with about 1550 Pg involved.

Concerns about imbalances in the global carbon cycle are not new; rapidly increasing amounts of CO_2 entering the atmosphere from human activities, including burning of fossil fuels, were first noted a century ago (Arrhenius, 1896). Since then, interest in the rates of flow of carbon, and amounts sequestered in various pools in the biosphere, has waxed and waned. For example, Plass (1956) expressed concern about the amounts of CO_2 being released by the burning of fossil fuels worldwide. An additional contribution to increased global CO_2 is the relatively large amounts of soil organic matter being "mined" by extensive cultivation throughout the major "breadbaskets" of the world. In several regions, for example, the North American Great Plains, the former Soviet Union, and Canada, the loss is quite large, perhaps up to 40% of the surface layers (Haas et al., 1957; Wilson, 1978; Coleman et al., 1984). Mann (1986) concluded in a survey of 625 soils studied pairwise, cultivated versus noncultivated on the same soil type, that 20% or more of carbon was lost over decadal time spans from soils with high amounts of carbon (ranging from 6 to 16 kilogram per square meter). Interestingly, she noted that modest gains occur in soils that are initially very low in soil organic carbon, such as very sandy-textured ones, if they are put into cultivation. A significant amount of carbon fixation, and subsequent movement into the SOM, in the surface-to-30-centimeter (cm) depth, will occur over several years' time span. If extensive application of fertilizers is required to achieve these gains, then overall the global carbon balance is still toward the positive side, in terms of carbon costs for fossil fuel–derived nitrogen, for example (Vitousek *et al.*, 2002).

In an extensive review of SOM models and global estimates of changes in soil organic carbon under a $2 \times CO_2$ climate, Post *et al.* (1996) ran simulations with the Rothamsted model over a 100-year time period. They found that global soil organic matter change was only about one-third of the way toward an eventual equilibrium under the enriched CO_2 regime. The change predicted with the United Kingdom Meteorological Office (UKMO) shows a small net sequestration of carbon in soil early in the climate transition period resulting from increases in tropical ecosystem soil carbon. However, this is followed by large carbon releases from arctic and boreal soils later in the century-long climate transition period. The largest net release of carbon from soil occurred at the end of the 100-year climate transition, after which the net releases decreased gradually as the soil carbon pools approached equilibrium under the double CO_2 regime (Post *et al.*, 1996).

Houghton (2003) noted that the carbon balance of the world's terrestrial ecosystems is uncertain. Several top-down (atmospheric) and bottom-up (forest inventory and land-use change) approaches are in use and difficult to compare, because they contain incomplete accounting inherent in their methods. After a brief discussion of the methods and their inherent limitations, we consider the possible resolution of the uncertainties arising from use of these methods. Of the top-down estimates, the first uses concentrations of oxygen (O_2) and CO_2 to partition atmospheric sinks of carbon between land and ocean. Using this assessment, terrestrial ecosystems were globally a net sink for carbon, averaging 0.2 (± 0.7) $Pg\,C\,yr^{-1}$ and 1.4 (± 0.7) $Pg\,C\,yr^{-1}$ in the 1980s and 1990s, respectively. The reason for the large increase between decades is unknown. A second top-down method is inverse modeling, which uses atmospheric transport models, together with spatial and temporal variations in atmospheric concentrations of CO_2 obtained through a network of flask air samples, to infer surface sources and sinks of carbon. The budget will not reflect accurately any changes in the amount of carbon on land or in the sea if some of the carbon fixed by terrestrial plants or used in weathering minerals is transported by rivers to the ocean and respired or released to the atmosphere there. The two top-down methods based on atmospheric measurements yield similar global estimates of a net terrestrial sink of about 0.7 (± 0.8) $Pg\,C\,yr^{-1}$ for the 1990s

(Houghton, 2003). Two bottom-up estimates have been used to estimate terrestrial sources and sinks over large regions: analyses of forest inventories and analyses of land-use change. One recent synthesis of forest inventories, which included converting wood volumes to total biomass and accounting for the fate of harvested products and changes in pools of woody debris, forest floor, and soils, found a net northern midlatitude terrestrial sink of from 0.6 to $0.7 \, \mathrm{Pg C yr^{-1}}$ for the years around 1990 (Goodale et al., 2002, cited in Houghton, 2003). The estimate is only one-third of that calculated from atmospheric data corrected for river transport. Houghton (2003) noted that accumulation of carbon belowground, not directly measured in forest inventories, was underestimated and might account for the difference in estimates. Because the few studies that have measured the accumulation of carbon in forest soils have consistently found soils account for only a small proportion (5–15%) of measured ecosystem sinks, Houghton (2003) concluded that, despite the fact that soils worldwide hold from two to three times more carbon than does biomass, there is no evidence as yet that they account for a significant terrestrial carbon sink. The second sort of bottom-up estimate, analyses of land-use change, calculated that globally, all factors of land-use change averaged 2.0 and $2.2 \, \mathrm{Pg C yr^{-1}}$ respectively, in the 1980s and 1990s (Houghton, 2003). In contrast to the unknown biases of atmospheric methods, analyses based on land-use change have deliberate biases built into them. These latter analyses consider only the changes in terrestrial carbon resulting directly from human activity. In other words, there may be other sources and sinks of carbon not related to land-use change, such as those caused by CO_2 fertilization or changes in climate, that are considered by other methods but ignored in analyses of land-use change. The terrestrial sources and sinks of carbon in petagrams of carbon per year as estimated by different methods are given in Table 8.2. (Houghton, 2003).

A major concern noted by Houghton (2003) was the unknown rate of turnover of carbon belowground. One of the major sources of carbon inputs in all terrestrial ecosystems has been attributed to fine roots. Uncertainties in estimates of root longevity have markedly hampered proper quantification of net primary production (NPP) and belowground carbon allocation, particularly in forests. In a comparison of fine root carbon inputs in two field sites, a hardwood forest (sweetgum Liquidambar styraciflua L.) in Tennessee and a loblolly pine (Pinus taeda) forest in North Carolina, Matamala et al. (2003) measured the carbon-13 (^{13}C) isotopic signatures of live roots before and after carbon enrichment was applied in Free Air Carbon Enrichment (FACE) experiments. There was a marked difference in tree species, with mean residence time (MRT) in roots between 1 and 2 millimeters (mm) being 5.7 years for pine, and only 3 years for sweetgum roots of the same diame-

TABLE 8.2. Terrestrial Sources (+) and Sinks (−) of Carbon (Pg C yr^{-1}) Estimated by Different Methods

Region	Inversions based on atmospheric data and models	Analysis of land-use change	Forest inventories
Globe	−1.4 (±0.8)	2.2 (±0.8)	
North	−2.4 (±0.8)	−0.03 (±0.5)	−0.65 (±0.05)
Tropics	1.2 (±1.2)	2.2 (±0.8)	
South	−0.2 (±0.6)	0.02 (±0.2)	

From Houghton, 2003.

ter. Matamala *et al.* (2003) note that any estimates of belowground carbon inputs need to pay careful attention to the species of vegetation involved, and also to study root turnover rates by at least two different methods, as we noted in Chapter 2 on studies of root turnover. The implications of these much longer mean residence times of forest tree fine roots, if proven to be a general occurrence, are quite large and deserving of further study by terrestrial ecologists.

PROBLEMS IN MODELING SOIL CARBON DYNAMICS

A more general problem yet faces soil ecologists. One of our current needs is to "model the measurable," rather than "measure the modellable" (Elliott, 1994). There are pools in models such as Century, mentioned earlier, that are more easily conceptualized than actually measured. A more readily measurable entity is the labile pool, consisting primarily of the microbial biomass. The intermediate and long-term pools, existing from decades to millennia, are very difficult to measure directly, and much work is under way to more effectively isolate and characterize these pools by a variety of methods (Six *et al.*, 2002a, b). This problem requires integration across several levels of resolution, dealing with numerous human activities in sociology and economics that have a direct impact on soil management. These include the concept of the effectiveness of management of carbon resources, which is inversely related to the cost of subsidizing the lost functions of organic matter (Fig. 8.4) (Woomer and Swift, 1995). The effectiveness of carbon resource management decreases with sequential loss of constituents and subsequent loss of function as land use intensifies without subsidizing lost organic matter. Elliott (1994) and colleagues urged soil ecologists to isolate functional soil organic matter fractions and determine their roles in soil processes in order to understand the mechanisms controlling soil processes. This includes the mechanisms and processes

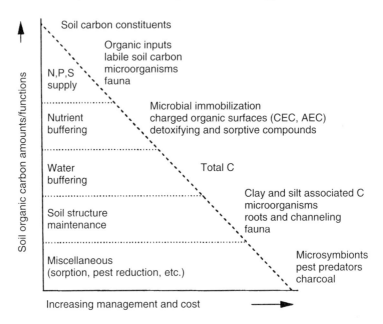

FIGURE 8.4. The functional role of soil organic matter within an ecosystem depends on the intensity with which that system is managed (from E.T. Elliott, personal communication; modified from Woomer and Swift, 1994).

involved in the formation and turnover of macro- and microaggregates in a wide range of soil types worldwide (see e.g., Beare *et al.*, 1994a, b; Six *et al.*, 1999, 2002a). Indeed, as noted in Chapter 3, chemical, microbiological, and macrobiological characterization of physically isolated fractions may provide the best opportunity for identifying functional pools of soil organic matter. For example, each major category of soil biota has a significant effect on one or more aspects of soil structure, including production of organic compounds that bind aggregates, and hyphal entanglement of soil particles (microflora), producing fecal pellets and creating biopores (meso- and macrofauna) (Hendrix *et al.*, 1990; Linden *et al.*, 1994). A complete list of influences of soil biota is given in Table 4.12 (Hendrix *et al.*, 1990).

Some recent developments have been made in conceptualizing SOM dynamics, which should have a considerable impact on the ways in which soils are viewed and managed for carbon sequestration. There are three principal mechanisms by which SOM is stabilized: (1) it is physically stabilized, or protected from decomposition through microaggregation; (2) SOM is closely associated with silt and clay particles; and (3) it is biochemically stabilized through the formation of recalcitrant SOM compounds. These stabilization mechanisms are strongly related to the ways in which SOM pools are protected (Six *et al.*, 2002a). The protective capacity of soil has been represented graphically, in an ascending series (Fig. 8.5) (Six, 2002a) showing silt and clay at an asymptotic maximum that is the maximum protection possible, because anything above that is considered nonprotected. A conceptual model of SOM dynamics with the aforementioned measurable pools follows a sequence beginning with above- and belowground inputs into unprotected soil carbon, moving either into microaggregate-associated soil carbon by aggregate turnover, or via adsorption/desorption into soil- and clay-associated soil carbon. It then moves subsequently into nonhydrolyzable soil carbon via condensation and complexation reactions, producing the biochemically protected soil carbon (Fig. 8.6) (Six *et al.*, 2002a). In contrast with earlier conceptual and simulation models, this scheme is indeed measurable, hence meeting the criterion of Elliott (1994) of "modeling the measurable." It also has the virtue of reflecting realities in lightly versus heavily weathered soils. The former, with a greater proportion of 2 : 1 clay mineral dominated soils, have a greater silt- and clay-protected carbon pool than the 1 : 1 clay mineral dominated soils. The latter minerals, for example, kaolinite and gibbsite, tend to dominate in the typically heavily weathered tropical soils (Theng *et al.*, 1989).

Six *et al.* (2002b) proceeded to test some of the assumptions in the conceptual model given above by performing a major synthesis of SOM dynamics in a wide range of temperate and tropical soils worldwide, comprising more than 32 sites. Six *et al.* (2002b) found a 1.8 times longer average MRT of carbon in the soil surface of temperate versus tropical soils (63 ± 7 versus 35 ± 6 years). This indicates that there is generally a faster carbon turnover in tropical than temperate soils. Interestingly, the range of MRT values was similar for both temperate and tropical soils, being 14–141 versus 13–108 years, respectively. The higher turnover rate for tropical soils is due primarily to faster turnover rates of the slow carbon pool in tropical soils (Feller and Beare, 1997, cited in Six *et al.*, 2002b).

BIOLOGICAL INTERACTIONS IN SOILS AND GLOBAL CHANGE

Perhaps the principal element of the global change scenario is the steadily increasing annual temperature, which rises about 0.1°C annually. As this increase occurs, there should be a perceptible increase in loss of soil carbon (Schimel *et al.*, 1994; Scharpenseel *et al.*, 1990; Jenkinson *et al.*, 1991). A countervailing tendency will exist with effects of CO_2 fertilization, enhancing plant primary production. However, several authors have noted the further constraints of other limiting

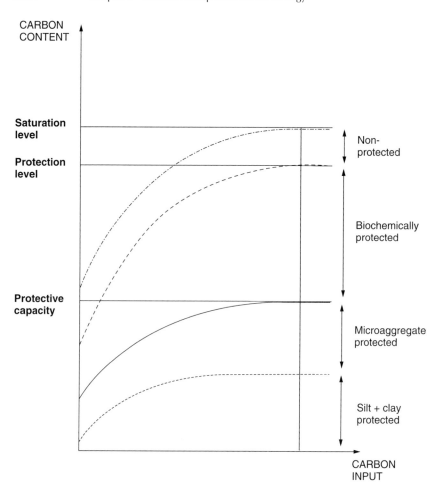

FIGURE 8.5. The protective capacity of soil (which governs the silt- and clay-protected carbon and microaggregate-protected carbon pools), the biochemically stabilized carbon pool, and the unprotected carbon pool define a maximum carbon content for soils. The pool size of each fraction is determined by its unique stabilizing mechanism (from Six *et al.*, 2002a).

resources to growth, such as mineral nutrients (Pastor and Post, 1988; Schimel, 1993; Hungate *et al.*, 2003). More holistic modeling efforts of changes in SOM, particularly ones that include soils, plants, herbivores, and detritivores together, are more realistic in their outcomes than those that model the plants, heterotrophs, and soil carbon pools separately (Schimel, 1993; Tinker *et al.*, 1996).

Global change effects on relationships to soil biota (aboveground to belowground) can be modeled as a nested set of control variables.

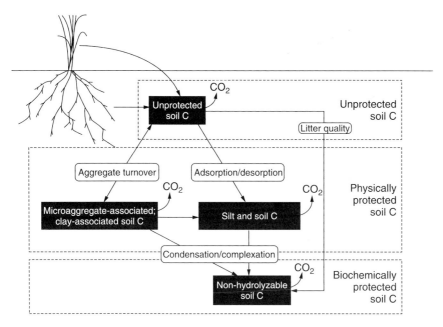

FIGURE 8.6. Conceptual model of soil organic matter (SOM) dynamics with measurable pools. The soil processes of aggregate formation/degradation, SOM adsorption/desorption and SOM condensation/complexation, and the litter quality of the SOM determine the SOM pool dynamics (from Six *et al.*, 2002a).

Morphological features of dominant life forms determine engineering activities at the ecosystem level, physicochemical properties of plant functional groups, modifying the provision of nutritional resources at the community level, and biological properties of individual species controlling direct interactions at the population level (Fig. 8.7) (Wolters *et al.*, 2000). Changes in ecosystem functions created by plant-induced alterations in the disturbance regime, to and by resource consumption rates of soil organisms, should be confined to situations where essential traits of the vegetation are drastically changed. Such a change is most likely when the strength of environmental change overrides all other factors controlling plant assemblage structure, when plants with key attributes or functions invade or become extinct, and when species-poor environments are affected.

Much of the experimental research on ecosystem responses has focused on individual species-level responses, and is seldom concerned with multifactor system-level responses. An ongoing study of ecosystem-scale manipulations in a California annual grassland has now progressed past 3 years' duration. Shaw *et al.* (2002) measured system-level

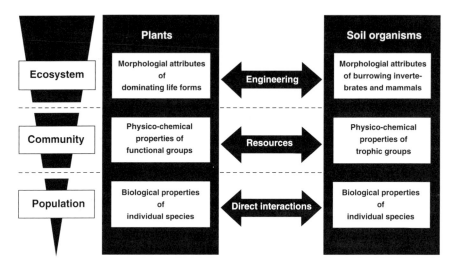

FIGURE 8.7.   Hierarchy of plant effects on soil biota and vice versa (from Wolters *et al.*, 2000).

responses in control and elevated CO_2 plots using the Jasper Ridge Global Change Experiment (JRGCE). The JRGCE imposed four global change factors at two levels: (1) CO_2 (ambient and 680 parts per million), (2) temperature (ambient and ambient plus 80 Watts (W) per square meter of thermal radiation), (3) precipitation (ambient and 50% above ambient plus 3-week-long growing season prolongation), and (4) N deposition (ambient and ambient plus 7 g of nitrate N per square meter per year), in a complete factorial design. In the third year of manipulations of the JRGCE, elevated CO_2 stimulated the production of aboveground biomass in the treatments with all of the other factors at ambient levels. Aboveground biomass increased more than 32%, which is comparable to that in other single-factor CO_2 enrichment experiments (e.g., the increase in North Carolina pine plantations was 25%, and the increase in an Arizona free-air CO_2 enrichment experiment [FACE] was 20–43% [Kimball *et al.*, 1995]). Interestingly, although each of the treatments involving increased temperature, nitrogen, or precipitation increased aboveground biomass and NPP, elevated CO_2 consistently shrank these increases. In fact, with the three other factors and ambient CO_2 increased NPP by 84%, but increased CO_2 more than halved this, down to 40%. Belowground NPP was even more suppressed, with an average effect across all treatments of minus 22%. This study by Shaw *et al.* (2002) is an object lesson in the need to pursue multifactorial studies over many years to ascertain the full effects of the manifold variables involved in global change phenomena.

A similar control hierarchy arises in the soil system as follows: attributes of burrowing macroinvertebrates and mammals establish different dynamic equilibria for soil at the ecosystem level; physicochemical properties of trophic groups affect the vegetation at the community level; and biological properties of individual species regulate direct interactions at the population level (Fig. 8.7) (Wolters *et al.*, 2000).

Land use change is probably one of the greatest agents of change in soil biology and ecology, and given the fact that it is so pervasive on all continents now, one can readily agree with Wolters *et al.* (2000) that land use change rapidly and persistently alters all levels of above- and belowground interactions and acts on a large scale.

Climate change, of the magnitude envisioned over the next century (an average global temperature increase of 2.5°C), will lead to a major shift in the boundaries of ecological systems. There will be climate-induced alteration in the makeup of many plant communities and changes in litter quality due to changes in species composition. For example, as mixed spruce–hardwood forests in the southern boreal region are replaced by hardwood due to global warming, the anticipated higher-quality litter will provide increased availability of resources to the soil organisms. Of course, in more northerly climes, as major climate changes occur, there will be significant alterations in ecosystem functions when it affects organisms that carry out functions performed by few other organisms. Schimel and Gulledge (1998) predicted that in areas where episodic drying and rewetting of soil associated with climate change becomes more severe, populations of cellulolytic and ligninolytic fungi may be reduced, resulting in a decrease in litter decomposition greater than would be predicted by considering only the changes in soil and litter moisture.

ECOLOGY OF INVASIVE SPECIES IN SOIL SYSTEMS: AN INCREASING PROBLEM IN SOIL ECOLOGY

One of the primary concerns of ecology in the 21st century has been the increasing numbers of invasive species in ecosystems around the world. The publicity concerning invasive species in aquatic systems has been extensive. Case studies of lampreys invading the Great Lakes of North America via the Welland Canal and later via the St. Lawrence Seaway, and the rapid spread of the zebra mussel in lakes and streams over much of the Western Hemisphere are noteworthy examples. There is a less obvious but growing literature documenting the effects of introduced plants and animals displacing or outcompeting native species in soils in numerous ecosystems of the world. Some examples follow.

Eastern deciduous forests in North America have been invaded by two species of plants that are often dominant in the understory vegetation. *Berberis thunbergii* is a woody shrub that often forms dense thickets. *Microstegium vimineum*, a C_4 grass, forms dense carpets. The two invasives co-occur often. In a series of laboratory and greenhouse experiments in New Jersey, Ehrenfeld *et al.* (2001) found that the soil under these plants was increased in available nitrate and had elevated pH as well. The two invasive plants have different mechanisms to achieve a similar end result. *Berberis* combines large biomasses of nitrogen-rich roots with nitrogen-rich leaf litter, whereas *Microstegium* clumps combine small biomasses of nitrogen-rich roots with small biomasses of nitrogen-poor litter that leave much of the surface soil with few roots. Changing key chemical characteristics of soil (e.g., changed nitrate and pH) undoubtedly represent only two of numerous ways in which invasive plant species alter the playing field in contesting for dominance of patches of soil.

In the same research sites that were used by Ehrenfeld *et al.* (2001), Kourtev *et al.* (2002) measured alteration of microbial community structure and function by exotic plant species (Japanese barberry [*Berberis thunbergii*] and Japanese stilt grass [*Microstegium vimineum*], compared to a co-occurring native species [blueberry—*Vaccinium* spp.]). They found in both bulk and rhizosphere soils that phospholipid fatty acid (PLFA) profiles, enzyme activities, and substrate-induced respiration (SIR) profiles of microbial communities were significantly altered under the two exotic species. The PLFA profiles provided only an index of community structure rather than specific information about what species were active. A correlation of structure (PLFA) and function, namely enzymes, showed that a particular set of species is associated with a particular pattern of enzyme activities but does not provide information about which of the species were responsible. Kourtev *et al.* (2002) found that profiles of enzymatic and catabolic capacity in the soil definitely differed with different microbial communities.

One of the more noted plant invasions of the past century was that of the annual grass *Bromus tectorum* L., which has a current range of 40,000,000 hectares, notably in wide regions of Washington, Oregon, Idaho, and Utah. Evans *et al.* (2001) measured litter biomass and carbon–nitrogen and lignin–nitrogen ratios to determine the effects on litter dynamics in a site that had been invaded in 1994. Long-term soil incubations (415 d) were used to measure potential soil microbial respiration and net nitrogen mineralization. Plant-available nitrogen was measured for 2 years with ion-exchange bags, and potential changes in rates of gaseous nitrogen losses were measured using denitrification enzyme activity. *Bromus* invasion significantly increased litter biomass, and its litter had significantly greater carbon–nitrogen and lignin–nitrogen ratios than did native species. The changes in litter quality and chemistry decreased potential rates of nitrogen mineralization in sites with *Bromus* by decreasing nitrogen available for microbial activity. Evans *et al.* (2001) suggest that *Bromus* may cause a short-term decrease in nitrogen loss by decreasing substrate availability and denitrification activity, but over the long term, nitrogen losses are likely to be greater in invaded sites because of increased fire frequency and greater nitrogen volatilization during fire. This mechanism, in conjunction with land use change, will set into play a set of positive feedbacks that will decrease nitrogen availability and alter species composition.

In a companion study to that of Evans *et al.* (2001), Belnap and Phillips (2001) studied the effects of invasion by *Bromus tectorum* in three study sites in the Canyonlands of southwestern Utah. They measured litter and soil changes in sites that had been dominated previously by *Hilaria jamesii*, a fall-active C_4 grass, and *Stipa comata* and *Stipa hymenoides*, predominantly spring-active C_3 species. Belnap and Phillips (2001) measured the abundances of a wide range of microbes, microarthropods, and macroarthropods under *Hilaria* and *Stipa* communities, as well as in those that had been invaded by *Bromus* in 1994 (Fig. 8.8). There were significant changes in numbers and diversity, due in part to changes in amounts and qualities of litter. In the *Bromus* invaded plots, litter quantity was 2.2 times higher in *Bromus* and *Hilaria* together than in *Hilaria* alone, contrasted with *Stipa* and *Bromus,* which was 2.8 times greater than in the *Stipa* alone. Soil biota responded generally in opposite manners in the plots that combined two perennials and an annual grass. Active bacteria decreased in *Hilaria* versus *Hilaria* with *Bromus*, and increased in *Stipa* versus *Stipa* with *Bromus.* Most higher trophic-level organisms increased in Hilaria plus Bromus relative to Hilaria alone, while decreasing in *Stipa* plus *Bromus* relative to *Stipa* alone. The soil and soil food web characteristics of the newly invaded sites included the following: (1) lower species richness and numbers of fungi and invertebrates; (2) greater numbers of active bacteria; (3) similar species of bacteria and fungi as those invaded more than 50 years previously; (4) higher levels of silt (hence greater water holding capacity and soil fertility); and (5) a more continuous cover of living and dead plant material. The authors note that food web architecture can vary widely from what had existed previously within the same vegetation type, depending on the reactions to the invasive species relative to the previous uninvaded condition. Addition of a common resource can shift conditions significantly, and careful attention to the effects of species by season by site is definitely warranted.

A much different example of a soil invasion is the movement of the predatory New Zealand flatworm, *Arthurdendyus triangulatus,* into

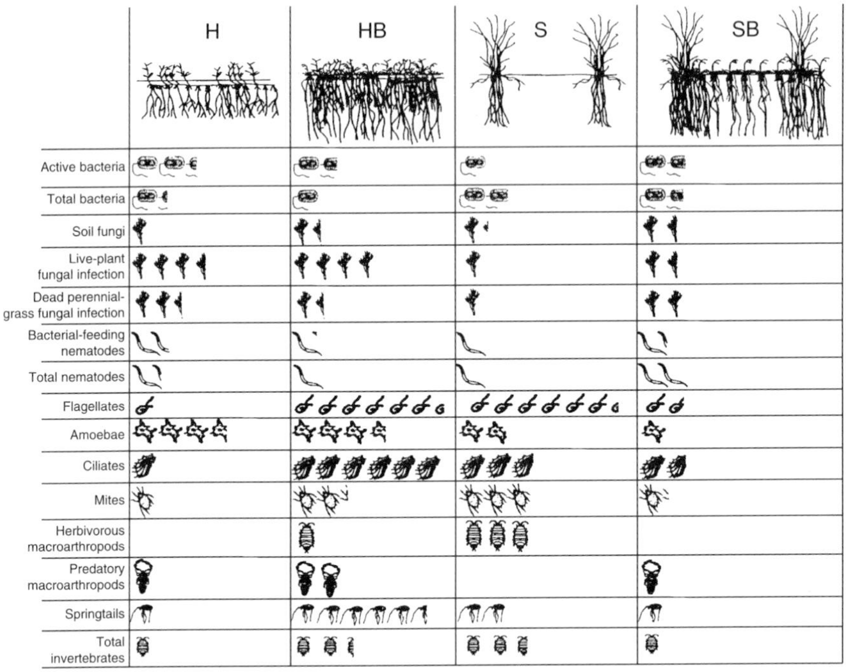

	H	HB	S	SB

FIGURE 8.8.   Diagram of soil food-web structure in the four different grassland communities, showing the relative abundance within a given functional group. The numbers of icons on a given line are relative to each other; thus twice as many icons indicates that organisms are twice as abundant. B = *Bromus*, H = *Hilaria*, S = *Stipa* (from Belnap and Phillips, 2001).

Western Europe, where it preys upon earthworms. It was established in Great Britain by unknown means, but probably on soil associated with nursery stock. Boag and Yeates (2001) suggest that the avenue of introduction was rather circuitous: initially the flatworms spread from botanic gardens to horticultural wholesalers, then to domestic gardens, and finally invaded agricultural lands. They note that it is one of twelve alien terrestrial planarians in Britain considered to be a pest, and hence seems to obey the "rule of tens" (*sensu* Williamson, 1996), in which only one in ten invasive species assumes an outbreak or pest status. The indigenous flatworms, including *Arthurdendyus*, are not a problem in New Zealand because New Zealand has a drier and warmer climate than in the west of Scotland, where the outbreaks are the most severe. Interestingly, under minimum tillage practice in New Zealand, where crop residues provide refuges (Yeates *et al.*, 1999), flatworms have the potential to reduce lumbricid earthworm populations. Invasions by exotic earthworm species are also becoming a problem in many temperate and tropical areas, as discussed further in Chapter 4 (Hendrix and Bohlen, 2002).

In general, numerous theoretical studies have usually supported Elton's (1958) biotic resistance hypothesis, in which more diverse communities better resist invading species (see Byers and Noonburg, 2003, and references cited therein). In a mathematical overview of more general aspects of biotic resistance to invasive species, Byers and Noonburg (2003) demonstrate that invasibility is influenced not only by the number of native species present, but also by the number of resources present in a given ecosystem. Building on a Lotka–Volterra competition model, Byers and Noonburg's model predicts that increasing invasibility with native diversity across large scales is the result of decreasing mean interaction strength as resources increase. The strength of the positive relationship between native and exotic species diversity and relative contribution of factors extrinsic to the community depend on whether niche breadth increases with the number of available resources. Interestingly, the same mechanism—the sum of interspecific competitive effects ($\Sigma \alpha_{ij} n_j$)—drives the opposite pattern of decreasing invasibility with native richness at small scales because resource numbers are held constant. As a consequence, Byers and Noonburg (2003) conclude that Elton's biotic resistance hypothesis, interpreted as a small-scale phenomenon, is consistent with large-scale patterns in exotic species diversity.

SOILS AND "GAIA": POSSIBLE MECHANISMS FOR EVOLUTION OF "THE FITNESS OF THE SOIL ENVIRONMENT?"

As was mentioned in Chapter 1, there are many positive feedback mechanisms in soils, in which organisms have arisen and/or evolved together. These include roots and arbuscular mycorrhiza (AM), and many of the genera and families of soil fauna. The following discussion is based on the very insightful and stimulating article by van Breemen (1993) entitled "Soils as biotic constructs favouring net primary productivity." Van Breemen asks the central question: Have soils merely been influenced by biota, or have biota created soils as natural bodies with properties favorable for terrestrial life? He presents five hypotheses or postulates related to the overarching theme: (1) there are soil properties "favorable" for terrestrial life in general; (2) biota, including plants and the soil dwelling organisms, are able to affect those soil properties; (3) on a scale of ecosystems and a global ("Gaian") scale, biotic action makes the outermost (1–100 cm) layer of the earth's crust more favorable for terrestrial life in general than it would have been in their absence; (4) at an ecosystem-level scale, biota tend to offset the effects of unfavorable properties of the soil or soil parent material by modifying those soil properties; and (5) modification of soil properties may play a role in species competition.

Following from the ideas of Odum and Biever (1984), there should be some positive or donor-recipient controls on interactions between primary producers and other biota, with the AM being a prominent example. As we have noted earlier, feeding on detrital organic matter in the soil is generally the principal energy flow in terrestrial ecosystems. Therefore feedback loops arising in the soil community (such as detrital food webs, see Chapter 6) should have a major effect on net primary productivity. Thus soil-biota interactions may be a most fruitful area to investigate and test hypotheses about positive effects of biota on the environment.

As far as favorable soil properties are concerned, changes and general improvements in soil porosities and aggregation as well as soil organic matter status are prime examples of general improvements in soil characteristics; these changes occur through cumulative interactions of the soil biota. This is not a simple linear progression however; there are examples of surface-feeding earthworms, which remove enough of the surface leaf litter material to cause a greater amount of soil erosion in their presence than in their absence (Johnson, 1990).

In the areas of soil texture and structure, as well as soil chemical properties, there are numerous examples of soil biotic interactions having a generally beneficial effect in the top meter of soil material. One example of this is provided by Gill and Abrol (1986), who described how planting *Eucalyptus teretocornis* and *Acacia nilotica* on an alkali soil (pH 10.5) markedly decreased pH and salinity within 3 to 6 years. These changes were probably caused by a suite of factors including increased water permeability, which followed the development of root channels and the accumulation of organic matter in the upper 20–50 cm of the soil profile. Other biota, notably termites, can promote higher salt content in soils, as detected by measurements in inhabited and abandoned termite hills compared to the surrounding soil (de Wit, 1978). Many of these processes tend to increase the amounts of heterogeneity within soil profiles, which has been well reviewed recently by Stark (1994).

At both ecosystem and global scales, there are significant effects of biota on rock and soil weathering. The early pioneering researches of Vernadsky (1944, 1998) and Volobuev (1964) in particular originated and made popular the concept of "organic weathering." The able partnership of roots and microbes in mineral translocation is noteworthy, for example, removing the interlayer K from phlogopite (vermiculitization) within the first 2 mm of the rhizosphere. For other references on biological impacts on mineral weathering, see Schlesinger (1996). As noted in Chapter 6, the soil physical effects of earthworms on soil structure, formation of heterogeneous pores, and high structural stability are hall-marks of soil-by-biota interactions over long time-intervals. There are several examples of transformations that counteract unfavorable soil properties. These arise principally from the influence of the biota on translocation and concentration of nutrients in the upper 1 meter of the earth's mantle, the living soil. In general, biota tend to invest more in increasing nutrient supply under nutrient-poor than in nutrient-rich conditions. Root production and activity, as a fraction of total net primary production, tends to be higher in the nutrient-poor conditions (Odum, 1971). This should be considered against a background of the generally slow growth rates and nutrient fluxes that occur in many wild plants on low-nutrient soils (Chapin, 1980). There is also an intriguing nutrient conservation process that occurs in many low-nutrient ecosystems. Development of mor humus types, characterized by thick organic horizons, is typical for "poor" (low productivity) sites, and may represent nutrient conservation brought about as a result of the slow-to-decompose litter formed in the surface layers (Vos and Stortelder, 1988). This in turn may lead to further inhibition of decomposition and net primary production, so is an example of a positive feedback effect, which may require occasional fires or other disturbances to act as a suitable "reset" over millennial time spans. Soil phosphorus, in its various inorganic and organic forms, is perhaps the most limiting element in terrestrial ecosystems (van Breemen, 1993). Storage of phosphorus by secondary iron and aluminum phases is partly under biotic control and may be regarded as part of tight biotic cycling of phosphorus for three reasons: (1) secondary iron and aluminum oxides result from biologically mediated weathering of primary minerals; (2) the oxides are often precipitated under the influence of iron oxidizing bacteria; and (3) the oxides can be kept in a mostly amorphous form by interaction with humic substances, from which phosphorus can be extracted by plants more efficiently than from crystalline oxides.

A further development in assessment of soil genesis and ecosystem condition is a quantitative assessment of forest humus forms, on a scale ranging from 1 (Eumull) to 7 (Dysmoder), which is called the *humus index* (Ponge *et al.*, 2002).

In the 72 sites studied, the humus forms were arranged as follows:

    1. Eumull (crumby A horizon, Oi horizon absent, Oe horizon absent, Oa horizon absent)

    2. Mesomull (crumby A horizon, Oi horizon present, Oe and Oa horizons absent)

    3. Oligomull (crumby A horizon, Oi horizon present, Oe horizon 0.5 cm thick, Oa horizon absent)

4. Dysmull (crumby A horizon, Oi horizon present, Oe horizon 1 cm or more thick, Oa horizon absent)

5(a). Amphimull (crumby A horizon; Oi, Oe, and Oa horizons present)

5(b). Hemimoder (compact A horizon, Oi and Oe horizons present, Oa horizon absent)

6. Eumoder (compact A horizon, Oi and Oe horizons present, Oa horizon 0.5 or 1 cm thick)

7. Dysmoder (compact A horizon, Oi and Oe horizons present, Oa horizon more than 1 cm thick)

This index is well correlated with several morphological and chemical variables describing forest floors and topsoil profiles: thickness of the Oe horizon, depth of the crumby mineral horizon, Munsell hue, pH_{KCl} and pH_{H_2O}, H and Al exchangeable acidity, percentage base saturation, cation-exchange capacity, exchangeable bases, carbon and nitrogen content, and available phosphorus of the A horizon (Fig. 8.9) (Ponge *et al.*, 2002). Used in concert with the Ponge (2003) concept of humus forms as a framework of soil biodiversity, this approach should provide a more general comparative tool for assessing the chemical and biological conditions of a wide range of soil systems worldwide.

All of the foregoing perhaps raises more questions than answers. However, the general trend is for the number of species and individuals with positive effects to increase, both in successional sequences and over evolutionary time. In essence, the property of an individual that improves the environment for that individual, or increases its reproductive success, will benefit both it and its competitors as well. The selective advantage for such a trait(s) is probably small, viewed in a classical Darwinian context. If viewed in more general contexts such as enhancement of site qualities, then this can be considered a more general application of community and ecosystem development. Van Breemen (1992) notes that development of a trait in an earthworm allowing it to better control the moisture content and CO_2/O_2 balance of its immediate surroundings would redound to the benefit of other organisms and site properties. If requirements of plants or a plant species happen to match those of the earthworm, then coevolution of the plant and worm might be possible too. Wilson (1980) suggested that one might envision further development and evolution of a community of microbes, which could coevolve with the earthworm, to better enhance nutrient cycling processes. This is an evolutionary example of significant processes at "hot spots," as noted in Chapter 6. The scenario is speculative, but serves as an example of where we may be expecting to see additional breakthroughs occurring in the cryptic and fascinating world of soil ecology.

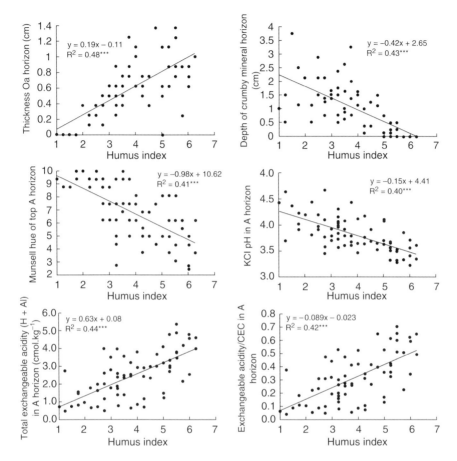

FIGURE 8.9. Correlation of the humus index with some variables measured in the topsoil profiles (means of four replicates) (from Ponge *et al.*, 2002).

SOIL ECOLOGY IN THE THIRD MILLENNIUM

We now come full circle to an issue that was raised in Chapter 1: rapidly increasing human population growth providing ever-increasing pressure on a finite base of natural resources. As noted by Daily (1997), the direct substitution cost of a hydroponic plant production system for one hectare of soil is the equivalent of $850,000, and still rising. When one adds to that the cost of cleansing and recycling, this is a sizable fraction of the more than $30 trillion dollar cost of annual goods and services provided by ecosystems globally (Costanza *et al.*, 1997).

In dealing with environmental remediation and environmental assessment in general, what is a "healthy" soil? Is there a simple one-sentence definition of "soil quality," *sensu* Doran (2002)? Is there a clean soil similar to either clean air or water? The short answer is no. The longer answer is quite complex, but informative, if one takes an ecosystem-level approach (Coleman *et al.*, 1998). A general definition of soil quality is "the capacity of a soil to function within ecosystem boundaries to sustain biological productivity, maintain environmental quality, and promote plant and animal health" (Doran and Parkin, 1994). This is a beginning, but the healthy activity of all organisms, including microorganisms, should be considered explicitly (Coleman *et al.*, 1998). As noted in Chapter 7, the state of our knowledge of microbial diversity, indeed that of a majority of the organisms active in soil, is still at a rudimentary stage. As a heuristic concept, soil quality has been useful for both education and assessment. These education and assessment tools encourage land managers to examine biological, chemical, and physical properties and processes occurring within their soil resources and to use that information as a framework for helping to make adaptive soil management decisions (Karlen *et al.*, 2001). Soil quality has not been embraced universally, because some soil scientists have been concerned that value-based decisions could supplant value-neutral science and thus lead to premature interpretations and assertions of soil quality before the concept has been thoroughly and analytically challenged (Sojka and Upchurch, 1999).

It is possible to examine the health of the litter-soil subsystem of terrestrial ecosystems by utilizing indicator indices. One example is the use of the ratio of microbial biomass carbon to soil organic carbon (C_{mic}/C_{org}). This index is related to soil carbon availability and the tendency for a soil to accumulate or lose organic matter. It has been used successfully in evaluating the status of restored ecosystems, for example, restored coal mine lands (Insam & Domsch, 1988).

A wide range of soil quality indices has been calculated, related to specific groups of microbes and fauna (Coleman *et al.*, 1998). These include nitrogen mineralization, soil respiration, respiration to microbial biomass ratios, faunal populations, and rates of litter decomposition (Knoepp *et al.*, 2000). Considerably less attention has been paid to ecosystem-level analyses. The following is an overview of several studies, undertaken in two agroecosystems in the Georgia Piedmont, in an aggrading forested ecosystem in western North Carolina, and in an agroecosystem in Nebraska.

In the agroecosystem study, a wide range of biological, chemical, and physical factors were measured in two field sites, in which alternative poultry-litter management practices were compared. Multivariate statistical techniques were used to determine the smallest set of chemical, physical, and biological indicators that accounted for at least 85% of the variability in the total data set at each site. This set was defined as the minimum data set (MDS) for evaluating soil quality (Andrews and Carroll, 2001). The efficacy of the chosen MDS was evaluated by performing multiple regressions of each MDS against numerical estimates of environmental and agricultural management sustainability goals (e.g., net revenues, phosphorus runoff potential, metal contamination, and amount of litter disposed of). Coefficients of determination ranged from 0.35 to 0.91, with an average R^2 of 0.71. Each MDS was then transformed and combined into an additive soil quality index (SQI). SQ indexes varied between the two sites, but Andrews and Carroll (2001) noted that this "designed SQI" enabled the indices to be tailored to local conditions.

In the forest ecosystem study, a combination of chemical and biological indices was used to measure soil quality in five watersheds arranged along an elevational gradient at the Coweeta Hydrologic Laboratory, in the southern Appalachian Mountains of western North Carolina. The selected characteristics of the elevation gradient stands are presented in Table 8.3 (Knoepp et al., 2000). The sites represented a gradient in vegetation and elevation and included xeric oak–pine (OP), cove hardwood (CH), mesic mixed-oak at low and high elevations (MO-L, MO-H), and mesic northern hardwood (NH) vegetation. The sites were then ranked on a range of soil chemical characteristics, nitrogen availability, litter decomposition rates, forest floor mass, coarse woody debris standing crop, soil oribatid mite populations as numbers and total species, and Shannon–Wiener biodiversity index (Table 8.4) (Knoepp et al., 2000), and then several measures of soil carbon availability: CO_2 flux, microbial carbon, qCO_2 ($\mu g\,C\,g\,soil^{-1}$), and qC_{mic} ($\mu g\,C_{mic}\,gC\,total^{-1}$). Note that all sites had approximately equal diversity of overstory tree species, with H values ranging between 1.93 for the NH, and 2.25 for OP.

The five sites were compared for overall soil/site quality, ranked using biological and chemical or physical quality and the aboveground indices of wood production, net primary productivity, and biodiversity. Overall, soil biological quality was highest for OP and MO-L, with the highest scores in nitrogen and carbon availability and fauna population indicators. Based on soil chemical and physical properties, NH ranked highest with the greatest cation and carbon and nitrogen concentration, and lowest bulk density. In sum, the highest-quality site is dependent on the goal desired for that site. In terms of wood production, MO-H was the highest-quality site. Both mixed-oak sites had the highest productivity using the total litterfall index. If one desired to maximize biodiversity, both aboveground and in the soil, all sites ranked highly (Knoepp et al., 2000). The overall take-home message is important for land use managers and ecologists in general: the site quality really depends on the objectives of the users and the context in which the sites (in this case, sit-

TABLE 8.3. Selected Characteristics of the Elevation Gradient Stands[a]

Site	OP	CH	MO-L	MO-H	NH
Elevation (m)	782	795	865	1001	1347
Aspect (deg)	180	340	15	75	20
Slope (deg)	34	21	34	33	33
Vegetation type	Oak–pine	Cove hardwoods	Mixed oak	Mixed oak	Northern hardwoods
Dominant species	*Kalmia latifolia, Quercus prinus, Q. rubra, Carya* spp.	*Liriodendron tulipifera, Quercus rubra, Tsuga candensis, Carya* spp.	*Rhododendron maximum, Quercus coccinea, Q. prinus*	*Rhododendron maximum, Quercus rubra, Q. prinus*	*Betula allegheniensis, Liriodendron tulipifera, Quercus rubra*
Moisture regime	xeric	mesic	mesic	mesic	mesic
Soil series and subgroup(s)	Evard/Cowee, Chandler, Edneyville/Chestnut, Typic Hapludults, Typic Dystrochrepts	Saunook, Tuckaseegee, Humic Hapludults, Typic Haplumbrepts	Trimont, Humic Hapludults	Chandler, Typic Dystrochrepts	Plott, Typic Haplumbrepts

[a]Data compiled from Coweeta Long-term Ecological Research Program records. From Knoepp, 2000.

TABLE 8.4. Soil Oribatid Mite Populations in Five Representative Sites in the Coweeta Hydrologic Laboratory Basin

Site / rank[a]	N[b]	# spp.[c]	J[d]	H[e]
OP/3	2237	78	0.752	3.28
MO-L/4	2234	96	0.746	3.41
MO-H/4	2192	92	0.761	3.44
CH/1	570	64	0.876	3.64
NH/2	1454	81	0.793	3.48

[a] Site rank for oribatid mite populations.
[b] Abundance of individuals collected.
[c] Total number of oribatid mite species identified.
[d] Pielou's evenness index.
[e] Shannon–Wiener biodiversity index.
After Lamoncha and Crossley, 1998; from Knoepp, 2000.

uated in two different watersheds) exist. There is ample room for further investigation in this important area in which scientists and managers from a variety of disciplines will collaborate. This is especially true when comparisons are made across wide continental gradients, for example, across ecoregions.

The foregoing examples involved very extensive sampling and analytical regimes that might preclude their wide adoption in soil quality studies. An innovative study in Nebraska employed the fact that differences in electromagnetic (EM) soil conductivity and available nitrogen levels over a growing season can be linked to feedlot manure/compost application and use of a green winter cover crop (Eigenberg *et al.*, 2002). A series of soil conductivity maps of a research cornfield were generated using global positioning system (GPS) and EM induction methods. The study was conducted over a 7-year period. Image processing techniques were used to establish EC treatment means for each of the growing season surveys. Sequential measurement of profile weighted soil electrical conductivity (EC_a) was effective in identifying the dynamic changes in available soil nitrogen as affected by animal manure and nitrogen fertilizer treatments during the corn-growing season. This real-time monitoring approach shows considerable promise in enabling farmers to more efficiently use nitrogen sources in cropping management systems and in minimizing nitrogen losses to the environment.

It is imperative to have a robust, quantitative, and universally applicable metric for soil quality. Considering the 4.5×10^9 hectares that are tropical soils, use of an updated fertility capability soil classification (FCC) system (Sanchez *et al.*, 2003) should be helpful for soil ecologists.

It employs quantitative topsoil attributes including percentage of total organic carbon saturation (van Noordwijk *et al.*, 1998) compared with undisturbed or productive site and soil taxonomy. The top three soil constraints in the tropics include moisture limitations, low nutrient capital reserves, and high erosion risk. Because many small farmers in tropical regions depend on organic sources for nutrient inputs to their crops, this becomes an ideal situation to practice sound organic agriculture. This approach has been promoted ably by the Tropical Soil Biology and Fertility Programme, which has a network of research sites throughout eastern and southern Africa, India, and southeast Asia (see van Noordwijk *et al.*, 1998, and Swift, 1999, cited in Sanchez *et al.*, 2003, and Palm *et al.*, 2001).

For those who are interested in pursuing practical, hands-on studies in soil ecology, Chapter 9 contains some selected field and laboratory exercises that should be of use in both research and teaching activities.

9 Laboratory and Field Exercises in Soil Ecology

INTRODUCTION

The following exercises are ones that we have found to be useful to better acquaint our students with techniques used to conduct field research in soil ecology. They provide a combination of process and taxonomic identification work, both of which are necessary to make some headway in our field. Further details of sampling and analytical methods can be found in Weaver *et al.* (1994) and Samner (2000).

MINIRHIZOTRON STUDIES

Principle

A video camera is used to record root densities at discrete time intervals to determine the growth and turnover of roots *in situ*. (See Chapter 2 for details.)

Description of a Minirhizotron

A clear polycarbonate tube 5 centimeters (cm) in diameter and 1.8 meters (m) in length is used as a minirhizotron. Each tube is marked with a groove and depths are stamped with reverse numbers externally at 10-cm intervals. A single groove etched along the length of the tube is used as a reference for orienting the video camera. The section of tube exposed above the soil surface is wrapped with black plastic tape and the top is plugged with a rubber stopper to keep light from entering the minirhizotron. The bottom end of each tube is sealed with a polycarbonate stopper.

Installation of the Minirhizotrons

This example is for an agroecasystem. Holes are drilled by an auger mounted on a tractor in the field plot at an angle of 20° from the vertical along the direction of the crop row. Holes are cleaned with a sharp-ended stainless steel tube of the same diameter as the outside of the polycarbonate tube. A minirhizotron is then driven into the hole to obtain an intimate tube–soil interface.

Enough minirhizotrons should be installed to enable one to obtain standard error estimates that are less than 10% of the mean. This usually requires between six and eight tubes to be installed per treatment (Cheng et al., 1990; Fahey et al., 1999; Pregitzer et al., 2002).

Observation and Recording

Observation of roots intersecting minirhizotrons is accomplished using a video camera, a small monitor, and a videocassette recorder (VCR) or writable compact disc (CD). The image produced by the camera is observed in the monitor and recorded simultaneously. During recording, the camera is moved from the bottom of the tube upward at a speed of approximately 0.3 cm per second or less, recording a 2-cm wide picture strip. This allows observation and counting of roots and minimizes field operation time.

Getting Data from the Videotape

The numbers of roots in the 2-cm wide strip observed are counted for each 10-cm length of the tube. Counts are independent of the length or the diameter of the root at the interface. If a root branches it receives one count for the main root and one for each branch. Whenever a root crosses the depth indication groove it receives one count in each depth interval.

Convert root counts to root length densities (RLD), as cm/cm^3, using the equation

$$RLD = (Nd) \div (Ad)$$

where N is the number of intersecting roots, d the outside diameter of the tube, and A the area of tube observed. (Upchurch and Ritchie, 1983; Upchurch and Taylor, 1990)

Tracing Technique

A high-quality, four- or five-head VCR is required. Frames of the root picture strip are frozen and traced on a sheet of clear plastic in sequence. Root length of each 10-cm interval is obtained by tracing roots on the clear plastic using a map measurer.

Root length can be converted to RLD using the equation

$$RLD = Lr \div CAD$$

where Lr is the root length, C the magnification factor, A the area of tube observed, and D the distance from the outside surface of the tube that can be observed in soil (usually assumed to be 0.2 or 0.3 cm in the literature) (Fahey *et al.*, 1999).

Automated Root Length Measures

More automated root length measures may be achieved using software packages such as "Roots" and WinRhizo," which are obtainable from a number of vendors.

SOIL RESPIRATION STUDIES

Principle

A dilute solution of alkali (typically 1 Molar [M] NaOH) is placed in an open glass jar above the soil surface. A metal cylinder (usually 10 cm in diameter, an irrigation pipe will do) is installed well into the A horizon, carbon dioxide evolution is measured by absorption in the alkali solution for 24 hours, and then the CO_2 amount absorbed is measured by back-titration of the excess alkali remaining. When expressed on a per unit area basis, the soil respiration data are comparable with literally hundreds of values from the literature about many different ecosystems.

Soil respiration is one of the most commonly used methods of determining metabolic activities of organisms in soil. As noted in Chapter 3, there is interest in determining the relative contributions of carbon dioxide evolved by the secondary consumers (microbes and fauna) as differentiated from that originating from respiring roots. For the purposes of this laboratory exercise, we will rely on comparisons of respiration from different ecosystems such as an arable field, forest, or grassland. Soil respiration, reflecting all of the biotic activity, is often measured to compare and contrast the side effects of chemicals such as pesticides and heavy metals. Soil respiration can be determined directly in the field. By measuring soil temperature and percentage of water in the soil, the relative contributions of these key abiotic variables can be calculated, which is useful in comparative ecosystem studies (e.g., Coleman, 1973). For more background on this method, consult also Alef and Nannipieri (1995).

Materials and Supplies Needed

Metal or plastic cylinders (30 cm long, 15 cm diameter) with one end beveled for ease in insertion into soil

Plastic lids to seal the cylinders, available from Sinclair & Rush, Inc., St. Louis, MO

Aluminum foil, to cover cylinders and lids, to minimize any heat loading

Screw-capped glass jars (4–6 cm diameter, 7–8 cm high)

Tripods made of metal or plastic

NaOH solution (0.5 or 1.0 M)

Barium chloride solution (1.0 M)

HCl solution, or definitive (determined) molarity, ideally 1.0 M

Thymolphthalein indicator: 1 gram (g) thymolphthalein is dissolved in 100 milliliters (mL) 95% ethanol. (Other references [e.g., Alef and Nannipieri, 1995] describe the more often used phenolphthalein, but it has a less-definitive end point, changing from dark to light pink at the end point. In contrast, the thymolphthalein changes from blue to colorless at the end point very rapidly, which gives the definitive end point so desired in this titration.)

Procedure

About a week after the respiration chambers are installed (pushed 15 cm into soil), the CO_2 flush from the soil disturbance caused by the installation should have diminished. It is now possible to measure the amount of CO_2 respired from a known surface area by trapping it into jars of 1 M NaOH during 24 hours.

Pipette 20 mL of NaOH solution into the glass jar, and place it on the tripod in the center of the selected cylinder. Immediately cover the cylinder with the plastic snug-fitting lid, and cover the entire setup with aluminum foil. Note time zero.

After 24-hour incubation, retrieve the jar, pipetting 20 mL $BaCl_2$ into it before covering with screw-on lid. Take the jars to the laboratory for titration, using the color indicator noted previously and the standard 1 M HCl.

Note that control treatments are performed by incubating sealed jars of NaOH solution in the field.

Calculations

CO_2 evolution rates are calculated as follows:

$$CO_2 - C \text{ (mg)} = (B - X)ME$$

where B is the HCl needed to titrate the NaOH solution from the blank; X is the HCl (mL) needed to titrate the NaOH solution in the experimental jars, exposed to the soil atmosphere; $M = 1.0$ (HCl molarity); and

E is the equivalent weight (22 for CO_2, and 6 for C). The data are thus expressed as milligrams of CO_2 or $CO_2 - C$ per square meter per day.

LITTER DECOMPOSITION STUDIES

Principle

Weight loss by bagged leaf litter has been a useful method for measuring leaf litter breakdown for nearly 50 years (Bocock and Gilbert, 1957). Known masses of litter enclosed in mesh bags or envelopes may be exposed in field sites and then retrieved at later times for remeasurement of mass. Disappearance or breakdown of bagged litter is a valuable means of comparing substrates such as leaf litter, twigs, or roots. Different habitats or geographic regions may also be compared. Litterbags have been successfully used in hardwood and conifer forests, deserts, agroecosystems, and arctic soil situations. In prairies, ingrowth of grass may present a problem. Litterbags have been used to sample a subset of soil microarthropods, in order to discover which species are active in a given stage of decomposition or to follow the development of microarthropod communities. Litterbags retrieved from field experiments are extracted with Berlese funnels before estimation of mass loss. Fractions of litterbag material may be removed also for nematode extraction. In this experimental context, different mesh sizes may be used to exclude macroarthropods, microarthropods, or microfauna. Breakdown rates are more rapid in bags with larger mesh and slowest in fine-mesh bags that admit only microfauna. Litterbags may be treated with insecticides or fungicides to manipulate specific groups of soil biota for studies of their effects on decomposition.

Litterbags consistently underestimate decomposition rates; thus they are properly used in a comparative context. Such underestimation is most extreme for rapidly decomposing substrates—leaves of *Cornus florida* or (in agricultural systems) leguminous foliage. For more recalcitrant litter types—*Rhododendron* or some *Quercus* species—mass loss from bagged litter more closely approximates that of unconfined leaves.

As an alternative method, a group of leaves secured by a nylon string attached to their petioles may be used in conjunction with litterbags. Loss of leaf area as well as mass loss may be measured (Hargrove and Crossley, 1988). Because entire fragments may be broken off and lost, this string method overestimates decomposition rates and is viewed as a comparative method. Decomposition of rapidly decaying substrates is characteristically overestimated. Some leaf litter species, such as sweetgum, tend to become detached from their petioles early in litter breakdown.

Another method of measuring leaf area loss was employed by Edwards and Heath (1963), who buried leaf disks of known area in litterbags. When they were retrieved the leaf area loss was estimated using a grid, and soil animals were enumerated. See Chapter 5 for a discussion of litter decomposition.

Litterbag Construction

Various types of netting have been used for litterbags. Mosquito netting, nylon drapery mesh, and plastic window screen are suitable materials. Cloth bags must be sewed together. Plastic screen bags may be constructed by using a soldering iron to melt and seal the edges of the bag. Cloth bags are more flexible, conform to the shape of the forest floor, and admit more fauna because the mesh openings are more flexible. Window screen bags are more rigid and openings are fixed, but they are much easier to construct. Size of bags has ranged from 0.25 square meter (m^2) down to 10 millimeters (mm); 10 cm by 10 cm (one decimeter) is a frequently used dimension. Actual dimensions of 12-by-12 cm yield an effective area of about 100 cm^2. A flap on the open end of the bag allows it to be closed with a safety pin.

Leaf litter substrates should be air dried if possible before insertion into the litterbags. We use contrasting litters with different palatabilities, carbon–nitrogen ratios, etc., such as dogwood, chestnut oak, or rhododendron. Decimeter litterbags will contain about 1.5 g of dry deciduous leaf material before breakage becomes a problem. After mass determination, an identifying label may be placed inside the litterbag. (Aluminum tags attached outside the bags tend to attract large mammals.)

Place the bags out in field sites such as old fields, forest floor, or agricultural fields. On each sample date, randomly select four replicate litter bags from each of several treatments. Upon retrieval, litterbags should be placed in plastic bags and returned to the laboratory as soon as possible. The bag may be opened and a small increment removed for nematode extraction (with suitable estimation of litter mass). The intact litterbag may then be extracted on a Tullgren funnel. Finally, the litter substrate is removed from the bag for mass determination.

Process the litter using the following steps:

1. Clean the outside of the litterbag (brush off any sand and litter particles).

2. Remove the metal label; record its number, the pick-up date, treatment code, and your initials on a new label.

3. Weigh a beaker (with tape label on it) and record on weighing sheet.

4. Above a piece of waxed paper, carefully take all the litter material out of the litterbag and put it into the beaker. If there was any soil intrusion into the litterbag, make sure you remove the soil before you put the litter in the beaker. Put the beaker in the 60°C oven and dry the litter for a minimum of 2 days.

5. After at least 2 days, record the dry weight (beaker + litter dry weight) and transfer the litter to a labeled paper bag.

Calculations

Express results as percentage of dry mass remaining:

$$\frac{M_0 - M_t}{M_0} \times 100$$

where

M_0 = initial dry mass at time zero
M_t = final dry mass at time t

Plot logarithm percentage mass remaining (y-axis) versus time (x-axis) to derive exponential decay curves and calculate decay constants $(-K)$ for statistical comparisons among treatments.

ANALYSES FOR SOIL MICROBIAL BIOMASS

The Chloroform-Fumigation K_2SO_4-Extraction Method

Principle

This procedure compares the amount of total organic carbon (TOC) in a chloroform-fumigated soil sample to that in a nonfumigated soil sample to determine soil microbial biomass. In the chloroform-fumigated sample, TOC will be higher because the sample contains the cell contents of lysed microbial cells. Hence the difference in extracted TOC between fumigated and nonfumigated samples will provide a measure of microbial biomass (Vance *et al.*, 1987). Note that you can only assume that this TOC is of microbial origin if the soil samples have been picked free of roots, litter, earthworms, etc. (the microfaunal contribution to TOC is less than 5%).

Fumigations are carried out for a period of 2 days in vacuum-desiccators with alcohol-free chloroform. Soil samples are extracted with 0.5 M K_2SO_4 and the filtrate is analyzed for TOC. Analysis results need to be adjusted to a TOC/g dry soil value. It is important to refrigerate soil samples until the fumigation and K_2SO_4 extractions are performed.

Preparation and Handling of Potassium Sulfate

Make 0.5 M K_2SO_4 by adding 87.13 g K_2SO_4 to a 1000-ml volumetric flask and bring to about 2/3 volume with deionized water. Place on a stir plate until K_2SO_4 is in solution and bring up to volume with deionized water. K_2SO_4 powder has a health rating of 1 (may cause irritation); goggles and gloves should be used when handling this chemical.

Sample Preparation

1. Remove all visible roots and leaf litter from freshly collected soil samples with a forceps. Roots may be dried and weighed to get a measure of root density of the soil sample.

2. Mix the soil sample thoroughly to make it "homogeneous." Label two 125-mL Erlenmeyer flasks (one NF—nonfumigated—and one F—fumigated); include, of course, the sample site and number on the label. **Note:** Labels must be in pencil on special white tape, because marker ink and often glue dissolve in chloroform fumes.

3. Weigh about 25 g soil (fresh weight; or 20 g dry weight) into each flask (water content should be in the 20–30% range for best fumigation results). Record weights for later correction to dry weight.

4. Include at least four blanks (2 F, 2 NF) (empty Erlenmeyer flasks).

Potassium Sulfate Extractions

1. Use a 50 mL dispenser to add 50 mL of 0.5 M K_2SO_4 to each NF flask. Cover flasks with parafilm and place them in a rotary shaker for 30 minutes (at about 200 rotations per minute [rpm]).

2. While samples are shaking, prepare your funnel and filter paper setup. Wearing powder-free gloves, fold Whatman 42 filter paper, (15-cm diameter), so that it forms a cone, and place it into a plastic funnel. Fill each funnel with deionized water to rinse out any traces of soluable carbon in the paper.

3. After 30 minutes, take flasks out of the shaker. For each flask: transfer white tape label to a 60-mL plastic bottle; place vial under funnel; gently shake extract to suspend the soil and pour it immediately onto filter paper. Do not try to get all the soil onto the filter. Store the extracts in a freezer until analysis on the TOC analyzer.

Chloroform Fumigation

1. Clean vacuum-desiccators and cover the bottom with water (about 1 cm). Place the 'F' flasks inside the desiccator. As many flasks as will fit without spilling the soil or blocking the opening of another flask may be placed in each desiccator (about 13 flasks). Include one blank (empty Erlenmeyer flask) in each desiccator.

2. Place a 100-mL beaker in each desiccator. Add a few glass beads and 40 of alcohol-free chloroform. Use gloves and work in the fume hood at all times when working with chloroform. Close the desiccator with the lid (use some Lubriseal or Vaseline for good sealing).

3. Start the water aspirator, check the vacuum suction, then attach the hose to the desiccator. This will create a vacuum in the desiccator, causing the chloroform to boil and evaporate, and saturate the desiccator atmosphere. Allow the chloroform to boil vigorously for 5 minutes. Then (first) close the vacuum route by sealing the desiccator (by turning the top), and then stop the water aspirator. The vacuum will suck water into the desiccator if you turn off the water before sealing the desiccator. Reapply the vacuum 3 times over the next hour. The chloroform may not boil after the first time. Allow the fumigation to proceed for 2 days.

Chloroform Removal

1. After 2 days, release the pressure in the vacuum-desiccators. Remove the beaker of chloroform, and store it in the back of the hood to let the remaining chloroform evaporate.

2. Because it is important that all residual chloroform is removed from the soil samples before proceeding with K_2SO_4 extractions, reapply the vacuum to the desiccator. However, do not seal it: simply detach the vacuum hose (while the tap is running) and allow the air to get back in and circulate. Repeat this procedure a few times to clear the chloroform from the flasks. Then place the flasks near the opening of the fume hood, and lower the window to increase the velocity of wind flowing over the flasks (the removal of chloroform will take about an hour).

3. Do the potassium sulfate extractions (see earlier section) for the F flasks.

Calculations

Microbial biomass carbon is calculated as:

(TOC [F] – TOC [NF]) ÷ K_C where $K_C = 0.38$ (Vance *et al.*, 1987).

NOTES:

Sample blanks should be run with all preparations and the results should be used to correct values above.

All glassware (volumetric flasks, Erlenmeyer flasks) and plasticware (Nalgene sample bottles) must be acid-washed (10% HCl) and dried before and after use.

The K_2SO_4 soil extracts can be used for several analyses other than TOC, for example, microbial nitrogen (NO_3 and NH_4) (see Cabrera and Beare, 1993; and Jenkinson, 1988) or microbial phosphorus and sulfur (see Page *et al.* 1982).

SAMPLING AND ENUMERATION OF NEMATODES

Principle

Using a procedure to determine the numbers of free-living nematodes in litter and soil samples, the abundances of these important mesofaunal organisms in terrestrial ecosystems can be estimated.

Sampling Considerations

When designing a sampling plan for nematodes, there are practical and theoretical considerations to be made, which are balanced with the amount of time and money available for the study.

• *Goal of the study and required accuracy:* The goal of the study will greatly determine the degree of accuracy required. For qualitative studies such as diagnosis of a plant disease or taxonomic work, the requirement is relatively low. In an ecological study of nematode communities more accuracy is needed, because rare species should be included in the sampling. Even more effort is needed in sampling for disease control, for example, to certify that a certain field is disease-free. In general, a higher accuracy costs more time and money.

• *Variation in space (horizontal and vertical distribution):* One of the biggest problems in sampling is that nematodes have a patchy distribution, that is, they are not randomly distributed in the soil. This has biological reasons (concentration around roots or in islands of organic debris), agronomic reasons (cropping history; planting diseased material creating "hot spots" of disease), as well as physical reasons (e.g., texture gradients). If any of these factors are known, sampling variability can be decreased by sampling and analyzing such areas separately (called *stratified* sampling). In terms of vertical variation, most nematodes (and other soil organisms) can be found in the topsoil (0–10 cm). In agricultural fields, when sampling roots, sample depth is often as deep as the plow depth (typically 15 cm) or deeper. Again, it depends on the purpose of the study.

• *Variation in time:* Seasonal cycles and life cycles will influence the results of the sampling. Seasonal fluctuations in moisture, temperature, and food availability will influence abundances of different species in different ways. For sampling plant parasitic nematodes, knowledge of the life cycle is most relevant. For population studies of an endoparasite, sampling soil when the crop is on the land does not make sense, because most nematodes will be inside the roots.

• *Statistical considerations:* The purpose of sampling is to *estimate,* as accurately as possible, nematode abundance in a site. The "sampling error" (expressed as variance or standard deviation) is the sum of *systematic* and *random* errors, introduced at any stage in the sampling process (taking cores, mixing soil, extracting soil, counting subsamples, etc.). Systematic errors can be reduced by improving methods and working as carefully as possible. Random errors can be reduced by taking more and bigger samples, as much as time and money allows. A rule of thumb is that variability *within* a sample (or plot) is usually smaller than *between* samples (or plots). Thus, it makes more sense to take many samples than to analyze many *sub*-samples (likewise, it is more efficient to sample many different plots than to take many samples within the same plot). Random sampling is best but not always most practical. Systematic sampling (e.g., taking samples along a transect, every few meters) is more often used. In that case, the starting point should be randomly chosen.

Sampling Tools and Precautions

Soil (and roots) can be sampled with any simple container such as a food can, a shovel, or special corers. **Notes:** When sampling to advise farmers on the presence/absence of plant parasitic nematodes, make sure the sampling tools are cleaned and disinfected after every soil sample. Always store the soil samples in plastic bags (in paper bags the soil will dry out), and keep them away from the sun (heat kills nematodes!). Handle the samples carefully: throwing them around, or any other mechanical violence (soil sieving, mixing), will kill nematodes as well. The sooner the samples are extracted, the better. For reliable results, the maximum storage time, at 4°C, is 1 week only.

Nematode Extraction: Baermann Funnel Method

Principle

The principle is very simple: nematodes move out of the substrate toward the water, and sink to the bottom of the funnel stem due to gravity forces. The method is suitable for extraction of mobile nematode from substrates like soils and sediments, plant material,[a] and litter.[a] Advantages and disadvantages of the method: it is cheap and simple; extraction efficiency is quite good if your sample is small relative to the funnel diameter (with only a thin layer of substrate on the sieve), but decreases very rapidly when your sample gets bigger than that. For alternatives, see Southey (1986).

Materials and Supplies Needed

Funnel with a piece of rubber tubing attached to the stem, closed with a pinch clamp
Funnel rack
Kimwipes
Clean tap water (not deionized water), centrifuge tube for sample collection
Small screen to support soil in funnel (e.g., aluminum window screen [avoid copper])

Procedure

1. Put the funnel in the rack. Make sure it is level. Fill it with clean water. Remove air bubbles by squeezing the rubber tube, or by draining water by opening the clamp.

2. **Carefully** mix the soil sample (excessive mixing will kill nematodes). Take a 10-g subsample, spread it out on a double layer of Kimwipe tissue. Record sample weight. Place the sample on the screen in the funnel. Do this carefully, so that small soil particles do not get through the tissues and obscure nematode samples.

3. The soil should be moist enough, but not totally submerged. Adjust the water level in the funnel using a spray bottle (don't spray *on* the sample, for same reasons as discussed in step two), or drain by opening the clamp. Cover the funnel with a petri dish or wax paper to avoid dust and evaporation.

4. The extraction time is 2 to 4 days. The longer the time, the more nematodes are harvested, but the question arises whether this increase is because of catching slow-moving nematodes, or due to hatchlings from eggs in the soil or larvae from fast-reproducing species—some have a life cycle of 2 days (especially at lab temperature). For that reason, 48 hours is an optimal time. Check after the first 24 hours if the soil is still moist enough.

---

[a]Nematode extraction from plant and litter material: Cut the material in small pieces before extraction.

5. Harvest: By opening the clamp, get approximately 10 mL in a centrifuge tube (e.g., plastic Corning, 15 mL with screw cap). Store immediately under refrigeration, for maximum of 1 week. To kill and preserve (using 5% formaldehyde) the nematodes, see article about identification of nematode feeding groups by Yeates *et al*. (1993).

Killing and Fixing Nematodes with Hot and Cold Formalin (5%)

(**Precautionary note:** Conduct this work in a fume hood.)

Using hot formalin, the nematodes are instantly killed and fixed. Immediately cooling down with cold formalin prevents overheating that could cause deformations. With this hot–cold formalin method, nematodes will be fixed in taxonomic groups according to characteristic body curvatures, which makes routine identification easier. Another advantage is that most species keep their body transparency. This is in contrast to the method of fixing nematodes with cold formalin only, where nematodes die slowly instead of instantly, which causes deformations and loss of transparency.

Materials and Supplies Needed

Water aspirator (vacuum pump) with fine pipette tip attached to the hose
Nematode suspension in plastic 15-mL centrifuge tube with formalin 5% (10.8 mL formaldehyde 37%, in 89.2 mL DI water) in a squeeze bottle

Procedure

1. Let the nematodes in the centrifuge tube settle to the bottom for at least 24 hours. Do this close to the water aspirator (vacuum pump) to avoid sample disturbance.
2. *Carefully* aspirate the "supernatant" water until 2 mL (the amount in the conical tip of the tube) is left.
3. Add (approximately) 4 mL of 90°C dilute formalin and *immediately* cool down with (approximately) 4 mL cold formalin from the squeeze bottle.
4. Make sure the cap is tightened so formalin can't evaporate.

SAMPLING AND ENUMERATION OF MICROARTHROPODS

Principle

Using heat and dryness, microarthropods are "induced" to move out of litter and soil samples, into a collecting fluid, for enumeration and identification. With flotation, differential buoyancy of the microarthropods is used to advantage with a flotation medium.

Methods for the Study of Microarthropods

Sampling

Mites and collembolans are collected by extracting them from a sample of their soil habitat. Extraction procedures may be either Berlese funnel extraction of soil cores or litterbags, or alternatively, flotation (Bater, 1996). Total microarthropod populations are then estimated by extrapolating from the size of the sample (weight or area) to field dimensions. For sampling protocols and considerations of sampling design, consult Hall (1996), Schinner *et al.* (1996), Coleman *et al.* (1999), and Larink (1997).

Soils vary greatly in structure, composition, pore size, moisture regime, and so forth; sampling methods need to be suited to the ecosystem under investigation. Most quantitative samples of microarthropods are taken from soil cores 5–10 cm in diameter by 5–10 cm deep. The smaller cores yield satisfactory results; a 5-by-5-cm core will contain several hundred mites and collembolans. A split core tool with a sharp beveled tip, designed to hold a sleeve for the soil sample, is preferable for most soils (Fig. 9.1). For many soils the great majority of microarthropods are found within 5 cm of the surface. In grassland soils and disturbed soils they may be distributed more deeply, and additional 5-cm increments may need to be extracted, to a depth of 15 cm or more. Sample cores should be extracted in a high-efficiency extractor (Fig. 9.2) as soon as possible; storage for any significant period of time will result in lower numbers of microarthropods extracted.

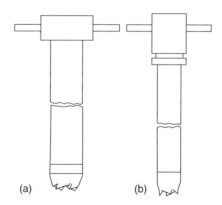

(a)　(b)

FIGURE 9.1. Soil coring device: (a) outer tube; (b) inner sleeve, typically made of aluminum. Effective core diameter: 5 cm (modified from Gorny and Grüm, 1993).

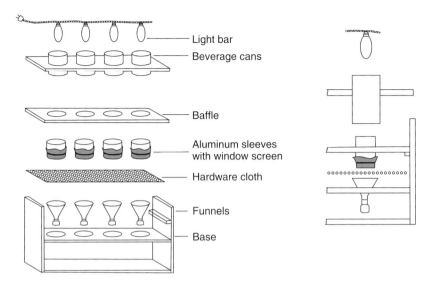

— Light bar
— Beverage cans

— Baffle

— Aluminum sleeves with window screen
— Hardware cloth

— Funnels

— Base

FIGURE 9.2.   Design and assembly of the high-efficiency microarthropod extractor (from Crossley and Blair, 1991).

Extraction of Microarthropods from Samples

There are many modifications of the basic Berlese funnel (Edwards, 1991) and most will yield satisfactory results. Heat is used to desiccate the sample, driving arthropods out and down into a collection fluid. Many designs are called Tullgren funnels, after the originator of the use of electric lights as heat sources. (The original Berlese funnels used steam as a heat source). Larger funnels (Fig. 9.3), used for extracting big samples of litter, can work effectively with small cores as well. Arrays of smaller funnels can handle more samples in a smaller space, and have become the most widely used piece of extraction equipment (Bater, 1996). Soil cores contained in their sleeves are extracted in an inverted position, surface layer down, so that arthropods can escape using natural channels in the soil. The upper portion (bottom) of the core should be moistened with water to improve extraction efficiency (T. R. Seastedt, personal communication). Seventy-percent ethyl alcohol is the usual collection fluid for the extracted arthropods. A ten-percent picric acid solution is preferred by some authors (e.g., Meyer, 1996). Care must be taken to keep mineral soil from falling into the sample (the berleseate), because samples contaminated with soil are hard to sort. To this end, a single layer of cheesecloth may be inserted between sample and funnel.

Litterbags are an alternative method to soil cores for sampling microarthropods. They offer several advantages: Microarthropods

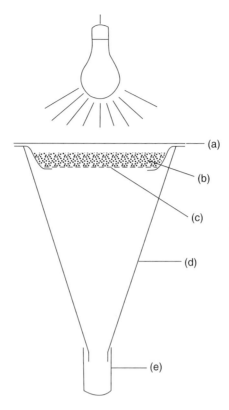

FIGURE 9.3. Schematic diagram of an extractor for soil macrofauna: (a) sample cover, (b) soil sample, (c) sample screen, (d) aluminum funnel, (e) collection container with 70% alcohol or similar collection fluid (from Gorny and Grüm, 1993).

using different substrates, such as different litter species, may be detected (Hansen, 2000). Different stages of litter decomposition may be compared, and time sequences described. Litterbags are readily extracted, intact, on large Tullgren funnels.

Flotation is another alternative method for extracting microarthropods from soil samples. Using organic solvents or saturated sugar solutions, arthropods may be separated from the soils, washed through a fine mesh screen, and thus recovered (see Fig. 4.9). In comparison with Tullgren extraction, flotation usually yields higher numbers of microarthropods. Some collembolans, such as members of the family Onychiuridae, respond poorly to Tullgren extraction; flotation is the method of choice for sampling these arthropods. The disadvantages of flotation are (1) the method is extremely laborious in comparison to Tullgren extraction and (2) it is not effective for samples with large amounts of organic matter. Finally, the use of organic solvents probably violates most laboratory health and safety considerations (Griffiths, 1996).

A more attractive procedure uses saturated sugar solutions instead of organic solvents (Snider and Snider, 1997). The method (Table 9.1) yielded much higher captures of microarthropods than did Tullgren extraction alone, in samples from two hardwood forest sites in upper Michigan. The method was extremely laborious; the authors report that 10 to 12 hours were required to sort a single sample. They note that financial constraints usually preclude labor-intensive procedures such as flotation, even though much more accurate population estimates were obtained.

Sample Sorting and Identification

A good dissecting microscope with magnification in the range of 10–40× is essential. A preliminary sorting will separate collembola, mites, and other microarthropods. The latter category includes the few tiny spiders, small beetles, and other insects (adults and larvae) that can usually be identified with the dissecting scope. Some may require slide mounts (see later discussion). Collembolan and mite specimens can be transferred to sorting dishes with a fine-tipped pipette (such as a Pasteur pipette), a camel's hair brush trimmed to 3–4 lashes, or a flattened, curved dissecting needle.

Collembolans: Identification of springtails almost always requires high magnification (400× or greater) of cleared specimens, using a good phase contrast microscope. Christiansen (1990) recommends clearing specimens and observing them in temporary mounts; heavily pigmented forms may require more clearing than the mounting medium provides. He suggests the following reagents:

TABLE 9.1. Procedure for Flotation Extraction of Microarthropods from Soil Samples, Using Saturated Sugar Solutions

Step one	Place the soil core in a plastic bag; crumble it gently.
Step two	Transfer the crumbled core into a 1-liter wide-mouth jar. Wet it with distilled water.
Step three	Add saturated sugar solution, leaving approximately 3 cm of headspace.
Step four	Cover the jar with a lid, shake it gently, and let it stand for 2 hours to allow organic matter to float to the surface.
Step five	Decant the solution through a number 200-mesh sleeve into a large bowl to trap the organic matter (do not include silt from the bottom of the jar).
Step six	Rinse the organic matter with distilled water, then wash it from the sieve into a sample jar containing 95% ethyl alcohol.
Step seven	Return the sugar solution to the bowl; return to step four.
Step eight	Repeat steps four to seven for three iterations, then combine all organic matter into one jar.

After Snider and Snider, 1997.

1. Potassium hydroxide, a 5% solution, for brief periods only.
2. Lactic acid, an excellent clearing agent; collembolans tolerate long exposure to it. May be mixed with an equal portion of glycerine.
3. André's fluid (40 cc chloral hydrate, 30 cc glacial acetic acid, 30 cc distilled water). Clears rapidly but may cause damage to specimens.
4. Bleach. A 5.35% solution of sodium hypochlorite will clear heavily pigmented specimens, but is destructive to cuticles.

Christiansen (1990) further recommends the use of depression slides for study of cleared specimens, because weight of the cover glass may crush the arthropod. Slide-mounted specimens are convenient for reexamination and for reference specimens (see later discussion about preparation of temporary and permanent mounts of microarthropods).

Mites: Preliminary sorting of mites into subgroups (Prostigmata, Mesostigmata, adult Oribatei, immature Oribatei, and Acaridida) can be accomplished successfully with experience. Even so, slide mounts will be necessary for confirmation of these identifications. In the sorting dish, the mites are separated into morphospecies and representatives are mounted on microscope slides, depending upon the group. Mesostigmata, Acaridida, and Prostigmata are usually mounted in Hoyer's medium (see description below), with preliminary clearing for large or heavily pigmented specimens. Oribatids require special consideration because their heavily pigmented and brittle exoskeleton is easily crushed by a coverslip.

Clearing agents for mites are similar to those used for collembolans. A popular one is lactophenol (Krantz, 1978):

Lactic acid	50 parts
Phenol crystals	25 parts
Distilled water	25 parts

Specimens may be left in lactophenol at room temperature for several days, or heated for more rapid action. Larger specimens may need to be punctured. After soaking, large mites such as trombidiids may be pressed with a flattened dissecting needle before mounting them. André's fluid (described previously) is also recommended for mites as well as collembolans. Nesbitt's fluid (40 g chloral hydrate, 25 mL distilled water, and 2.5 mL concentrated acetic acid) is useful for specimens that do not respond to milder clearing agents. Oribatids stored in lactic acid for a few weeks are usually satisfactorily cleared for study.

Most permanent or semipermanent mounting media used for mites (and collembolans as well) are aqueous, in that they contain, or are soluble in, water (Krantz, 1978). Gum arabic and chloral hydrate are the principal ingredients. Hoyer's medium is one of the most popular:

Distilled water	50 ml
Gum Arabic	30 g
Chloral hydrate	200 g
Glycerine	20 ml

Clear crystals of gum arabic are preferred. Powdered gum arabic is difficult to wet but may be dissolved in alcohol, which is then allowed to evaporate (R. A. Norton, personal communication). If slide mounted specimens are given gentle heat (40°C) for a few days, considerable clearing of the specimen will take place. Slide mounts using Hoyer's medium, if ringed, will last for some years but eventually deteriorate. Canada Balsam is not satisfactory for mites or collembolans because the refractive index of the medium is so similar to that of the cuticle (Christiansen, 1990). Permount is suitable but requires that specimens be dehydrated and mounted from xylene (Adl, 2003). Generally speaking, mite and collembolan specimens should be archived in 70% alcohol, although "constant vigilance" is necessary to guard against evaporation of the preservative (Christiansen, 1990).

As noted above, many specimens of oribatid mites cannot be mounted on slides in the usual manner without crushing them and thus obscuring their features. Following clearing, specimens may be examined in depression slides, partially covered with a cover slip, in lactic acid or glycerin and manipulated with a fine needle. R. A. Norton (personal communication) recommends slide mounts using a procedure with a small cavity drilled into a microscope slide. A small drop of fluid is placed next to the tiny hole and allowed to flow into it. The mite is positioned and allowed to partially dry; then mounting medium and coverslip may be added. Norton further recommends a 50–50 mixture of Hoyer's medium and Nesbitt's fluid for preliminary clearing of oribatids.

SAMPLING AND ENUMERATION OF MACROARTHROPODS

Principle

See comments about sampling in Chapter 4.

Methods for Sampling Macroarthropods

Sampling

Macroarthropods are a diverse group, with representatives in several classes and orders of the Arthropoda. Most are visible to the unaided eye and hand collecting and sorting is a reasonable procedure for some of them. Many, such as the cryptozoa, are crepuscular or nocturnal, and even hand collecting may require special procedures such as flashlights, black lights, or baits.

Berlese or Tullgren Extraction

Large Tullgren funnels will extract microarthropods from samples of forest floor, but sample size is necessarily limited to 0.1–2.5 m^2 collections. Larger samples will be necessary for those taxa whose densities are smaller than 1–2 per m^2, and careful examination and hand sorting may be the preferable means of processing them. Forest floor materials can be removed from a measured area and then crumbled over a coarse screen atop a white enameled pan. Arthropods are captured as they are released from the sample. An alternate procedure involves crumbling the sample gently into a container of water and collecting arthropods that float to the surface (Bater, 1996).

Flotation

In most terrestrial ecosystems, macroarthropods are inhabitants of the mineral soil layers, and sampling them requires the use of coring tools. Soil cores of 5–25 mm diameter and 10–25 mm deep will recover the majority of macroarthropods in most systems. These monoliths may be crumbled and hand sorted, or washed through a set of sieves with running water (Edwards, 1991). The flotation procedure of Behre (1987) involves washing the soil through such a set of sieves of decreasing mesh sizes, with collecting bowls arranged in steps and connected to an overflow. The magnesium sulfate solution used for washing is thus collected and reused.

Emergence Traps

Transient soil inhabitants emerge and may be trapped to estimate their densities. Emergence traps made from screen wire and covering a known area can be fitted against the soil to collect adult macroarthropods when they appear (Callaham *et al.*, 2003).

Pitfall Trapping

Can traps are a useful, inexpensive, and rapid method for assessing communities of macroarthropods. Pitfall traps (Fig. 9.4) have limited usefulness for assessing population sizes (Coleman *et al.*, 1999), because catches reflect both density and mobility of arthropods. Still, pitfall traps are a valuable method for comparing habitats, assessing seasonal shifts in macroarthropod communities, and evaluating species richness. Species-area curves constructed from a series of trappings (Fig. 9.5) revealed the difference in species richness of a ground beetle community in Iowa (Larsen *et al.*, 2003).

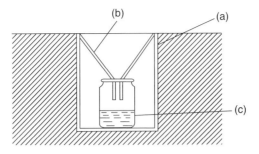

FIGURE 9.4. Double cylinder for the capture of epigeic fauna: (a) outer cylinder, permanently set in the soil, (b) funnel, (c) removable inner jar with collection/preserving fluid (from Gorny and Grüm, 1993).

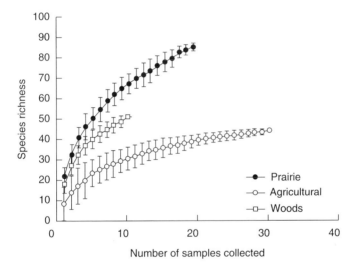

FIGURE 9.5. Species accumulation curves for ground beetles in tallgrass prairie, agricultural, and woodland habitats in northeastern Iowa based on annual pitfall trap samples collected over 5 years from each site and habitat (from Larsen *et al.*, 2003).

Pitfall traps consist of cans or jars, 5–25 mm in diameter, set flush with the soil surface. Arthropods blundering into the traps are directed by a funnel into a vial with preservative (Fig. 9.4). Alcohol and propylene glycol have been used as preservatives (Larsen *et al.*, 2003). Propylene glycol is not subject to evaporative loss, but is poisonous to vertebrates and thus not recommended. If specimens are to be used for chemical analyses, a dry killing agent such as naphthalene or para-dichlorobenzene may be substituted. Pitfall traps should be emptied daily. Heavy rain may ruin the samples. Raised covers may be used to protect the pitfall traps and offer some protection from flooding by rainfall.

SAMPLING AND ENUMERATION OF EARTHWORMS

Principle

Earthworms may be sampled in a variety of ways, depending on behavioral traits and habitat preferences. Different groups of earthworms inhabit different portions of a soil volume and may be sampled accordingly. For example, some epigeic species (e.g., *Bimastos* spp., *Amynthas* spp.) may be collected by hand in the litter layer or in and under logs. Deep burrowing, anecic species such as *Lumbricus terrestris* may be best sampled with chemical irritants that cause the worms to emerge to the soil surface where they may be hand collected. Many endogeic and epigeic species can be quantitatively collected by digging and hand sorting a known volume of soil. In general, the appropriate methods depend on the purpose of sampling (e.g., quantitative sampling versus qualitative biodiversity surveys). Collection and enumeration methods are reviewed in detail by Lee (1985), Edwards and Bohlen (1996), and Hendrix (2000).

Collection of Earthworms

Collection techniques can be classified as passive, behavioral, and indirect (Table 9.2).

Passive Techniques

Hand digging and sorting, which is the most commonly used method for quantitative sampling of earthworms, involves digging pits of known volume (e.g., 25 by 25 by 25 cm), breaking the soil by hand, and collecting all earthworms and cocoons found. Collected specimens are immediately preserved in 70% ethanol or 5% formalin for later counting and identification, or they may be kept alive in cool, moist media for use in experiments. Washing and sieving is an elaboration of hand sorting: the soil is dispersed in water, poured through a sieve, and the earthworms and cocoons hand picked from the sieve contents. Bouché and Beugnot (1972) describe mechanical approaches to washing and sieving. Flotation of sieve contents in a high-density solution, such as 1.16–1.20 specific gravity $MgSO_4$, is another means of separating earthworms and other soil fauna.

Behavioral Techniques

Several approaches have been taken to extracting earthworms from soil based on their behavioral response to certain stimuli. A number of

TABLE 9.2. Descriptions of Methods for Collecting Earthworms

Method	Description	Advantages	Disadvantages
Passive			
Hand sorting	Known volume of soil cut with spade or corer, broken apart and worms removed by hand	Simple, reliable in the field; low cost	Laborious; may not collect deep-burrowing species, small earthworms, and cocoons
Washing and sieving	Known volume of soil cut with spade or corer, soaked in dispersant/preservative, and washed through sieve(s) by hand or mechanical device	Higher recovery of cocoons and small individuals	Laborious; may not collect deep-burrowing species
Flotation	Material from hand sorting or washing/sieving floated in high-density solution (e.g., $MgSO_4$)	Separates earthworms from soil and plant debris; cocoons and small individuals collected	Laborious; may not collect deep-burrowing species
Behavioral			
Chemical extraction	Soil saturated with chemical irritant (e.g., 0.2% formalin) causing earthworms to emerge onto soil surface	Simple; effective on deep-burrowing anecic species	Not effective on all species, in all soils or under all conditions
Heat extraction	Soil blocks or cores suspended under heat lamps in water into which earthworms migrate	Effective on dense root mats	Not effective on all species; inconvenient for field use
Electrical extraction	Metal rods inserted into soil and connected to AC electrical source	Useful for selective or comparative sampling	Highly variable; not convenient in the field; dangerous
Mechanical vibration	Stake or rod inserted into soil and vibrated with bow or flat iron	Simple, useful for selective or comparative sampling	Not effective on all species
Trapping	Pitfall or baited traps placed in soil and sampled at desired intervals	Simple, useful for selective or comparative sampling	Not effective on all species
Mark–recapture	Individuals tagged, released, and population sampled at intervals	Useful for estimating population density, dispersal, and mortality	Laborious
Indirect			
Cast counting	Surface castings enumerated and identified	Simple	Not a quantitative estimate of population density

Summarized from Lee (1985) and Edwards and Bohlen (1996); reproduced from Hendrix (2000).

chemical irritants have been used, including $HgCl_2$, $KMnO_4$, mustard, and formalin. Aqueous solutions of 0.165–0.550% formalin are most commonly used and have been shown to be effective on *L. terrestris* when applied in three sequential doses totaling $18 \, L \cdot m^{-2}$; but formalin may be less effective on other species (Satchell, 1969; Callaham and Hendrix, 1997). Chemical extraction with aqueous mustard powder solution has been shown to be as effective as formalin in some cases; this method avoids the use of toxic formaldehyde. Effectiveness varies with earthworm species and activity, soil water content, porosity, and temperature. Comparisons with hand sorting should be done before adopting extraction techniques for quantitative sampling.

Heat extraction is a modification of that used for enchytraeids (discussed in next section). Soil cores or blocks are placed in pans of water and exposed to heat from overhead light bulbs; earthworms are collected from the water after several hours. This technique was more effective than hand sorting or formalin extraction on small earthworms in dense root mats (Satchell, 1969). As with hand sorting, it is not effective on deep-burrowing, anecic species such as *L. terrestris*.

Mechanical vibration employs a rod or stake driven into the soil, vibration for a few minutes with a bow or flat piece of metal such as an automobile leaf spring, and collection of earthworms that emerge onto the soil surface. Some megascolecid species have been sampled with this technique (Reynolds, 1973; Hendrix *et al.*, 1994), but it is not effective on lumbricids and probably only useful for selective or comparative sampling of certain populations.

Electrical extraction of earthworms involves inserting metal rods into the soil, connecting them to a source of alternating current, and collecting earthworms that come to the soil surface. Different voltages and amperages have been used with varying degrees of success; effectiveness of the technique is highly dependent on soil water content, electrolyte concentration, and temperature. As with mechanical vibration, the soil volume sampled is not known and therefore this method is best suited for qualitative or comparative sampling. However, a commercially available electrical sampler ("octet" device developed by Thielemann [1986]) was evaluated by Schmidt (2001) and found to be highly effective for quantitative sampling of lumbricid species in pastures. Electrical extraction methods are potentially very dangerous and should only be used with extreme caution.

Two earthworm-trapping techniques have been described. Pitfall traps (open-top containers buried level with the soil surface and containing a fixative solution such as picric acid) may be useful for sampling surface-active species in diurnal or seasonal studies. Arrays of traps are installed and sampled at 12-hour, 24-hour, or longer intervals. Baited traps, such as perforated clay pots containing manure or other attractants and inserted into the soil, may also be useful for collecting certain species. As with other behavioral methods, trapping is probably highly selective and best suited for qualitative or comparative sampling.

Mark, release, and recapture techniques have been widely used to study population dynamics of animals including earthworms. Large numbers of individuals of desired species are collected, marked (e.g., with brands or nontoxic dyes), and released into the population of interest. Sampling over time and distance from the target site, and enumeration of tagged relative to untagged individuals, yields information on dispersal, mortality, and population density. Radioisotope and, more recently, immunofluorescent antibody techniques have been employed in earthworm mark–recapture studies.

Indirect Techniques

For earthworm species that cast on the soil surface, such as *Aporrectodea longa,* numbers and identity of castings may be a useful index of population activity. Because casting is dependent on soil temperature and moisture, this technique is highly variable and not a quantitative estimate of population density.

In summary, digging and hand sorting or washing are probably the most reliable means of sampling earthworms. However, no single method will be adequate to sample earthworm populations in all situations. Combinations of methods will probably achieve reasonable results. For example, formalin or mustard solution can be applied to the bottom of pits previously excavated for hand sorting, to extract deep burrowing anecic forms not sampled by digging (Edwards and Bohlen, 1996). Combinations of various methods may be useful in other situations.

Identification of Earthworms

A useful taxonomic key to lumbricid species is found in Schwert (1990). More detailed keys to non-lumbricids (based on interval anatomy) are Fender (1990), James (1990), and Sims and Gerard (1985).

SAMPLING AND ENUMERATION OF ENCHYTRAEIDS

Principle

Because of their much smaller size than earthworms, enchytraeids are not effectively sampled by hand sorting from soil. Instead, behavioral methods such as heat extraction are often employed. Van Vliet (2000) reviews methods for sampling and extracting enchytraeids.

Collection of Enchytraeids

Quantitative sampling of enchytraeids is often done with cylindrical core samplers that keep the soil intact. Optimum sampler size is 5–7.5 cm in diameter (van Vliet, 2000). Because enchytraeids show clumped distributions, sufficient replicates need to be taken to estimate population density and composition. However, the size and number of the sampling units is mostly chosen as a compromise between accuracy of the abundance estimates and the amount of work involved (Didden, 1993).

The most commonly used method to extract enchytraeids from soil is the wet funnel method (similar to the Baermann funnel used for nematodes, described previously), using an extractor such as that shown in Figure 9.6 (O'Connor, 1955). In this method, a thin soil sample is placed on a sieve in a funnel filled with water and exposed to light and heat. After about 3 hours, the light intensity is increased gradually until the soil surface has reached a temperature of 45°C. Enchytraeids respond by moving downward away from the heat and pass through the sieve

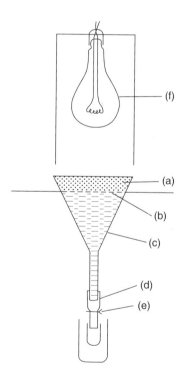

FIGURE 9.6.   Modified Baermann funnel to extract enchytraeids: (a) soil sample, (b) wire-gauze sieve, (c) funnel, (d) rubber tube, (e) funnel outlet with spring clip, (f) 25-watt light bulb (from Gorny and Grüm, 1993).

into the water below. A modified extraction method was described by Graefe (1973) and Schauermann (1983) in which enchytraeids are extracted from soil without heat. Extraction time is extended to several days for soils rich in organic matter and up to 2 weeks for mineral soils. The time of extraction is limited by the possibility of an oxygen deficit in the water, which may kill larger organisms.

Didden *et al.* (1995) compared the two methods and found more efficient extraction without heat. The length of the cold extraction period and the total extraction time had a significant positive influence on the extraction efficiency. The Graefe (1973) method is less expensive and easier to set up, but the long extraction time makes it more difficult to handle large numbers of samples. The O'Connor (1955) method is faster and can be modified with a longer initialization period before heat is applied.

Identification of Enchytraeids

Taxonomic identification of enchytraeids is difficult. The key by Dash (1990) is useful and more detailed monographs can be found in Nielsen and Christensen (1959, 1961, 1963).

References

Aber, J., McDowell, W., Nadelhoffer, K., Magill, A., Berntson, G., Kamakea, M., McNulty, S., Currie, W., Rustad, L., and Fernandez, I. (1998). Nitrogen saturation in temperate forest ecosystems. Hypotheses Revisited. *BioScience* 48, 921–934.

Abuzinadah, R. A., and Read, D. J. (1989). The role of proteins in the nitrogen nutrition of ectomycorrhizal plants. V. Nitrogen transfer in birch (*Betula pendula*) grown in association with mycorrhizal and non-mycorrhizal fungi. *New Phytol.* 112, 61–68.

Adejuyigbe, C. O., Tian, G., and Adeoye, G. O. (1999). Soil microarthropod populations under natural and planted fallows in southwestern Nigeria. *Agroforestry Systems* 47, 263–272.

Adl, S. M. (2003). "The Ecology of Soil Decomposition." CABI Publishing, Wallingford, U.K.

Albers, B. P., Beese, F., and Hartmann, A. (1995). Flow-microcalorimetry measurements of aerobic and anaerobic soil microbial activity. *Biol. Fertil. Soils* 19, 203–208.

Aldrete, A. N. G. (1990). Insecta: Psocoptera. *In* "Soil Biology Guide" (D. L. Dindal, ed.), pp. 1033–1052. Wiley, New York.

Alef, K. (1995). Heat output. *In* "Methods in Applied Soil Microbiology and Biochemistry" (K. Alef and P. Nannipieri, eds.), pp. 223–224. Academic Press, London.

Alef, K., and Nannipieri, P. (1995a). Microbial biomass. *In* "Methods in Applied Soil Microbiology and Biochemistry" (K. Alef and P. Nannipieri, eds.), pp. 375–417. Academic Press, London.

Alef, K., and Nannipieri, P. (1995b). Chapter 7, Enzyme activities. *In* "Methods in Applied Soil Microbiology and Biochemistry" (K. Alef and P. Nannipieri, eds.), pp. 311–373. Academic Press, London.

Allen, E. B., Allen, M. F., Helm, D. Y., Trappe, J. M., Molina, R., and Rincon, E. (1995). Patterns and regulation of mycorrhizal plant and fungal diversity. *Plant & Soil* 70, 47–62.

Allen, M. F. (1991). "The Ecology of Mycorrhizae." Cambridge University Press, Cambridge.

Allen, M. F. (ed.) (1992). "Mycorrhizal Functioning: An Integrative Plant–Fungal Process." Chapman and Hall, London.

Allen, T. F. H., and Starr, T. B. (1982). "Hierarchy. Perspectives for Ecological Complexity." University of Chicago Press, Chicago.

Allen-Morley, C. R., and Coleman, D. C. (1989). Resilience of soil biota in various food webs to freezing perturbations. *Ecology* 70, 1127–1141.

Alphei, J., Bonkowski, M., and Scheu, S. (1995). Application of the selective inhibition method to determine bacterial: fungal ratios in three beechwood soils rich in carbon-optimization of inhibitor concentrations. *Biol. Fertil. Soils* 19, 173–176.

Anderson, D. W. (1988). The effect of parent material and soil development on nutrient cycling in temperate ecosystems. *Biogeochemistry* 5, 71–97.

Anderson, D. W., and Coleman, D. C. (1985). The dynamics of organic matter in grassland soils. *J. Soil Water Conserv.* 40, 211–216.

Anderson, J. M. (1973). Carbon dioxide evolution from two temperate, deciduous woodland soils. *J. Appl. Ecol.* 10, 361–378.

Anderson, J. M. (1975). Succession, diversity and trophic relationships of some soil animals in decomposing leaf litter. *J. Anim. Ecol.* 44, 475–495.

Anderson, J. M. (1975). The enigma of soil animal species diversity. In "Progress in Soil Zoology" (J. Vanek, ed.), pp. 51–58. Academia, Prague.

Anderson, J. M. (1992). Responses of soils to climate change. *Adv. Ecol. Res.* 22, 163–210.

Anderson, J. M. (2000). Food web functioning and ecosystem processes: problems and perceptions of scaling. In "Invertebrates as Webmasters in Ecosystems" (D. C. Coleman and P. F. Hendrix, eds.), pp. 3–24. CAB International, Wallingford, U.K.

Anderson, J. M., Huish, S. A., Ineson, P., Leonard, M. A., and Splatt, P. R. (1985). Interactions of invertebrates, micro-organisms, and tree roots in nitrogen and mineral element fluxes in deciduous woodland soils. In "Ecological Interactions in Soil: Plants, Microbes and Animals." (A. H. Fitter, D. Atkinson, D. J. Read, and M. B. Usher, eds.), pp. 377–392. Blackwell Scientific, Oxford.

Anderson, J. P. E., and Domsch, K. H. (1978). A physiological method for the quantitative measurement of microbial biomass in soils. *Soil Biol. Biochem.* 10, 215–221.

Anderson, R. V., Coleman, D. C., and Cole, C. V. (1981). Effects of saprotrophic grazing on net mineralization. Terrestrial nitrogen cycles. *Ecol. Bull. (Stockholm)* 33, 201–216.

Anderson, R. V., Coleman, D. C., Cole, C. V., and Elliott, E. T. (1981b). Effect of the nematodes *Acrobeloides* sp. and *Mesodiplogaster Iheritieri* on substrate utilization and nitrogen and phosphorus mineralization in soil. *Ecology* 62, 549–555.

Anderson, R. V., Gould, W. D., Woods, L. E., Cambardella, C., Ingham, R. E., and Coleman, D. C. (1983). Organic and inorganic nitrogenous losses by microbivorous nematodes in soil. *Oikos* 40, 75–80.

Anderson, T.-H., and Domsch, K. H. (1993). The metabolic quotient for CO_2 (qCO_2) as a specific-activity parameter to assess the effects of environmental conditions, such as pH, on the microbial biomass of forest soils. *Soil Biol. Biochem.* 25, 393–395.

Anderson, T.-H. (1994). Physiological analysis of microbial communities in soil: applications and limitations. In "Beyond the Biomass" (K. Ritz, J. Dighton, and K. E. Giller, eds.), pp. 67–76. Wiley-Sayce, Chichester.

Andrade, G., Linderman, R. G., and Bethlenfalvay, G. J. (1998). Bacterial associations with the mycorrhizosphere and hyphosphere of the arbuscular mycorrhizal fungus *Glomus mosseae*. *Plant & Soil* 202, 79–87.

André, H. M., Ducarme, X., Anderson, J., Crossley, D. A., Jr., Koehler, H., Paoletti, M., Walter, D., and Lebrun, P. (2001). Skilled eyes are needed to go on studying the richness for the soil. *Nature* 409, 761.

André, H. M., Ducarme, X., and Lebrun, P. (2002). Soil biodiversity: myth, reality or conning? *Oikos* 96, 3–24.

Andrén, O., Lindberg, T., Paustian, K., and Rosswall, T. (eds.) (1990). "Ecology of Arable Land. Organisms, Carbon and Nitrogen Cycling." Munksgaard International, Copenhagen.

Andrén, O., Brussaard, L., and Clarholm, M. (1999). Soil organism influence on ecosystem-level process—bypassing the ecological hierarchy. *Appl. Soil Ecol.* 11, 177–188.

Andrew, J. A., Harrison, K. G., Matamala, R., and Schlesinger, W. H. (1999). Separation of root respiration from soil respiration using carbon-13 labeling during free-air carbon dioxide enrichment (FACE). *Soil Sci. Soc. Am. J.* 63, 1429–1435.

Andrews, S. S., and Carroll, C. R. (2001). Designing a soil quality assessment tool for sustainable agroecosystem management. *Ecol. Appls.* 11, 1573–1585.

Appelhoff, M., Fenton, M. F., and Harris, B. L. (1993). "Worms Eat Our Garbage: Classroom Activities for a Better Environment." Flowerfield Press, Kalamazoo, MI.

Arlian, L. G., and Woolley, T. A. (1970). Observations on the biology of *Liacarus cidarus* (Acari: Cryptostigmata, Liacaridae). *J. Kansas Ent. Soc.* 43, 297–301.

Arnett, R. H., Jr. (ed.) (1993). "American Insects: A Handbook of the Insects of America North of Mexico." Sandhill Crane Press, Inc., Gainesville, Florida.

Arrhenius, S. (1896). On the influence of carbonic acid in the air upon the temperature of the ground. *Philos. Mag.* 41, 237–257.

Artursson, V., and Jansson, J. K. (2003). Use of bromodeoxyuridine immunocapture to identify active bacteria associated with arbuscular mycorrhizal hyphae. *Appl. Environ. Microbiol.* 69, 6208–6215.

Auerbach, S. I. (1958). The soil ecosystem and radioactive waste disposal to the ground. *Ecology* 39, 522–529.

Azcón-Aguilar, C., and Barea, J. M. (1995). Arbuscular mycorrhizas and biological control of soil-borne plant pathogens—an overview of the mechanisms involved. *Mycorrhiza* 6, 457–464.

Bååth, E., Lohm, U., Lundgren, B., Rosswall, T., Soderstrom, B., and Sohlenius, B. (1981). Impact of microbial-feeding animals on total soil activity and nitrogen dynamics: a soil microcosm experiment. *Oikos* 37, 257–264.

Baker, D. D., and Schwintzer, C. R. (1990). Introduction. *In* "The Biology of *Frankia* and Actinorhizal Plants." pp. 1–13. Academic Press, San Diego.

Baker, E. W., Camin, J. H., Cunliffe, E., Woolley, T. A., and Yunker, C. E. (1958). "Guide to the Families of Mites." Institute of Acarology, University of Maryland, College Park.

Baker, G. H. (1998). The ecology, management and benefits of earthworms in agricultural soils, with particular reference to Southern Australia. *In* "Earthworm Ecology" (C. A. Edwards, ed.), pp. 229–258. Lewis Publishers, Boca Raton, FL.

Bal, L. (1982). "Zoological ripening of soils." Pudoc, Wageningen.

Baldock, J. A. (2002). Interactions of organic materials and microorganisms with minerals in the stabilization of soil structure. *In* "Interactions between Soil Particles and Microorganisms" (P. M. Huang, J.-M. Bollag, and N. Senesi, eds.), pp. 85–131. Wiley, Chichester.

Bamforth, S. S. (1980). Terrestrial protozoa. *J. Protozool.* 27, 33–36.

Bamforth, S. S. (1997). Protozoa: recyclers and indicators of agroecosystem quality. *In* "Fauna in Soil Ecosystems" (G. Benckiser, ed.), pp. 63–84. Marcel Dekker, New York.

Banerjee, B. (1970). A mathematical model on sampling diplopods using pitfall traps. *Oecologia* 4, 102–105.

Barber, D. A., and Martin, J. K. (1976). The release of organic substances by cereal roots into soil. *New Phytol.* 76, 69–80.

Barker, G. M. (2001). Gastropods on land: phylogeny, diversity and adaptive morphology. *In* "The Biology of Terrestrial Molluscs" (G. M. Barker, ed.), pp. 1–146. CAB International, Wallingford, U.K.

Barois, I. (1999). Ecology of earthworm species with large environmental tolerance and/or extended distributions. *In* "Earthworm Management in Tropical Agroecosystems" (P. Lavelle, L. Brussaard, and P. Hendrix, eds.), pp. 57–85. CABI Publishing, New York.

Barra, J. A., and Christiansen, K. (1975). Experimental study of aggregation during the development of *Pseudosinella impediens* (Collembola, Entomobryidae). *Pedobiologia* 15, 343–347.

Bater, J. E. (1996). Micro- and macro-arthropods. *In* "Methods for the Examination of Organismal Diversity in Soils and Sediments" (G. S. Hall, ed.), pp. 163–174. CAB International, New York.

Battley, E. H. (1987). "Energetics of Microbial Growth." Wiley-Interscience, New York.

Beare, M. H., Neely, C. L., Coleman, D. C., and Hargrove, W. L. (1990). A substrate-induced respiration (SIR) method for measurement of fungal and bacterial biomass on plant residues. *Soil Biol. Biochem.* 22, 585–594.

Beare, M. H., Neely, C. L., Coleman, D. C., and Hargrove, W. L. (1991). Characterizations of a substrate-induced respiration method for measuring fungal, bacterial and total microbial biomass on plant residues. *Agric. Ecosys. Environ.* 34, 65–73.

Beare, M. H., Parmelee, R. W., Hendrix, P. F., Cheng, W., Coleman, D. C., and Crossley, D. A., Jr. (1992). Microbial and faunal interactions and effects on litter nitrogen and decomposition in agroecosystems. *Ecol. Monogr.* 62, 569–591.

Beare, M. H., Hendrix, P. F., and Coleman, D. C. (1994a). Water-stable aggregates and organic matter fractions in conventional and no-tillage soils. *Soil Sci. Soc. Am. J.* 58, 777–786.

Beare, M. H., Cabrera, M. L., Hendrix, P. F., and Coleman, D. C. (1994b). Aggregate-protected and unprotected pools of organic matter in conventional and no-tillage Ultisols. *Soil Sci. Soc. Am. J.* 58, 787–795.

Behan, V. M., and Hill, S. B. (1978). Feeding habits and spore dispersal of oribatid mites in the North American arctic. *Rev. Écol. Biol. Sol* 15, 497–516.

Behan-Pelletier, V. M., and Hill, S. B. (1983). Feeding habits of sixteen species of Oribatei (Acari) from an acid peat bog, Glenamoy, Ireland. *Rev. Écol. Biol. Sol* 20, 221–267.

Behan-Pelletier, V. M., and Norton, R. A. (1983). Epidamaeus (Acari: Damaeidae) of arctic western North America and extreme northeast U.S.S.R. *Can. Entomol.* 115, 1253–1289.

Behan-Pelletier, V. M. (1993). Eremaeidae (Acari:Oribatida) of North America. Memoirs. Ent. Soc. Canada, No. 168, 193 pp.

Behan-Pelletier, V. M., and Bissett, B. (1993). Biodiversity of nearctic soil arthropods. *Can. Biodiversity* 2, 5–14.

Behan-Pelletier, V., and Newton, G. (1999). Linking soil biodiversity and ecosystem function—the taxonomic dilemma. *BioScience* 49, 149–215.

Behan-Pelletier, V. M., and Walter, D. E. (2000). Biodiversity of oribatid mites (Acari: Oribatida) in tree canopies and litter. *In* "Invertebrates as Webmasters in Ecosystems" (D. C. Coleman and P. F. Hendrix, eds.), pp. 187–202. CABI Publishing, Wallingford, U.K.

Behre, G. F. (1987). Die Sieb–Flotations–Methode. Bau und Erprobung eines Ökologischen Arbeitsgerätes zur mechanischen Auslese von Bodenarthropoden. *Ber. Naturw. Ver. Wuppertal* 40, 52–55.

Belnap, J. (2002). Nitrogen fixation in biological soil crusts from Southeastern Utah, USA. *Biol. Fertil. Soils* 35, 128–135.

Belnap, J., and Phillips, S. L. (2001). Soil biota in an ungrazed grassland: response to annual grass (*Bromus tectorum*) invasion. *Ecol. Appls.* 11, 1261–1275.

Bending, G. D., and Read, D. J. (1996). Nitrogen mobilization from protein–polyphenol complex by ericoid and ectomycorrhizal fungi. *Soil Biol. Biochem.* 28, 1603–1612.

Bengtsson, G., Hedlund, K., and Rundgren, S. (1994). Food- and density-dependent dispersal: evidence from a soil Collembolan. *J. Anim. Ecol.* 63, 513–520.

Bengtsson, J. (1998). Which species? What kind of diversity? Which ecosystem function? Some problems in studies of relations between biodiversity and ecosystem function. *Appl. Soil Ecol.* 10, 191–199.

Bentham, H., Harris, J. A., Birch, P., and Short, K. C. (1992). Habitat classification and soil restoration assessment using analysis of soil microbiological and physico-chemical characteristics. *J. Appl. Ecol.* 29, 711–718.

Berg, B. (1986). Nutrient release from litter and humus in coniferous forest soils—a mini review. *Scand. J. For. Res.* 1, 359–369.

Berg, B., and Staaf, H. (1981). Leaching, accumulation and release of nitrogen in decomposing forest litter. *Ecol. Bull. Stockholm* 33, 163–178.

Bernard, E. C. (1985). Two new species of *Protura* (Insecta) from North America. *Proc. Biol. Soc. Washington* 98, 72–80.

Bever, J. D. (1999). Dynamics within mutualism and the maintenance of diversity: inference from a model of interguild frequency dependence. *Ecology Letters* 2, 52–61.

Bever, J. D., Schultz, P. A., Pringle, A., and Morton, J. B. (2001). Arbuscular mycorrhizal fungi: more diverse than meets the eye, and the ecological tale of why. *BioScience* 51, 923–931.

Bignell, D. E. (1984). The arthropod gut as an environment for microorganisms. *In* "Invertebrate–Microbial Interactions" (J. M. Anderson, A. D. M. Rayner, and D. W. H. Walton, eds.), pp. 205–227. Cambridge University Press, Cambridge.

Bignell, D. E. (2000). Introduction to symbiosis. *In* "Termites: Evolution, Sociality, Symbioses, Ecology" (T. Abe, D. E. Bignell, and M. Higashi, eds.), pp. 189–208. Kluwer Academic, Dordrecht, The Netherlands.

Bignell, D. E., and Eggleton, P. (2000). Termites in ecosystems. *In* "Termites: Evolution, Sociality, Symbioses, Ecology" (T. Abe, D. E. Bignell, and M. Higashi, eds.), pp. 363–387. Kluwer Academic, Dordrecht, The Netherlands.

Binkley, D. (2002). Ten-year decomposition in a loblolly pine forest. *Can. J. For. Res.* 32, 2231–2235.

Bintrim, S., Donohue, T., Handelsman, J., Roberts, G., and Goodman, R. (1997). Molecular phylogeny of Archaea from soil. *Proc. Nat. Acad. Sci. USA* 94, 277–282.

Blair, J. M. (1988a). Nitrogen, sulfur and phosphorus dynamics in decomposing deciduous leaf litter in the southern Appalachians. *Soil Biol. Biochem.* 20, 693–701.

Blair, J. M. (1988b). Nutrient release from decomposing foliar litter of three tree species with special reference to calcium, magnesium and potassium dynamics. *Plant & Soil* 110, 49–55.

Blair, J. M., and Crossley, D. A., Jr. (1988). Litter decomposition, nitrogen dynamics and litter microarthropods in a southern Appalachian hardwood forest 8 years following clearcutting. *J. Appl. Ecol.* 25, 683–698.

Blair, J. M., Crossley, D. A., Jr., and Callaham, L. C. (1992). Effects of litter quality and microarthropods on N dynamics and retention of exogenous ^{15}N in decomposing litter. *Biol. Fertil. Soils* 12, 241–252.

Blair, W. F. (1977). "Big Biology: The US/IBP." Dowden, Hutchinson and Ross, Stroudsburg, PA.

Blakemore, R. J. (2002). "Cosmopolitan Earthworms: An Eco-taxonomic Guide to the Peregrine Species of the World" (first CD edition). VermEcology, PO Box 414 Kippax, ACT 2615, Australia (426 pp. and 80 figs.).

Bloem, J., de Ruiter, P., and Bouwman, L. (1997). Soil food webs and nutrient cycling in agroecosystems. *In* "Modern Soil Microbiology" (J. D. Van Elsas, J. T. Trevors, and E. M. H. Wellington, eds.), pp. 245–278. Marcel Dekker, New York.

Blum, M. S., and Edgar, A. L. (1971). 4-Methyl-3-heptanone: identification and role in opilionid exocrine secretions. *Insect Biochem.* 1, 181–188.

Blumberg, A. Y., and Crossley, D. A., Jr. (1983). Comparison of soil surface arthropod populations in conventional tillage, no-tillage and old field systems. *Agroecosystems* 8, 247–253.

Boag, B., and Yeates, G. W. (2001). The potential impact of the New Zealand flatworm, a predator of earthworms, in Western Europe. *Ecol. Appls.* 11, 1276–1286.

Bocock, K. L., and Gilbert, O. J. W. (1957). The disappearance of leaf litter under different woodland conditions. *Plant & Soil* 9, 179–185.

Bohlen, P. J., Edwards, C. A., Zhang, Q., Parmelee, R. W., and Allen, M. (2002). Indirect effects of earthworms on microbial assimilation of labile carbon. *Appl. Soil Ecol.* 20, 255–261.

Böhm, W. (1979). "Methods of Studying Root Systems." Springer, Berlin.

Bolger, T. M., Heneghan, L. J., and Neville, P. (2000). Invertebrates and nutrient cycling in coniferous forest ecosystems: spatial heterogeneity and conditionality. *In* "Invertebrates as Webmasters in Ecosystems" (D. C. Coleman and P. F. Hendrix, eds.), pp. 161–184. CABI Publishing, Wallingford, U.K.

Bolger, T. (2001). The functional value of species biodiversity—a review. *Proc. Irish Roy. Acad.* 101B, 199–204.

Bolton, B. (1994). "Identification Guide to the Ant Genera of the World." Harvard University Press, Cambridge.

Bomberg, M., Jurgens, G., Saana, A., Sen, R., and Timonen, S. (2003). Nested PCR detection of Archaea in defined compartments of pine mycorrhizospheres developed in boreal forest humus microcosms. *FEMS Microbiol. Ecol.* 43, 163–171.

Bongers, T. (1990). The maturity index: an ecological measure of environmental disturbance based on nematode species composition. *Oecologia* 83, 14–19.

Bonkowski, M., Cheng, W., Griffiths, B. S., Alphei, J., and Scheu, S. (2000). Microbial–faunal interactions in the rhizosphere and effects on plant growth. *Eur. J. Soil Biol.* 36, 135–147.

Bonkowski, M., Griffiths, B., and Scrimgeour, C. (2000). Substrate heterogeneity and microfauna in soil organic "hotspots" as determinants of nitrogen capture and growth of ryegrass. *Appl. Soil Ecol.* 14, 37–53.

Bornebusch, C. H. (1930). "The Fauna of Forest Soil." Forst. ForsVaes. Danm. Copenhagen.

Borneman, J., and Triplett, E. W. (1997). Molecular microbial diversity in soils from Eastern Amazonia: evidence for unusual microorganisms and microbial population shifts associated with deforestation. *Appl. Environ. Microbiol.* 63, 2647–2653.

Borror, D. J., DeLong, D. M., and Triplehorn, C. A. (1981). "An Introduction to the Study of Insects" (5th edition). Saunders College Publishing, Philadelphia.

Borror, D. J., Triplehorn, C. A., and Johnson, N. F. (1989). "An Introduction to the Study of Insects" (6th edition). Saunders, Philadelphia.

Bouché, M. B., and Beugnot, M. (1972). Contribution à l'approache méthodologique de l'étude des Biocenoses. II. L'extraction des macroéléments du sol par lavage-tamisage. *Ann. Zool. Ecol. Anim.* 4, 537–544.

Bouché, M. B. (1975). Action de la faune sur les etats de la matiére organique dans les ecosystemes. *In* "Biodégradation et Humification" (G. Kilbertus, O. Reisinger, A. Mourey, and J. A. Cancela da Fonseca, eds.), pp. 157–168. Pierron, Sarrugemines.

Bouché, M. B. (1977). Stratégies lombriciennes. In: Soil organisms as components of ecosystems. (U. Lohm and T. Persson, eds.). *Ecol. Bull. (Stockholm)* 25, 122–133.

Bouché, M. B. (1983). The establishment of earthworm communities. In: "Earthworm Ecology from Darwin to Vermiculture" (J. E. Satchell, ed.), pp. 431–448. Chapman and Hall, London.

Boudreaux, H. B. (1970). "Arthropod Phylogeny with Special Reference to Insects." Wiley, New York.

Boutton, T. W., and Yamasaki, S. (eds.) (1996). "Mass Spectrometry of Soils." Marcel Dekker, New York.

Bouwman, A. F. (ed.) (1990). "Soils and the Greenhouse Effect." Wiley, Chichester.

Bowen, H. J. M. (1979). "Environmental Chemistry of the Elements." Academic Press, London, New York.

Box, J. E., Jr., and Johnson, J. W. (1987). Minirhizotron rooting comparisons of three wheat cultivars. *In* "Minirhizotron Observation Tubes: Methods and Applications for Measuring Rhizosphere Dynamics" (H. M. Taylor, ed.), pp. 123–130. Am. Soc. Agron. Special Publ. No. 50, Madison, Wisconsin.

Box, J. E., Jr., and Hammond, L. C. (1990). "Rhizosphere Dynamics." Westview Press, Boulder, Colorado.

Bradford, M. A., Jones, T. H., Bardgett, R. D., Black, H. I. J., Boag, B., Bonkowski, M., Cook, R., Eggers, T., Gange, A. C., Grayston, S. J., Kandeler, E., McCaig, A. E., Newington, J. E., Prosser, J. I., Setälä, H., Staddon, P. L., Tordoff, G. M., Tscherko, D., and Lawton, J. H. (2002). Impacts of soil faunal community composition on model grassland ecosystems. *Science* 298, 615–618.

Brady, N. C. (1974). "The Nature and Properties of Soils" (8th edition). MacMillan, New York.

Brady, N. C., and Weil, R. R. (2000). "Elements of the Nature and Properties of Soils." Prentice-Hall, Upper Saddle River, New Jersey.

Brauman, A., Bignell, D. E., and Tayasu, I. (2000). Soil-feeding termites: biology, microbial associations and digestive mechanisms. *In* "Termites: Evolution, Sociality, Symbioses, Ecology" (T. Abe, D. E. Bignell, and M. Higashi, eds.), pp. 233–259. Kluwer Academic, Dordrecht, The Netherlands.

Breznak, J. A. (1984). Biochemical aspects of symbiosis between termites and their intestinal microbiota. *In* "Invertebrate–Microbial Interactions" (J. M. Anderson, A. D. M. Rayner, and D. W. H. Walton, eds.), pp. 173–203. Cambridge University Press, Cambridge.

Breznak, J. (2000). Ecology of prokaryotic microbes in the guts of wood- and litter-feeding termites. *In* "Termites: Evolution, Sociality, Symbioses, Ecology" (T. Abe, D. E. Bignell, and M. Higashi, eds.), pp. 209–231. Kluwer Academic, Dordrecht, The Netherlands.

Brinkhurst, R. O., and Cook, D. G. (eds.) (1980). "Aquatic Oligochaete Biology." Plenum Press, New York.

Brockmeyer, V., Schmid, R., and Westheide, W. (1990). Quantitative investigations of the food of two terrestrial enchytraeid species (Oligochaeta). *Pedobiologia* 34, 151–184.

Brown, G. G. (1995). How do earthworms affect microfloral and faunal community diversity? *Plant & Soil* 170, 247–269.

Bruno, J. F., Stachowicz, J. J., and Bertness, M. D. (2003). Inclusion of facilitation into ecological theory. *Trends Ecol. Evol.* 18, 119–125.

Brussaard, L., Bouwman, L. A., Geurs, M., Hassink, J., and Zwart, K. B. (1990). Biomass, composition and temporal dynamics of soil organisms of a silt loam soil under conventional and integrated management. *Neth. J. Agric. Sci.* 38, 283–302.

Brussaard, L., and Kooistra, M. J. (eds.) (1993). "Soil Structure/Soil Biota Interrelationships." Elsevier, Amsterdam.

Bryant, R. J., Woods, L. E., Coleman, D. C., Fairbanks, B. C., McClellan, J. F., and Cole, C. V. (1982). Interactions of bacterial and amoebal populations in soil microcosms with fluctuating moisture content. *Appl. Environ. Microbiol.* 43, 7447–752.

Buckley, D., Graber, J., and Schmidt, T. (1998). Phylogenetic analysis of nonthermophilic members of the kingdom Crenarchaeota and their diversity and abundance in soils. *Appl. Environ. Microbiol.* 64, 4333–4339.

Buckman, H. O., and Brady, N. C. (1970). "The Nature and Properties of Soils" (7th edition). Macmillan, New York.

Bundt, M., Widmer, F., Pesaro, M., Zeyer, J., and Blaser, P. (2001). Preferential flow paths: biological "hot spots" in soils. *Soil Biol. Biochem.* 33, 729–738.

Burch, J. B., and Pearce, T. A. (1990). Terrestrial Gastropoda. *In* "Soil Biology Guide" (D. Dindal, ed.), pp. 201–309. Wiley, New York.

Butcher, J. W., Snider, R., and Snider, R. J. (1971). Bioecology of edaphic Collembola and Acarina. *Ann. Rev. Entomol.* 16, 249–288.

Byers, J. E., and Noonburg, E. G. (2003). Scale dependent effects of biotic resistance to biological invasion. *Ecology* 84, 1428–1433.

Byers, R. A., Barratt, B. I. P., and Calvin, D. (1989). Comparison between defined-area traps and refuge traps for sampling slugs in conservation-tillage crop environments. *In* "Slugs and Snails in World Agriculture" (I. Henderson, ed.), pp. 187–192. British Crop Protection Council Monograph No. 41, Thornton Heath, U.K.

Cabrera, M. L., and Beare, M. H. (1993). Alkaline persulfate oxidation for determining total nitrogen in microbial biomass extracts. *Soil Sci. Soc. Am. J.* 57, 1007–1012.

Caldwell, M. M., and Camp, L. B. (1974). Belowground productivity of two cool desert communities. *Oecologia* 17, 123–130.

Callaham, M. A., Jr., and Hendrix, P. F. (1997). Relative abundance and seasonal activity of earthworms (Lumbricidae and Megascolecidae) as determined by hand-sorting and formalin extraction in forest soils on the southern Appalachian Piedmont. *Soil Biol. Biochem.* 29, 317–322.

Callaham, M. A., Jr., Whiles, M. R., Meyer, K. C., Brock, B. L. and Charlton, R. E. (2002). Feeding ecology and emergence production of annual cicadas (Homoptera, Cicadidae) in tallgrass prairie. *Oecologia* 123, 535–542.

Callaham, M. A., Jr., Whiles, M. R., and Blair, J. M. (2003). Annual fire, mowing and fertilization effects on two cicada species (Homoptera: Cicadidae) in tallgrass prairie. *Amer. Midl. Nat.* 148, 90–101.

Cambardella, C. A., and Elliott, E. T. (1994). Carbon and nitrogen dynamics of soil organic matter fractions from cultivated grassland soils. *Soil Sci. Soc. Am. J.* 58, 123–130.

Catts, E. P., and Haskell, N. H. (1990). "Entomology and Death: A Procedural Guide." Joyce's Print Shop, Clemson, South Carolina.

Chakraborty, S., Old, K. M., and Warcup, J. H. (1983). Amoebae from a take-all suppressive soil which feed on *Gaeumannomyces graminis tritici* and other soil fungi. *Soil Biol. Biochem.* 15, 17–24.

Chakraborty, S., and Warcup, J. H. (1983). Soil amoebae and saprophytic survival of *Gaeumannomyces graminis tritici* in a suppressive pasture soil. *Soil Biol. Biochem.* 15, 181–185.

Chalot, M., and Brun, A. (1998). Physiology of organic nitrogen acquisition by ectomycorrhizal fungi and ectomycorrhizas. *FEMS MicroBiol. Revs.* 22, 21–44.

Chapin, F. S., III (1980). The mineral nutrition of wild plants. *Ann. Rev. Ecol. System.* 11, 233–260.

Chapin, F. S., III, Zavaleta, E. S., Eviner, V. T., Naylor, R. L., Vitousek, P. M., Reynolds, H. L., Hooper, D. U., Lavorel, S., Sala, O. E., Hobbie, S. E., Mack, M. C., and Diaz, S. (2000). Consequences of changing biodiversity. *Nature* 405, 234–242.

Cheng, W. (1996). Measurement of rhizosphere respiration and organic matter decomposition using natural ^{13}C. *Plant & Soil* 183, 163–168.

Cheng, W., and Coleman, D. C. (1990). Effect of living roots on soil organic matter decomposition. *Soil Biol. Biochem.* 22, 781–787.

Cheng, W., Coleman, D. C., and Box, J. E., Jr. (1990). Root dynamics, production and distribution in agroecosystems on the Georgia Piedmont using minirhizotrons. *J. Appl. Ecol.* 27, 592–604.

Cheng, W., Coleman, D. C., Carroll, C., and Hoffman, C. A. (1993). In situ measurement of root respiration and soluble carbon concentrations in the rhizosphere. *Soil Biol. Biochem.* 25, 1189–1196.

Cheshire, M. (1979). "Soil Carbohydrates." Academic Press, London.

Cheshire, M. V., Sparling, G. P., and Mundie, C. M. (1984). Influence of soil type, crop and air drying on residual carbohydrate content and aggregate stability after treatment with periodate and tetraborate. *Plant & Soil* 76, 339–347.

Chiariello, N., Hickman, J. C., and Mooney, H. A. (1982). Endomycorrhizal role for interspecific transfer of phosphorus in a community of annual plants. *Science* 217, 941–943.

Christiansen, K. (1970). Experimental studies on the aggregation and dispersion of Collembola. *Pedobiologia* 10, 180–198.

Christiansen, K. A. (1990). Insecta: Collembola. *In* "Soil Biology Guide" (D. L. Dindal ed.), pp. 965–995. Wiley, New York.

Christiansen, K. A. (1992). Springtails. *Kansas School Nat.* 39, 1–16.

Christiansen, K. A., and Bellinger, P. F. (1980–81). "The Collembola of North America North of the Rio Grande." Grinnell College, Grinnell, Iowa.

Christiansen, K. A., and Bellinger, P. F. (1998). "The Collembola of North America North of the Rio Grande" (2nd edition). Grinnell College, Grinnell, Iowa.

Christiansen, K., Doyle, M., Kahlert, M., and Gobaleza, D. (1992). Interspecific interactions between collembolan populations in culture. *Pedobiologia* 36, 274–286.

Clarholm, M. (1981). Protozoan grazing of bacteria in soil-impact and importance. *Microb. Ecol.* 7, 343–350.

Clarholm, M. (1985). Possible roles for roots, bacteria, protozoa and fungi in supplying nitrogen to plants. *In* "Ecological Interactions in Soil: Plants, Microbes and Animals" (A. H. Fitter, D. Atkinson, D. J. Read, and M. B. Usher, eds.), pp. 355–365. Blackwell, Oxford.

Clarholm, M. (1994). The microbial loop in soil. *In* "Beyond the Biomass" (K. Ritz, J. Dighton, and K. E. Giller, eds.), pp. 221–230. Wiley-Sayce, Chichester.

Coineau, Y. (1974). "Introduction a l'Étude des microarthropodes du Sol et de ses Anexes. Documents pour l'Enseignement Practique de l'Écologie." Doin, Paris.

Cole, L. C. (1946). The cryptozoa of an Illinois woodland. *Ecol. Monogr.* 16, 49–86.

Coleman, D. C. (1973). Soil carbon balance in a successional grassland. *Oikos* 24, 195–199.

Coleman, D. C. (1976). A review of root production processes and their influence on soil biota in terrestrial ecosystems. *In* "The role of terrestrial and aquatic organisms in decomposition processes" (J. M. Anderson and A. Macfadyen, eds.), pp. 417–434. Blackwell, Oxford.

Coleman, D. C. (1985). Through a ped darkly: an ecological assessment of root–soil–microbial–faunal interactions. *In* "Ecological Interactions in Soil: Plants, Microbes and Animals" (A. H. Fitter, D. Atkinson, D. J. Read, and M. B. Usher, eds.), pp. 1–21. Blackwell, Oxford.

Coleman, D. C. (1994a). Compositional analysis of microbial communities. *In* "Beyond the Biomass" (K. Ritz, J. Dighton, and K. Giller, eds.), pp. 201–220. Wiley-Sayce, Chichester.

Coleman, D. C. (1994b). The microbial loop concept as used in terrestrial soil ecology studies. *Microb. Ecol.* 28, 245–250.

Coleman, D. C. (2001). Soil Biota, Soil Systems and Processes. *In* "Encyclopedia of Biodiversity," vol. 5 (S. Levin, ed.), pp. 305–314. Academic Press, San Diego, CA.

Coleman, D. C., Andrews, R., Ellis, J. E., and Singh, J. S. (1976). Energy flow and partitioning in selected man-managed and natural ecosystems. *Agro-Ecosystems* 3, 45–54.

Coleman, D. C., Cole, C. V., Anderson, R. V., Blaha, M., Campion, M. K., Clarholm, M., Elliott, E. T., Hunt, H. W., Schaefer, B., and Sinclair, J. (1977). Analysis of rhizosphere–saprophage interactions in terrestrial ecosystems. *In* "Soil Organisms as Components of Ecosystems" (U. Lohm and T. Persson, eds.), pp. 299–309. Ecol. Bull. 25, Stockholm.

Coleman, D. C., and Sasson, A. (1980). Decomposers subsystem, Chapter 7. *In* "Grasslands, Systems Analysis, and Man. IBP Synthesis, vol. 19" (A. Breymeyer and G. van Dyne, eds.), pp. 609–655. Cambridge University Press, London.

Coleman, D. C., Reid, C. P. P., and Cole, C. V. (1983). Biological strategies of nutrient cycling in soil systems. *Adv. Ecol. Res.* 13, 1–55.

Coleman, D. C., Cole, C. V., and Elliott, E. T. (1984). Decomposition, organic matter turnover and nutrient dynamics in agroecosystems. *In* "Agricultural Ecosystems—Unifying Concepts" (R. Lowrance, B. R. Stinner, and G. J. House, eds.), Wiley-Interscience, New York.

Coleman, D. C., Oades, J. M., and Uehara, G. (eds.) (1989). "Dynamics of Soil Organic Matter in Tropical Ecosystems." University of Hawaii Press, Honolulu.

Coleman, D. C., Ingham, E. R., Hunt, H. W., Elliott, E. T., Reid, C. P. P., and Moore, J. C. (1990). Seasonal and faunal effects on decomposition in semiarid prairie, meadow and lodgepole pine forest. *Pedobiologia* 34, 207–219.

Coleman, D. C., Odum, E. P., and Crossley, D. A., Jr. (1992). Soil biology, soil ecology, and global change. *Biol. Fertil. Soils* 14, 104–111.

Coleman, D. C., Hendrix, P. F., Beare, M. H., Cheng, W., and Crossley, D. A., Jr. (1993). Microbial and faunal dynamics as they affect soil organic matter dynamics in subtropical Agroecosystems. *In* "Soil Biota and Nutrient Cycling Farming Systems" (M. G. Paoletti, W. Foissner, and D. C. Coleman, eds.), pp. 1–14. Lewis Publishing Company, Chelsea, Michigan.

Coleman, D. C., and Schoute, J. F. T. (1993). Translation of soil features across levels of spatial resolution—introduction to round table discussion. *Geoderma* 57, 171–181.

Coleman, D. C., Hendrix, P. F., Beare, M. H., Crossley, D. A., Jr., Hu, S., and van Vliet, P. C. J. (1994a). The impacts of management and biota on nutrient dynamics and soil structure in sub-tropical agroecosystems: impacts on detritus food webs. *In* "Soil Biota Management in Sustainable Farming Systems" (C. E. Pankhurst, B. M. Doube, V. V. S. R. Gupta, and P. R. Grace, eds.), pp. 133–143. CSIRO, Melbourne, Australia.

Coleman, D. C., Dighton, J., Ritz, K., and Giller, K. E. (1994b). Perspectives on the compositional and functional analysis of soil communities. *In* "Beyond the Biomass" (K. Ritz, J. Dighton, and K. E. Giller, eds.), pp. 261–271. Wiley-Sayce, Chichester.

Coleman, D. C., Hendrix, P. F., and Odum, E. P. (1998). Ecosystem health: an overview. *In* "Soil Chemistry and Ecosystem Health" (P. M. Huang, ed.), pp. 1–20. Soil Science Society of America Special Publication No. 52, Madison, Wisconsin.

Coleman, D. C., Blair, J. M., Elliott, E. T., and Wall, D. H. (1999). Soil invertebrates. *In* "Standard Soil Methods For Long–Term Ecological Research" (G. P. Robertson, D. C. Coleman, C. S. Bledsoe, and P. Sollins, eds.), Oxford University Press, New York.

Coleman, D. C., and Hendrix, P. F. (eds.) (2000). "Invertebrates as Webmasters in Ecosystems." CABI Publishing, Wallingford, U.K.

Connell, J. H., and Lowman, M. D. (1989). Low diversity tropical rain forests: some possible mechanisms for their existence. *Am. Nat.* 122, 661–696.

Copeland, T. P., and Imadaté, G. (1990). Insecta: Protura. *In* "Soil Biology Guide" (D. L. Dindal, ed.) pp. 911–933. Wiley, New York.

Cornelissen, J. H. C., Aerts, R., Cerabolini, B., Werger, M. J. A., and van der Heijden, M. G. A. (2001). Carbon cycling traits of plant species are linked with mycorrhizal strategy. *Oecologia* 129, 611–619.

Costanza, R., d'Arge, R., de Groot, R., Farber, S., Grasso, M., Hannon, B., Limburg, K., Naeem, S., O'Neill, R. V., Paruelo, J., Raskin, R. G., Sutton, P., and van den Belt, M. (1997). The value of the world's ecosystem services and natural capital. *Nature* 387, 253–260.

Coûteaux, M.-M. (1972). Distribution des thécamoebiens de la litiére et de l'humus de deux sols forestier d'humus brut. *Pedobiologia* 12, 237–243.

Coûteaux, M.-M. (1985). Relation entre la densité apparente d'un humus et l'aptitude a la croissance de ses ciliés. *Pedobiologia* 28, 289–303.

Coûteaux, M.-M., Mousseau, M., Celerier, M. L., and Bottner, P. (1991). Increased atmospheric CO_2 and litter quality: decomposition of sweet chestnut leaf litter with animal food webs of different complexities. *Oikos* 61, 54–64.

Crawford, C. S. (1981). "Biology of desert invertebrates." Springer-Verlag, New York.

Crawford, C. S. (1990). Scorpiones, Solfugae, and associated desert taxa. *In* "Soil Biology Guide" (D. L. Dindal, ed.), pp. 421–475. Wiley, New York.

Crocker, R. L. (1952). Soil genesis and the pedogenic factors. *Quart. Rev. Biol.* 27, 139–168.

Cromack, K., Jr. (1973). "Litter production and litter decomposition in a mixed hard-wood watershed and in a white pine watershed at Coweeta Hydrologic Station, North Carolina." Ph.D. Dissertation, University of Georgia.

Cromack, K., Jr., Fichter, B. L., Moldenke, A. M., Entry, J. A., and Ingham, E. R. (1988). Interactions between soil animals and ectomycorrhizal fungal mats. *Agric. Ecosys. Environ.* 24, 161–168.

Crossley, D. A., Jr. (1960). Comparative external morphology and taxonomy of nymphs of the Trombiculidae (Acarina). *Univ. Kansas Sci. Bull.* 40, 135–321.

Crossley, D. A., Jr., and Hoglund, M. P. (1962). A litter-bag method for the study of microarthropods inhabiting leaf litter. *Ecology* 43, 571–573.

Crossley, D. A., Jr., and Blair, J. M. (1991). A high-efficiency, "low-technology" Tullgren-type extractor for soil microarthropods. *Agric. Ecosys. Environ.* 34, 187–192.

Crossley, D. A., Jr., Mueller, B. R., and Perdue, J. C. (1992). Biodiversity of microarthropods in agricultural soils: relations to functions. *Agric. Ecosys. Environ.* 40, 37–46.

Crossley, D. A., Jr., and Coleman, D. C. (1999). Microarthropods. *In* "The Handbook of Soil Science" (M. Sumner, ed.) pp. C59–C65. CRC Press, Boca Raton, FL.

Crossley, D. A., Jr., and Coleman, D. C. (1999). Macroarthropods. *In* "The Handbook of Soil Science" (M. Sumner, ed.) pp. C65–C70. CRC Press, Boca Raton, FL.

Crowe, J. H. (1975). The physiology of cryptobiosis in tardigrades. *Mem. Ist. Ital. Idrobiol. (Suppl.)* 32, 37–59.

Crowe, J. H., and Cooper, A. F., Jr. (1971). Cryptobiosis. *Sci. Am.* 225, 30–36.

Curl, E. A. (1979). Effects of mycophagous collembola on *Rhizoctonia solani* and cotton-seedling disease. *In* "Soil-Borne Plant Pathogens" (B. Schippers and W. Gams, eds.), pp. 253–269. Academic Press, London.

Curl, E. A., and Truelove, B. (1986). "The Rhizosphere." Springer-Verlag, Berlin, New York.

Currie, W. S. (2003). Relationships between carbon turnover and bioavailable energy fluxes in two temperate forest soils. *Global Change Biol.* 9, 919–929.

Curry, J. P. (1994). "Grassland Invertebrates. Ecology, Influence on Soil Fertility and Effects on Plant Growth." Chapman & Hall, London.

Curry, J. P., and Good, J. A. (1992). Soil faunal degradation and restoration. *Adv. Soil Sci.* 17, 171–215.

Curry, J. P., Byrne, D., and Boyle, K. E. (1995). The earthworm population of a winter cereal field and its effects on soil and nitrogen turnover. *Biol. Fertil. Soils* 19, 166–172.

Cutler, D. W. (1920). A method for estimating the number of active protozoa in the soil. *J. Agric. Sci.* 10, 135–143.

Cutler, D. W. (1923). The action of protozoa on bacteria when inoculated into sterile soil. *Ann. Appl. Biol.* 10, 137–141.

Cutler, D. W., Crump, L. M., and Sandon, H. (1923). A quantitative investigation of the bacterial and protozoan population of the soil. *Phil. Trans. Roy. Soc. (B) Biol. Sci.* 211, 317–350.

Daily, G. (ed.) (1997). "Nature's Services. Societal Dependence on Natural Ecosystems." Island Press, Washington, D. C.

Daniel, O., and Anderson, J. M. (1992). Microbial biomass and activity in contrasting soil materials after passage through the gut of the earthworm *Lumbricus rubellus* Hoffmeister. *Soil Biol. Biochem.* 24, 465–470.

Darbyshire, J. F. (ed.) (1994). "Soil Protozoa." CAB International, Wallingford.

Darbyshire, J. F., and Greaves, M. (1967). Protozoa and bacteria in the rhizosphere of *Sinapis alba* (L), *Trifolium repens* (L.), and *Lolium perenne* (L). *Can. J. Microbiol.* 13, 1057–1068.

Darwin, C. (1837). On the formation of mould. *Proc. Geol. Soc.* 2, 574–576.

Darwin, C. (1881). "The formation of vegetable mould, through the action of worms, with observations on their habits." John Murray, London.

Dash, M. C. (1990). Oligochaeta: Enchytraeidae. *In* "Soil Biology Guide" (D. L. Dindal, ed.), pp. 311–340. Wiley, New York.

David, J.-F., and Gillon, D. (2002). Annual feeding rate of the millipede *Glomeris marginata* on holm oak (*Quercus ilex*) leaf litter under Mediterranean conditions. *Pedobiologia* 46, 42–52.

Davidson, D. A., Bruneau, P. M. C., Grieve, I. C., and Young, I. M. (2002). Impacts of fauna on an upland grassland soil as determined by micromorphological analysis. *Appl. Soil Ecol.* 20, 133–143.

De Angelis, D. L. (1992). "Dynamics of Nutrient Cycling and Food Webs." Chapman & Hall, London.

De Ruiter, P. C., Moore, J. C., Zwart, K. B., Bouwman, L. A., Hassink, J., Bloem, J., De Vos, J. A., Marinissen, J. C. Y., Didden, W. A. M., Lebbink, G., and Brussaard, L. (1993). Simulation of nitrogen mineralization in the below-ground food webs of two winter wheat fields. *J. Appl. Ecol.* 30, 95–106.

De Ruiter, P. C., Neutel, A.-M., and Moore, J. C. (1995). Energetics and stability in below-ground food webs. *In* "Food Webs: Integration of Patterns and Dynamics" (G. A. Polis and K. O. Winemiller, eds.), Chapman and Hall, New York.

De Ruiter, P. C., Neutel, A.-M., and Moore, J. C. (1998). Biodiversity in soil ecosystems: the role of energy flow and community stability. *Appl. Soil Ecol.* 10, 217–228.

De Wit, H. A. (1978). "Soils and grassland types of the Serengeti plain (Tanzania)." Ph.D. Dissertation, Agricultural University, Wageningen, The Netherlands.

Denef, K., Six, J., Bossuyt, H., Frey, S. D., Elliott, E. T., Merckx, R., and Paustian, K. (2001). Influence of dry–wet cycles on the interrelationship between aggregate, particulate organic matter, and microbial community dynamics. *Soil Biol. Biochem.* 33, 1599–1611.

Diaz, S., and Cabido, M. (2001). Vive la différence: plant functional diversity matters to ecosystem processes. *Trends Ecol. Evol.* 16, 646–655.

Didden, W. A. M. (1990). Involvement of Enchytraeidae (Oligochaeta) in soil structure evolution in agricultural fields. *Biol. Fertil. Soils* 9, 152–158.

Didden, W. A. M. (1993). Ecology of Enchytraeidae. *Pedobiologia* 37, 2–29.

Didden, W. A. M. (1995). The effect of nitrogen deposition on enchytraeid-mediated decomposition and mobilization—a laboratory experiment. *Acta Zool. Fennica* 196, 60–64.

Dighton, J. (2003). "Fungi in Ecosystem Processes." Dekker, New York.

Dighton, J., and Coleman, D. C. (1992). Phosphorus relations of roots and mycorrhizas of *Rhododendron maximum* L. in the Southern Appalachians, N. Carolina. *Mycorrhiza* 1, 175–184.

Dillon, E. S., and Dillon, L. S. (1961). "A Manual of Common Beetles of Eastern North America." Row, Peterson, Evanston, IL.

Dindal, D. L. (ed.) (1990). "Soil Biology Guide." John Wiley & Sons, New York.

Donner, J. (1966). "Rotifers." Warne, London.

Doran, J. W. (1980a). Microbial changes associated with residue management with reduced tillage. *Soil Sci. Soc. Am. J.* 44, 518–524.

Doran, J. W. (1980b). Soil microbial and biochemical changes associated with reduced tillage. *Soil Sci. Soc. Am. J.* 44, 765–771.

Doran, J. W. (2002). Soil health and global sustainability: translating science into practice. *Agric. Ecosys. Environ.* 88, 119–127.

Doran, J. W., Coleman, D. C., Bezdicek, D. F., and Stewart, B. S. (eds.) (1994). "Defining Soil Quality for a Sustainable Environment. SSSA Special Publication." ASA, Madison, WI.

Doran, J. W., and Parkin, T. B. (1994). Defining and assessing soil quality. *In* "Defining Soil Quality for a Sustainable Environment" (J. W. Doran, D. C. Coleman, D. F. Bezdicek, and B. A. Stewart, eds.), pp. 3–21. SSSA Special Publ. 35. ASA, Madison, WI.

Dornbush, M. E., Isenhart, T. M., and Raich, J. W. (2002). Quantifying fine-root decomposition: an alternative to buried litterbags. *Ecology* 83, 2985–2990.

Dósza-Farkas, K. (1996). Reproduction strategies in some enchytraeid species. *In* "Newsletter on Enchytraeidae," No. 5 (K. Dósza-Farkas, ed.), pp. 25–33. Eötvös Loránd University, Budapest.

Draney, M. S. (1997). Ground-layer spiders (Araneae) of a Georgia Piedmont floodplain agroecosystem: species list, phenology, and habitat selection. *J. Arachnol.* 25, 333–351.

Dunger, W. (1983). "Fauna in Soils." Ziemsen Verlag, Wittenberg Lutherstadt, Germany.

Eash, N. S., Karlen, D. L., and Parkin, T. B. (1994). Fungal contributions to soil aggregation and soil quality. *In* "Defining Soil Quality for a Sustainable Environment" (J. Doran, D. C. Coleman, D. F. Bezdicek, and B. A. Stewart, eds.), pp. 221–228. SSSA Special Publication No. 35, Madison, WI.

Edgar, A. L. (1990). Opiliones (Phalangida). *In* "Soil Biology Guide" (D. L. Dindal, ed.), pp. 529–581. Wiley, New York.

Edwards, C. A. (1959). The ecology of Symphyla. II. Seasonal soil migrations. *Entomol. Exp. Appl.* 2, 257–267.

Edwards, C. A. (1990). Symphyla. *In* "Soil Biology Guide" (D. L. Dindal, ed.), pp. 891–910. Wiley, New York.

Edwards, C. A. (1991). The assessment of populations of soil-inhabiting invertebrates. *Agric. Ecosys. Environ.* 34, 145–176.

Edwards, C. A. (1998). "Earthworm Ecology." St. Lucie Press, Boca Raton.

Edwards, C. A. (2000). Soil invertebrate controls and microbial interactions in nutrient and organic matter dynamics in natural and agroecosystems. *In* "Arthropods as Webmasters in Ecosystems" (D. C. Coleman and P. F. Hendrix, eds.), pp. 141–159. CABI Publishing, Wallingford, U.K.

Edwards, C. A. (ed.) (2004). "Earthworm Ecology" (2nd edition). St. Lucie Press, Boca Raton.

Edwards, C. A., and Heath, G. W. (1963). The role of soil animals in breakdown of leaf material. *In* "Soil Organisms" (J. Doeksen and J. van der Drift, eds.), pp. 76–80. North Holland Publishing Co., Amsterdam.

Edwards, C. A., and Lofty, J. R. (1977). "Biology of Earthworms" (2nd edition). Chapman and Hall, London.

Edwards, C. A., and Bohlen, P. J. (1996). "Earthworm Biology and Ecology" (3rd edition). Chapman and Hall, London.

Edwards, N. T. (1991). Root and soil respiration responses to ozone in *Pinus taeda L.* seedlings. *New Phytol.* 18, 315–321.

Edwards, W. M., and Shipitalo, M. J. (2004). Consequences of earthworms in agricultural soils: aggregation and porosity. *In* "Earthworm Ecology" (C. A. Edwards, ed.), pp. 147–161. Lewis Publishers, Boca Raton.

Ehrenfeld, J. G., Kourtev, P., and Huang, W. (2001). Changes in soil functions following invasions of understory plants in deciduous forests. *Ecol. Appls.* 11, 1287–1300.

Eigenberg, R. A., Doran, J. W., Nienaber, J. A., Ferguson, R. B., and Woodbury, B. L. (2002). Electrical conductivity monitoring of soil condition and available N with animal manure and a cover crop. *Agric. Ecosys. Environ.* 88, 183–193.

Eisenbeis, G., and Wichard, W. (1987). "Atlas on the Biology of Soil Arthropods." Springer-Verlag, Stuttgart.

Elliott, E. T. (1986). Hierarchic aggregate structure and organic C, N, and P in native and cultivated grassland soils. *Soil Sci. Soc. Am. J.* 50, 627–633.

Elliott, E. T., and Coleman, D. C. (1977). Soil protozoan dynamics in a shortgrass prairie. *Soil Biol. Biochem.* 9, 113–118.

Elliott, E. T., Anderson, R. V., Coleman, D. C., and Cole, C. V. (1980). Habitable pore space and microbial trophic interactions. *Oikos* 35, 327–335.

Elliott, E. T., Horton, K., Moore, J. C., Coleman, D. C., and Cole, C. V. (1984). Mineralization dynamics in fallow dryland wheat plots, Colorado. *Plant & Soil* 76, 149–155.

Elliott, E. T., and Coleman, D. C. (1988). Let the soil work for us. *Ecol. Bull. (Copenhagen)* 39, 23–32.

Elliott, E. T., Janzen, H. H., Campbell, C. A., Cole, C. V., and Myers, R. J. K. (1994). Principles of ecosystem analysis and their application to integrated nutrient management and assessment of sustainability, Proc. Sustainable Land Management for the 21st Century. Vol. 2: Plenary Papers. ISSS, Acapulco, Mexico.

Elton, C. S. (1927). "Animal Ecology." Methuen, London.

Elton, C. S. (1958). "The Ecology of Invasions by Animals and Plants." Methuen-Wiley, London, New York.

Emerson, A. E. (1956). Regenerative behavior and social homeostasis of termites. *Ecology* 27, 248–258.

Entry, J. A., Rose, C. L., and Cromack, K., Jr. (1991). Litter decomposition and nutrient release in ectomycorrhizal mat soils of a Douglas fir ecosystem. *Soil Biol. Biochem.* 23, 285–290.

Entry, J. A., Rose, C. L., and Cromack, K., Jr. (1992). Microbial biomass and nutrient concentrations in hyphal mats of the ectomycorrhizal fungus *Hysterangium setchellii* in a coniferous forest soil. *Soil Biol. Biochem.* 24, 447–453.

Ettema, C. H., and Bongers, T. (1993). Characterization of nematode colonization and succession in disturbed soil using the Maturity Index. *Biol. Fertil. Soils* 16, 79–85.

Ettema, C. H., and Yeates, G. W. (2003). Nested spatial biodiversity patterns of nematode genera in a New Zealand forest and pasture soil. *Soil Biol. Biochem.* 35, 339–342.

Evans, R. D., and Belnap, J. (1999). Long-term consequences of disturbance on nitrogen dynamics in an arid ecosystem. *Ecology* 80, 150–160.

Evans, R. D., Rimer, R., Sperry, L., and Belnap, J. (2001). Exotic plant invasion alters nitrogen dynamics in an arid grassland. *Ecol. Appls.* 11, 1301–1310.

FAO (United Nations Food and Agriculture Organization) (1990). "Soilless Culture for Horticultural Crop Production." FAO, Rome.

Fahey, T. J., Bledsoe, C. S., Day, F. P., Ruess, R. W., and Smucker, A. J. M. (1999). Fine root production and demography. *In* "Standard Soil Methods for Long-Term Ecological Research" (G. P. Robertson, D. C. Coleman, C. Bledsoe, and P. Sollins, eds.), pp. 437–455. Oxford University Press, New York.

Farrah, S. R., and Bitton, G. (1990). Viruses in the soil environment. *In* "Soil Biochemistry," vol. 6 (J.-M. Bollag and G. Stotzky, eds.), pp. 529–556. Marcel Dekker, New York.

Farrar, J., Hawes, M., Jones, D., and Lindow, S. (2003). How roots control the flux of carbon to the rhizosphere. *Ecology* 84, 827–837.

Feller, C. (1997). The concept of soil humus in the past three centuries. *In* "History of Soil Science—International Perspectives." (D. H. Yaalon and S. Berkowicz, eds.), pp. 15–46. Catena Verlag, Reiskirchen, Germany.

Feller, C., and Beare, M. H. (1997). Physical control of soil organic matter dynamics in the tropics. *Geoderma* 79, 69–116.

Fender, W. M. (1995). Native earthworms of the Pacific Northwest: an ecological overview. *In* "Ecology and Biogeography of Earthworms in North America" (P. F. Hendrix, ed.), pp. 53–66. Lewis Publishers, Boca Raton, FL.

Fender, W. M., and McKey-Fender, D. (1990). Oligochaeta: Megascolecidae and other earthworms from western North America. *In* "Soil Biology Guide" (D. Dindal, ed.), pp. 357–378. Wiley, New York.

Fenster, C. R., and Peterson, G. A. (1979). Effects of no-tillage fallow as compared to conventional tillage in a wheat-fallow system. Research Bulletin 289. Agricultural Experiment Station, University of Nebraska, Lincoln.

Ferguson, L. M. (1990a). Insecta: Diplura. *In* "Soil Biology Guide" (D. L. Dindal, ed.), pp. 951–963. Wiley, New York.

Ferguson, L. M. (1990b). Insecta: Microcoryphia and Thysanura. *In* "Soil Biology Guide" (D. L. Dindal, ed.), pp. 935–949. Wiley, New York.

Ferris, H., Bongers, T., and de Goede, R. G. M. (2001). A framework for soil foodweb diagnostics: extension of the nematode faunal analysis concept. *Appl. Soil Ecol.* 18, 13–29.

Fierer, N., Schimel, J. P., and Holden, P. A. (2003). Variations in microbial community composition through two soil depth profiles. *Soil Biol. Biochem.* 35, 167–176.

Filser, J. (2002). The role of Collembola in carbon and nitrogen cycling in soil. *Pedobiologia* 46, 234–245.

Finlay, R. D., Frostegard, A., and Sonnerfeldt, A. M. (1992). Utilization of organic and inorganic nitrogen sources by ectomycorrhizal fungi in pure culture and in symbiosis with *Pinus contorta* Dougl. Ex Loud. *New Phytol.* 120, 105–115.

Fitter, A. H. (1985). Functional significance of root morphology and root system architecture. *In* "Ecological Interactions in Soil; Plants, Microbes and Animals" (A. H. Fitter, D. Atkinson, D. J. Read, and M. B. Usher, eds.), pp. 87–106. Blackwell, Oxford.

Fitter, A. H. (1991). The ecological significance of root system architecture: an economic approach. *In* "Plant Root Growth: An Ecological Perspective" (D. Atkinson, ed.), pp. 229–243. Blackwell, Oxford.

FitzPatrick, E. A. (1984). "Micromorphology of Soils." Chapman and Hall, London.

Foelix, R. F. (1996). "Biology of Spiders" (2nd edition). Oxford University Press, Oxford.

Fogel, R. (1985). Roots as primary producers in below-ground ecosystems. *In* "Ecological Interactions in Soil: Plants, Microbes and Animals" (A. H. Fitter, D. Atkinson, D. J. Read, and M. B. Usher, eds.), pp. 23–36. Blackwell, Oxford.

Fogel, R. (1991). Root system demography and production in forest ecosystems. *In* "Plant Root Growth: An Ecological Perspective" (D. Atkinson, ed.), pp. 89–101. Blackwell, Oxford.

Fogel, R., and Lussenhop, J. (1991). The University of Michigan Soil Biotron: a platform for soil biology research in a natural forest. *In* "Plant Root Growth: An Ecological Perspective" (D. Atkinson, ed.), pp. 61–73. Blackwell, Oxford.

Foissner, W. (1987a). Soil protozoa: Fundamental problems, ecological significance, adaptations in ciliates and testaceans, bioindicators, and guide to the literature. *Progress in Protistology* 2, 69–212.

Foissner, W. (1987b). Global soil ciliate (Protozoa, Ciliophora) diversity: a probability based approach using large sample collections from Africa, Australia and Antarctica. *Biodiversity and Conservation* 6, 1627–1638.

Foissner, W. (1994). Soil protozoa as bioindicators in ecosystems under human influence. *In* "Soil Protozoa" (J. F. Darbyshire, ed.), pp. 147–193. CABI Publishing, Wallingford.

Folke, C., Holling, C. S., and Perrings, C. (1996). Biological diversity, ecosystems and the human scale. *Ecol. Appls.* 6, 1018–1024.

Foster, R. C. (1985). In situ localization of organic matter in soils. *Quaest. Entomol.* 21, 609–633.

Foster, R. C. (1988). Microenvironments of soil microorganisms. *Biol. Fertil. Soils* 6, 189–203.

Foster, R. C. (1994). Microorganisms and soil aggregates. *In* "Soil Biota" (C. E. Pankhurst, B. M. Doube, V. V. S. R. Gupta, and P. R. Grace, eds.), pp. 144–155. CSIRO, Melbourne.

Foster, R. C., Rovira, A. D., and Cock, T. W. (1983). "Ultrastructure of the Root Soil Interface." Amer. Phytopathology Society, St. Paul, MN.

Foster, R. C., and Dormaar, J. F. (1991). Bacteria-grazing amoebae in situ in the rhizosphere. *Biol. Fertil. Soils* 11, 83–87.

Foth, H. D. (1990). "Fundamentals of Soil Science." Wiley, New York.

Fragoso, C., James, S. W., and Borges, S. (1995). Native earthworms of the north neotropical region: current status and controversies. In "Ecology and Biogeography of Earthworms in North America" (P. F. Hendrix, ed.), pp. 67–103. Lewis Publishers, Boca Raton.

Fragoso, C., Kanyonyo, J., Moreno, A., Senapati, B. K., Blanchart, E., and Rodriguez, C. (1999). A survey of tropical earthworms: taxonomy, biogeography and environmental plasticity. In "Earthworm Management in Tropical Agroecosystems" (P. Lavelle, L. Brussaard, and P. Hendrix, eds.), pp. 1–26. CABI Publishing, Wallingford, U.K.

Francé, R. H. (1921). "Das Edaphon." Arb. Biol. Inst. Muenchen, Stuttgart.

Frank, D. A., Kuns, M. M., and Guido, D. R. (2002). Consumer control of grassland plant production. Ecology 83, 602–606.

Freckman, D. W. (1994). "Life in the Soil/Soil Biodiversity: Its Importance to Ecosystem Processes. Report on a Workshop Held at the Natural History Museum, London." NREL, Colorado State University, Ft. Collins.

Freckman, D. W., and Virginia, R. A. (1989). Plant-feeding nematodes in deep-rooting desert ecosystems. Ecology 70, 1665–1678.

Freckman, D. W., and Ettema, C. H. (1993). Assessing nematode communities in agroecosystems of varying human intervention. Agric. Ecosys. Environ. 45, 239–261.

Frost, S. W. (1942). "General Entomology." McGraw-Hill, New York.

Fu, S., Kisselle, K. W., Coleman, D. C., Hendrix, P. F., and Crossley, D. A., Jr. (2001). Short-term impacts of aboveground herbivory (grasshopper) on the abundance and ^{14}C activity of soil nematodes in conventional tillage and no till agroecosystems. Soil Biol. Biochem. 33, 1253–1258.

Furlong, M. A., Singleton, D. R., Coleman, D. C., and Whitman, W. B. (2002). Molecular and culture-based analyses of prokaryotic communities from an agricultural soil and the burrows and casts of the earthworm Lumbricus rubellus. Appl. Environ. Microbiol. 68, 1265–1279.

Gadgil, R. L., and Gadgil, P. D. (1975). Suppression of litter decomposition by mycorrhizal roots of Pinus radiata. N.Z. J. For. Sci. 5, 33–41.

Garbaye, J. (1991). Biological interactions in the mycorrhizosphere. Experientia 47, 370–375.

Garey, J. R. (2001). Ecdysozoa: The relationship between Cycloneuralia and Panarthropoda. Zool. Anzeiger 240, 321–330.

Garrett, C. J., Crossley, D. A., Jr., Coleman, D. C., Hendrix, P. F., Kisselle, K. W., and Potter, R. L. (2001). Impact of the rhizosphere on soil microarthropods in agroecosystems on the Georgia piedmont. Appl. Soil Ecol. 16, 141–148.

Gates, G. E. (1967). On the earthworm fauna of the Great American Desert and adjacent areas. The Great Basin Naturalist 27, 142–1761.

Gaudinski, J. B., Trumbore, S. E., Davidson, E. A., and Zheng, S. (2000). Soil carbon cycling in a temperate forest: radiocarbon-based estimates of residence times, sequestration rates and partitioning of fluxes. Biogeochemistry 51, 33–69.

Gaudinski, J. B., Trumbore, S. E., Davidson, E. A., Cook, A. C., Markewitz, D., and Richter, D. D. (2001). The age of fine-root carbon in three forests of the eastern United States measured by radiocarbon. Oecologia 129, 420–429.

Gerson, U., Smiley, R. L., and Ochoa, R. (2003). "Mites (Acari) for Pest Control." Blackwell Science, Oxford.

Gilbert, G. S. (2002). Evolutionary ecology of plant diseases in natural ecosystems. Ann. Rev. Phytopath. 40, 13–43.

Gill, H. S., and Abrol, I. P. (1986). Salt affected soils and their amelioration through afforestation. In "Amelioration of Soil by Trees" (R. T. Prinsley and M. J. Swift, eds.), pp. 43–53. Commonwealth Science Council, London.

Gillard, O. (1967). Coprophagous beetles in pasture ecosystems. *J. Aust. Inst. Agric. Sci.* 33, 30–34.

Giller, K. E. (2001). "Nitrogen fixation in tropical cropping systems" (2nd edition). CABI Publishing, Wallingford, U.K.

Gilmore, S. K. (1972). Collembola predation on nematodes. *Search Agric.* 1, 1–12.

Gilmore, S. K., and Potter, D. A. (1993). Potential role of Collembola as biotic mortality agents for entomopathogenic fungi nematodes. *Pedobiologia* 37, 30–38.

Gisin, H. (1962). Sur la fauna européen des Collemboles IV. *Rev. Suisse Zool.* 69, 1–23.

Gisin, H. (1963). Collemboles d'Europe V. *Rev. Suisse Zool.* 70, 77–101.

Gisin, H. (1964). Collemboles d'Europe VII. *Rev. Suisse Zool.* 71, 649–78.

Gist, C. S., and Crossley, D. A., Jr. (1973). A method for quantifying pitfall trapping. *Environ. Entomol.* 2, 951–952.

Gist, C., Crossley, D. A., Jr., and Merchant, V. A. (1974). An analysis of life tables for *Sinella curviseta* (Collembola). *Environ. Entomol.* 3, 840–844.

Gist, C. S., and Crossley, D. A., Jr. (1975). A model of mineral cycling for an arthropod food web in a southeastern hardwood forest litter community. *In* "Mineral Cycling in Southeastern Ecosystems" (F. G. Howell and M. H. Smith, eds.), pp. 84–106. Energy Research and Development Administration, Washington, D. C.

Gjelstrup, P., and Petersen, P. (1987). Jordbundens mider og springhaler (Mites and springtails in the soil). *Natur. Hist. Museum, Århus* 26, 1–76.

Glinka, K. D. (1927). Dokuchaiev's ideas in the development of pedology and cognate sciences. USSR Acad. Sci. Russian Pedological Investigations, I.

Golley, F. B. (1993). "History of the ecosystem concept in ecology: more than the sum of the parts." Yale University Press, New Haven.

González, G., Ley, R. L., Schmidt, S. K., Zou, X., and Seastedt, T. R. (2001). Soil ecological interactions: comparisons between tropical and subalpine forests. *Oecologia* 128, 549–556.

González, G., and Seastedt, T. R. (2001). Soil fauna and plant litter decomposition in tropical and subalpine forests. *Ecology* 82, 955–964.

Goodnight, C. J., and Goodnight, M. L. (1960). Speciation among cave opilionids of the United States. *Am. Midl. Nat.* 64, 34–38.

Górny, M., and Grüm, L. (1993). "Methods in Soil Zoology." Polish Scientific Publishers, Warsaw.

Graefe, U. (1973). Systematische Untersuchungen and der Gattung *Achaeta* (Enchytraeidae, Oligochaeta). Diplomarbeit, Universität Hamburg, Hamburg, Germany.

Green, R. N., Trowbridge, R. L., and Klinka, K. (1993). Towards a taxonomic classification of humus forms. *Forest Sci. Monogrs.* 29, 1–49.

Greenslade, P. (1964). Pitfall trapping as a method for studying populations of Carabidae (Coleoptera). *J. Anim. Ecol.* 33, 301–310.

Greenslade, P. J. N. (1985). Pterygote insects and the soil: their diversity, their effects on soils and the problem of species identification. *Quaest. Entomol.* 21, 571–585.

Griffiths, B. S. (1994). Soil nutrient flow. *In* "Soil Protozoa" (J. F. Darbyshire, ed.), pp. 65–91. CABI Publishing, Wallingford, U.K.

Griffiths, B. S., and Caul, S. (1993). Migration of bacterial-feeding nematodes, but not protozoa, to decomposing grass residues. *Biol. Fertil. Soils* 15, 201–207.

Griffiths, B. S., Bonkowski, M., Dobson, G., and Caul, S. (1999). Changes in soil microbial community structure in the presence of microbial-feeding nematodes and protozoa. *Pedobiologia* 43, 297–304.

Griffiths, B. S., Ritz, K., Bardgett, R. D., Cook, R., Christensen, S., Ekelund, F., Sorensen, S., Bååth, E., Bloem, J., de Ruiter, P. C., Dolfing, J., and Nicolardot, B. (2000). Ecosystem response of pasture soil communities to fumigation-induced microbial diversity reductions: an examination of the biodiversity—ecosystem function relationship. *Oikos* 90, 279–294.

Griffiths, B. S., Ritz, K., Wheatley, R., Kuan, H. L., Boag, B., Christensen, S., Ekelund, F., Sorensen, S. J., Muller, S., and Bloem, J. (2001). An examination of the biodiversity—ecosystem function relationship in arable soil microbial communities. *Soil Biol. Biochem.* 33, 1713–1722.

Griffiths, D. A. (1996). Mites. *In* "Methods for the Examination of Organismal Diversity in Soils and Sediments" (G. S. Hall, ed.), pp. 175–185. CAB International, Wallingford, U.K.

Griffiths, E. (1965). Micro-organisms and soil structure. *Biol. Revs.* 40, 129–142.

Grodzinski, W., and Yorks, T. P. (1981). Species and ecosystem-level bioindicators of airborne pollution: an analysis of two major studies. *Water Air Soil Pollut.* 16, 33–53.

Groffman, P. M., and Bohlen, P. J. (1999). Soil and sediment biodiversity. *BioScience* 49, 139–148.

Guggenberger, G., and Kaiser, K. (2003). Dissolved organic matter in soil: challenging the paradigm of sorptive preservation. *Geoderma* 113, 293–310.

Guillet, B. (1990). Le vieillissement des matiéres organiques et des associations organo-minérales des andosols et des podsols. *Science du Sol* 28, 285–299.

Gunn, A., and Cherrett, J. M. (1993). The exploitation of food resources by soil meso- and macroinvertebrates. *Pedobiologia* 37, 303–320.

Gupta, V. V. S. R. (1989). "Microbial biomass sulfur and biochemical mineralization of sulfur in soils." Ph.D. Dissertation, University of Saskatchewan, Saskatoon.

Gupta, V. V. S. R., and Germida, J. J. (1988). Populations of predatory protozoa in field soils after 5 years of elemental S fertilizer application. *Soil Biol. Biochem.* 20, 787–791.

Gupta, V. V. S. R., and Germida, J. J. (1989). Influence of bacterial—amoebal interactions on sulfur transformations in soil. *Soil Biol. Biochem.* 21, 921–930.

Gupta, V. V. S. R., and Yeates, G. W. (1997). Soil microfauna as bioindicators of soil health. *In* "Biological Indicators of Soil Health" (C. Pankhurst, B. M. Doube, and V. V. S. R. Gupta, eds.), pp. 201–233. CABI Publishing, Wallingford, U.K.

Haas, H. J., Evans, C. E., and Miles, E. F. (1957). "Nitrogen and carbon changes in Great Plains soils as influenced by cropping and soil treatments." Technical Bulletin No. 1164, USDA, Washington, D.C.

Hadas, A. (1979). Heat capacity. *In* "The Encyclopedia of Soil Science, Part 1: Physics, Chemistry, Biology, Fertility, and Technology" (R.W. Fairbridge and C.W. Finkl, Jr., eds.), p. 189. Dowden, Hutchinson & Ross, Stroudsburg, PA.

Haimi, J., and Einbork, M. (1992). Effects of endogeic earthworms on soil processes and plant growth in coniferous forest soil. *Biol. Fertil. Soils* 13, 6–10.

Haines, B. L., and Swank, W. T. (1988). Acid precipitation effects on forest processes. *In* "Forest Hydrology and Ecology at Coweeta" (W. T. Swank and D. A. Crossley, Jr., eds.), pp. 359–366. Springer-Verlag, New York.

Hairston, N. G., Jr., and Hairston, N. G., Sr. (1993). Cause-effect relationships in energy flow, trophic structure, and interspecific interactions. *Am. Nat.* 142, 379–411.

Hall, G. S. (ed.) (1996). "Methods for the Examination of Organismal Diversity in Soils and Sediments." CAB International, New York.

Hall, S. J., and Raffaelli, D. G. (1993). Food webs: theory and reality. *Adv. Ecol. Res.* 24, 187–239.

Hallsworth, E. G., and Crawford, D. V. (1965). "Experimental Pedology." Butterworths, London.

Hansen, R. A. (1999). Red oak litter promotes a microarthropod functional group that accelerates its decomposition. *Plant & Soil* 209, 37–45.

Hansen, R. A. (2000). Diversity in the decomposing landscape. *In* "Invertebrates as Webmasters in Ecosystems" (D. C. Coleman and P. F. Hendrix, eds.), pp. 203–219. CABI Publishing, Wallingford, U.K.

Hansen, R. A., and Coleman, D. C. (1998). Litter complexity and composition are determinants of the diversity and species composition of oribatid mites (Acari: Oribatida) in litterbags. *Appl. Soil Ecol.* 9, 17–23.

Hanson, P. J., Edwards, N. T., Garten, C. T., and Andrews, J. A. (2000). Separating root and soil microbial contributions to soil respiration. *Biogeochemistry* 48, 115–146.

Hansson, A., Andrén, O., and Steen, E. (1991). Root production of four arable crops in Sweden and its effect on abundance of soil organisms. *In* "Plant Root Growth: An Ecological Perspective" (D. Atkinson, ed.), pp. 247–266. Blackwell, Oxford.

Hanula, J. L. (1995). Relationship of wood-feeding insects and coarse woody debris. *In* "Biodiversity and Coarse Woody Debris in Southern Forests" (J. W. McMinn and D. A. Crossley, Jr., eds.), pp. 55–81. USDA Forest Service, Asheville, N.C.

Hargrove, W. W., and Crossley, D. A., Jr. (1988). Video digitizer for the rapid measurement of leaf area lost to herbivorous insects. *Ann. Ent. Soc. Amer.* 81, 593–598.

Harmon, M. D., and Chen, H. (1991). Coarse woody debris dynamics in two old growth ecosystems. *BioScience* 41, 604–610.

Harris, R. F., Chesters, G., Allen, O. N., and Attoe, O. J. (1964). Mechanisms involved in soil aggregate stabilization by fungi and bacteria. *Soil Sci. Soc. Amer. Proc.* 28, 529–532.

Harris, W. F., Kinerson, R. S., Jr., and Edwards, N. T. (1977). Comparison of belowground biomass of natural deciduous forest and loblolly pine plantations. *Pedobiologia* 7, 369–381.

Harry, M., Jusseaume, N., Gambier, B., and Garnier-Sillam, E. (2001). Use of RAPD markers for the study of microbial community similarity from termite mounds and tropical soils. *Soil Biol. Biochem.* 33, 417–427.

Hartnett, D. C., and Wilson, G. W. T. (1999). Mycorrhizae influence plant community structure and diversity in tallgrass prairie. *Ecology* 80, 1187–1195.

Harvey, R. W., Kinner, N. E., Bunn, A., MacDonald, D., and Metge, D. (1995). Transport behavior of groundwater protozoa and protozoan-sized micro-spheres in sandy aquifer sediments. *Appl. Environ. Microbiol.* 61, 209–271.

Hattori, T. (1994). Soil micro environment. *In* "Soil Protozoa" (J. F. Darbyshire, ed.), pp. 43–64. CABI Publishing, Wallingford, U.K.

Hawksworth, D. L. (1991a). "The Biodiversity of Microorganisms and Invertebrates." CABI Publishing, Wallingford, U.K.

Hawksworth, D. L. (1991b). The fungal dimension of biodiversity: magnitude, significance and conservation. *Mycol. Res.* 95, 641–655.

Hawksworth, D. L. (2001). The magnitude of fungal diversity: the 1.5 millions species estimate revisited. *Mycol. Res.* 105, 1422–1432.

Heal, O. W., and Dighton, J. (1985). Resource quality and trophic structure in the soil system. *In* "Ecological Interactions in Soil: Plants, Microbes and Animals" (A. H. Fitter, D. Atkinson, D. J. Read, and M. B. Usher, eds.), pp. 339–354. Blackwell Scientific, Oxford.

Heal, O. W., Anderson, J. M., and Swift, M. J. (1997). Plant litter quality and decomposition: an historical overview. *In* "Driven by Nature: Plant Litter Quality and Decomposition" (G. Cadisch and K. E. Giller, eds.), pp. 3–30. CABI Publishing, Wallingford, U.K.

Hedley, M. J., and Stewart, J. W. B. (1982). Method to measure microbial phosphate in soil. *Soil Biol. Biochem.* 14, 377–385.

Helal, H. M., and Sauerbeck, D. (1991). Short-term determination of the actual respiration rate of intact plant roots. *In* "Plant Roots and Their Environment" (B. L. Michael and H. Persson, eds.), pp. 88–92. Elsevier Science Publishers, Amsterdam.

Henderson, L. J. (1913). "The Fitness of the Environment." Beacon Hill Press, Boston.

Henderson, L. S. (1952). Household insects. In "Insects, the Yearbook of Agriculture 1952" (A. Stefferud, ed.), pp. 469–475. U. S. Government Printing Office, Washington, D. C.

Hendrick, R. L., and Pregitzer, K. S. (1992). The demography of fine roots in a Northern hardwood forest. *Ecology* 73, 1094–1104.

Hendrix, P. F. (ed.) (1995). "Earthworm Ecology and Biogeography in North America." Lewis Publishers, Boca Raton, Florida.

Hendrix, P. F. (2000). Earthworms. *In* "Handbook of Soil Science" (M. E. Sumner, ed.), pp. C77–C85, CRC Press, Boca Raton, FL.

Hendrix, P. F., Parmelee, R. W., Crossley, D. A., Jr., Coleman, D. C., Odum, E. P., and Groffman, P. (1986). Detritus food webs in conventional and no-tillage agroecosystems. *Bioscience* 36, 374–380.

Hendrix, P. F., Crossley, D. A., Jr., Coleman, D. C., Parmelee, R. W., and Beare, M. H. (1987). Carbon dynamics in soil microbes and fauna in conventional and no-tillage agroecosystems. *INTECOL Bulletin* 15, 59–63.

Hendrix, P. F., Crossley, D. A., Jr., Blair, J. M., and Coleman, D. C. (1990). Soil biota as components of sustainable agroecosystems. *In* "Sustainable Agricultural Systems" (C. A. Edwards, R. Lal, P. Madden, R. H. Miller, and G. House, eds.), pp. 637–654. Soil & Water Conserv. Soc., Ankeny, Iowa.

Hendrix, P. F., Coleman, D. C., and Crossley, D. A., Jr. (1992). Using knowledge of soil nutrient cycling processes to design sustainable agriculture. *J. Sust. Agric.* 2, 63–82.

Hendrix, P. F., Callaham, M. A., Jr., and Kirn, L. (1994). Ecological studies of nearctic earthworms in the southern USA: II. Effects of bait harvesting on *Diplocardia* populations in Apalachicola National Forest in north Florida. *Megadrilogica* 5, 73–76.

Hendrix, P. F., and Bohlen, P. J. (2002). Exotic earthworm invasions in North America: ecological and policy implications. *BioScience* 52, 801–811.

Heneghan, L., Coleman, D. C., Zou, X., Crossley, D. A., Jr., and Haines, B. L. (1998). Soil microarthropod community structure and litter decomposition dynamics: a study of tropical and temperate sites. *Appl. Soil Ecol.* 9, 33–38.

Heneghan, L., Coleman, D. C., Zou, X., Crossley, D. A., Jr., and Haines, B. L. (1999). Soil microarthropod contributions to decomposition dynamics: tropical—temperate comparisons of a single substrate. *Ecology* 80, 1873–1882.

Hijii, N. (1987). Seasonal changes in abundance and spatial distribution of the soil arthropods in a Japanese cedar (*Cryptomeria japonica* D. Don) plantation, with special reference to Collembola and Acarina. *Ecol. Res.* 2, 159–173.

Hillel, D. J. (1991). "Out of the Earth: Civilization and the Life of the Soil." Free Press, New York.

Hillel, D. J. (1998). "Environmental Soil Physics." Academic Press, San Diego.

Hiltner, L. (1904). Über neuere erfahrungen und probleme auf dem gebiet der bodenbakteriologie und unter besonderer berücksichtigung der gründüngung und brache. *Arb. Dtsch. Landwirt. Ges.* 98, 59–78.

Hobbie, E. A., Weber, N. S., Trappe, J. M., and van Klinken, G. J. (2002). Using radiocarbon to determine the mycorrhizal status of fungi. *New Phytol.* 156, 129–136.

Hoffman, R. L. (1990). Diplopoda. *In* "Soil Biology Guide" (D. L. Dindal, ed.), pp. 835–860. Wiley, New York.

Hoffman, R. L. (1999). "Checklist of the Millipedes of North and Middle America." Special Publication Number 8, Virginia Museum of Natural History, Martinsville.

Holland, E. A., and Coleman, D. C. (1987). Litter placement effects on microbial and organic matter dynamics in an agroecosystem. *Ecology* 68, 425–433.

Hölldobler, B., and Wilson, E. O. (1990). "The Ants." Belknap Press, Harvard University, Cambridge.

Hominick, W. M. (2002). Biogeography. *In* "Entomopathogenic Nematology" (R. Gaugler, ed.), pp. 115–143. CABI Publishing, Wallingford, U.K.

Hongoh, Y., Ohkuma, M., and Kudo, T. (2003). Molecular analysis of bacterial microbiota in the gut of the termite *Reticulitermes speratus* (Isoptera: Rhinotermitidae). *FEMS Microbiol. Ecol.* 44, 231–242.

Hooper, D. U., Bignell, D. E., Brown, V. K., Brussaard, L., Dangerfield, J. M., Wall, D. H., Wardle, D. A., Coleman, D. C., Giller, K. E., Lavelle, P., van der Putten, W. H., de Ruiter, P. C., Rusek, J., Silver, W. L., Tiedje, J. M., and Wolters, W. (2000). Interactions between aboveground and belowground biodiversity in terrestrial ecosystems: patterns, mechanisms, and feedbacks. *BioScience* 50, 1049–1061.

Hoover, C., and Crossley, D. A., Jr. (1995). Leaf litter decomposition and microarthropod abundance along an altitudinal gradient. *In* " The Significance and Regulation of Soil Biodiversity" (H. P. Collins, G. P. Robertson and M. J. Klug, eds.), pp. 287–292. Kluwer Academic, Dordrecht, The Netherlands.

Hopkin, S. P. (1997). "The Biology of Springtails (Insecta: Collembola)." Oxford University Press, Oxford.

Hopkin, S. P., and Read, H. J. (1992). "The Biology of Millipedes." Oxford University Press, Oxford.

Horwath, W. R., Pregitzer, K. S., and Paul, E. A. (1994). ^{14}C allocation in tree—soil systems. *Tree Physiol.* 14, 163–176.

Hotopp, K. P. (2002). Land snails and soil calcium in central Appalachian mountain forest. *Southeastern Nat.* 1, 27–44.

Houghton, R. A. (1990). The global effects of deforestation. *Environ. Sci. Technol.* 24, 414–422.

Houghton, R. A. (2003). Why are estimates of the terrestrial carbon balance so different? *Global Change Biology* 9, 500–509.

Houghton, R. A., Hobbie, J. E., Melillo, J. M., Moore, B., Peterson, B. J., Shaver, G. R., and Woodwell, G. M. (1983). Changes in the carbon content of terrestrial biota and soils between 1860 and 1980: a net release of CO_2 to the atmosphere. *Ecol. Monogr.* 53, 235–262.

Houghton, R. A., Boone, R. D., Fruci, J. R., Hobbie, J. E., Melillo, J. M., and Palm, C. A. (1987). The flux of carbon from terrestrial ecosystems to the atmosphere in 1980 due to changes in land use: geographic distribution of the global flux. *Tellus* 39B, 122–139.

House, G. J., Stinner, B. R., and Crossley, D. A., Jr. (1984). Nitrogen cycling in conventional and no-tillage agroecosystems: analysis of pathways and processes. *J. Appl. Ecol.* 21, 991–1012.

House, G. J., Worsham, A. D., Sheets, T. J., and Stinner, R. E. (1987). Herbicide effects on soil arthropod dynamics and wheat straw decomposition in a North Carolina no-tillage agroecosystem. *Biol. Fertil. Soils* 4, 109–114.

Hugenholtz, P., Goebel, B. M., and Pace, N. R. (1998). Impact of culture-independent studies on the emerging phylogenetic view of bacterial diversity. *J. Bacteriol.* 180, 4765–4774.

Huhta, V., Wright, D. H., and Coleman, D. C. (1989). Characteristics of defaunated soil. I. A comparison of three techniques applied to two different forest soils. *Pedobiologia* 33, 417–426.

Humphreys, W. F. (1979). Production and respiration in animal populations. *J. Anim. Ecol.* 48, 427–453.

Hungate, B. A., Dukes, J. S., Shaw, M. R., Lou, Y., and Field, C. B. (2003). Nitrogen and climate change. *Science* 302, 1512–1513.

Hunt, G. S., Norton, R. A., Kelly, J. P. H., Collof, M. J., and Lindsay, S. M. (1998). "An interactive glossary of oribatid mites." CD-ROM, CSIRO Publishers, Melbourne.

Hunt, H. W., Coleman, D. C., Ingham, E. R., Ingham, R. E., Elliott, E. T., Moore, J. C., Rose, S. L., Reid, C. P. P., and Morley, C. R. (1987). The detrital food web in a shortgrass prairie. *Biol. Fertil. Soils* 3, 57–68.

Hunt, H. W., and Wall, D. H. (2002). Modelling the effects of loss of soil biodiversity on ecosystem function. *Global Change Biology* 8, 33–50.

Hunter, M. D., and Price, P. W. (1992). Playing chutes and ladders: heterogeneity and the relative roles of bottom-up and top-down forces in natural communities. *Ecology* 73, 724–732.

Hunter, P. E., and Rosario, R. M. T. (1988). Associations of Mesostigmata with other arthropods. *Annu. Rev. Entomol.* 33, 393–413.

Husband, R., Herre, E. A., Turner, S. L., Gallery, R., and Young, J. P. W. (2002). Molecular diversity of arbuscular mycorrhizal fungi and patterns of host association over time and space in a tropical forest. *Molecular Ecology* 11, 2669–2678.

Hutchinson, G. E. (1957). Concluding remarks. *In* "Cold Spring Harbor Symposium on Quantitative Biology," vol. 22, pp. 415–427. Cold Spring, N.Y.

Ingham, E. R., and Klein, D. A. (1982). Relationship between fluorescein diacetate—stained hyphae and oxygen utilization, glucose utilization, and biomass of sub-merged fungal batch cultures. *Appl. Environ. Microbiol.* 44, 363–370.

Ingham, E. R., Trofymow, J. A., Ames, R. N., Hunt, H. W., Morley, C. R., Moore, J. C., and Coleman, D. C. (1986a). Trophic interactions and nitrogen cycling in a semiarid grassland soil. I. Seasonal dynamics of the natural populations, their interactions and effects on nitrogen cycling. *J. Appl. Ecol.* 23, 597–614.

Ingham, E. R., Trofymow, J. A., Ames, R. N., Hunt, H. W., Morley, C. R., Moore, J. C., and Coleman, D. C. (1986b). Trophic interactions and nitrogen cycling in a semiarid grassland soil II. System responses to removal of different groups of soil microbes or fauna. *J. Appl. Ecol.* 23, 615–630.

Ingham, E. R., and Horton, K. A. (1987). Bacterial, fungal and protozoan responses to chloroform fumigation in stored soil. *Soil Biol. Biochem.* 19, 545–550.

Ingham, E. R., Coleman, D. C., and Moore, J. C. (1989). An analysis of food web structure and function in a shortgrass prairie, mountain meadow and lodgepole pine forest. *Biol. Fertil. Soils* 8, 29–37.

Ingham, R. E., Trofymow, J. A., Ingham, E. R. and Coleman, D. C. (1985). Interactions of bacteria, fungi, and their nematode grazers: Effects on nutrient cycling and plant growth. *Ecol. Monogr.* 55, 119–140.

Insam, H. (1990). Are the soil microbial biomass and basal respiration governed by the climatic regime? *Soil Biol. Biochem.* 22, 525–532.

Insam, H., and Domsch, K. H. (1988). Relationship between soil organic carbon and microbial biomass on chronosequences of reclamation sites. *Microb. Ecol.* 15, 177–188.

Ito, M., and Abe, W. (2001). Micro-distribution of soil inhabiting tardigrades (Tardigrada) in a sub-alpine coniferous forest of Japan. *Zool. Anzeiger* 240, 403–407.

Jackman, J. A. (1997). "A Field Guide to Spiders and Scorpions of Texas." Gulf Pub., Houston.

Jackson, R. R., and Willey, M. B. (1994). The comparative study of the behavior of *Myrmarachne*, ant-like jumping spiders (Araneae: Salticidae). *Zool J. Linnean Soc.* 110, 77–91.

Jacot, A. P. (1936). Soil structure and soil biology. *Ecology* 17, 359–379.

James, S. W. (1990). Oligochaeta: Megascolecidae and other earthworms from southern and midwestern North America. *In* "Soil Biology Guide" (D. L. Dindal, ed.), p. 379–386. Wiley, New York.

James, S. W. (1995). Systematics, biogeography and ecology of nearctic earthworms from eastern, central, southern and southwestern United States. *In* "Ecology and Biogeography of Earthworms in North America" (P. F. Hendrix, ed.), pp. 29–52. Lewis Publisher, Boca Raton, FL.

Jamieson, B. G. M. (1988). On the phylogeny and higher classification of the Oligochaeta. *Cladistics* 4, 367–401.

Jastrow, J. D. and Miller, R. M. (1991). Methods for assessing the effects of biota on soil structure. *Agric. Ecosys. Environ.* 34, 279–303.

Jastrow, J. D., Miller, R. M., and Lussenhop, J. (1998). Contributions of interacting biological mechanisms to soil aggregate stabilization in restored prairie. *Soil Biol. Biochem.* 30, 905–916.

Jenkinson, D. S. (1966). Studies on the decomposition of plant material in soil, II. Partial sterilization of soil and the soil biomass. *J. Soil Sci.* 17, 280–302.

Jenkinson, D. S. (1988). Determination of microbial biomass carbon and nitrogen in soil. *In* "Advances in Nitrogen Cycling in Agricultural Ecosystems" (J. R. Wilson, ed.), CABI Publishing, Wallingford, U.K.

Jenkinson, D. S., and Powlson, D. S. (1976). The effects of biocidal treatments on metabolism in soil. V. A method for measuring soil biomass. *Soil Biol. Biochem.* 8, 209–213.

Jenkinson, D. S., and Parry, L. C. (1989). The nitrogen cycle in the Broadbalk wheat experiment: a model for the turnover of nitrogen through the soil microbial biomass. *Soil Biol. Biochem.* 21, 535–541.

Jenkinson, D. S., Adams, D. E., and Wild, A. (1991). Model estimates of CO_2 emissions from soil in response to global warming. *Nature* 351, 304–306.

Jenkinson, D. S., Brookes, P. C., and Powlson, D. S. (2004). Measuring microbial biomass. *Soil Biol. Biochem.* 36, 5–7.

Jennings, T. J., and Barkham, J. P. (1975). Food of slugs in mixed deciduous woodland. *Oecologia* 26, 211–221.

Jenny, H. (1941). "Factors of Soil Formation." McGraw-Hill, New York.

Jenny, H. (1980). "The soil resource: origin and behavior." *Ecological Studies* 37 Springer-Verlag, New York.

Jenny, H., and Grossenbacher, K. (1963). Root—soil boundary zones as seen in the electron microscope. *Soil Sci. Soc. Amer. Proc.* 27, 273–277.

Joergensen, R. G., Anderson, T.-H., and Wolters, V. (1995). Carbon and nitrogen relationships in the microbial biomass of soils in beech (*Fagus sylvatica* L) forests. *Biol. Fertil. Soils* 19, 141–147.

Johnson, D. L. (1990). Biomantle evolution and the redistribution of earth materials and artifacts. *Soil Science* 149, 84–102.

Johnson, D., Leake, J. R., Ostle, N., Ineson, P., and Read, D. J. (2002). *In situ* $^{13}CO_2$ pulse-labelling of upland grassland demonstrates a rapid pathway of carbon flux from arbuscular mycorrhizal mycelia to the soil. *New Phytol.* 153, 327–334.

Jones, C. G., Lawton, J. H., and Shachak, M. (1994). Organisms as ecosystem engineers. *Oikos* 69, 373–386.

Jones, D. L., Farrar, J., and Giller, K. E. (2003). Associative nitrogen fixation and root exudation——what is theoretically possible in the rhizosphere? *Symbiosis* 35, 19–38.

Jones, P. C. T., and Mollison, J. E. (1948). A technique for the quantitative estimation of soil microorganisms. *J. Gen. Microbiol.* 2, 54–69.

Jongerius, A. (1964). "Soil Micromorphology." Elsevier, Amsterdam.

Jurgens, G., Lindström, K., and Saano, A. (1997). Novel group within the kingdom Crenarchaeota from boreal forest soil. *Appl. Environ. Microbiol.* 63, 803–805.

Kaczmarek, M. (1993). Apparatus and tools for the extraction of animals from the soil. *In* "Methods in Soil Zoology" (M. Górny and L. Grüm, eds.), pp. 112–141. Polish Scientific Publishers, Warsaw.

Kalisz, P. J., and Wood, H. B. (1995). Native and exotic earthworms in wildland ecosystems. *In* "Earthworm Ecology and Biogeography in North America" (P. Hendrix, ed.), pp. 117–126. Lewis Publishers, Boca Raton, FL.

Kaneko, N. (1988). Feeding habits and cheliceral size of oribatid mites in cool temperate forest soils in Japan. *Rev. Écol. Biol. Sol* 25, 353–363.

Karg, W. (1982). Investigations of habitat requirements, geographical distributions and origin of predatory mite genera in the Cohort Gamasina for use as bioindicators. *Pedobiologia* 24, 241–247.

Karlen, D. L., Andrews, S. S., and Doran, J. W. (2001). Soil quality: current concepts and applications. *Adv. Agron.* 74, 1–40.

Kasprzak, K. (1982). Review of enchytraeid (Oligochaeta, Enchytraeidae) community structure and function in agricultural ecosystems. *Pedobiologia* 23, 217–232.

Kaston, B. J. (1978). "How To Know the Spiders." McGraw Hill, Boston.

Kaya, H. K., Whipple, A. V., Child, A. L., Kraig, S., Bondonno, M., Dyer, K., and Maron, L. L. (1996). Entomopathogenic nematodes: natural enemies of root-feeding caterpillars on bush lupine. *Oecologia* 108, 167–173.

Keilin, D. (1959). The problem of anabiosis or latent life: history and current concept. *Proc. Roy. Soc. London, B* 150, 149–191.

Kemmpf, W. W. (1964). On the number of species of ants in the neotropical region. *Studia Entomologica* 8, 161–200.

Kent, A. D., and Triplett, E. W. (2002). Microbial communities and their interactions in soil and rhizosphere systems. *Ann. Revs. Microbiol.* 56, 211–236.

Kerley, S. J., and Read, D. J. (1995). The biology of mycorrhiza in the Ericaceae. 18. Chitin degradation by *Hymenoscyphus ericae* and transfer of chitin-nitrogen to the host-plant. *New Phytol.* 131, 369–375.

Kethley, J. (1990). Acarina: Prostigmata (Actenidida). *In* "Soil Biology Guide" (D. L. Dindal, ed.), pp. 667–756. Wiley, New York.

Kevan, D. K. M., and Scudder, G. G. E. (1989). "Illustrated keys to the families of terrestrial arthropods of Canada. 1. Myriapods (Millipedes, Centipedes, etc.)." Biological Survey of Canada (Terrestrial Arthropods), Ottawa.

Kilbertus, G. (1980). Études des microhabitats contenus dans les agrégats du sol leur relation avec la biomasse bacterienne et la taille des procaryotes presents. *Rev. Écol. Biol. Sol.* 17, 543–557.

Kilbertus, G., and Vannier, G. (1981). Relations microflore—microfaune dans la grotte de Sainte-Catherine (Pyrenées ariegeoises). II. Le regime alimentaire de *Tomocerus minor* (Lubbock) et *Tomocerus problematicus* Cassagnau (Insectes Collemboles). *Rev. Écol. Biol. Sol* 18, 319–338.

Kimball, B. A., Pinter, P. J., Jr., Garcia, R. L., LaMorte, R. L., Wall, G. W., Hunsaker, D. J., Wechsung, G., Wechsung, F., and Kartschall, T. (1995). Productivity and water use of wheat under free-air CO_2 enrichment (FACE). *Global Change Biology* 1, 42.

Kimble, J. M., and Levine, E. R. (1994). The Nairobi Conference: topics, results, and research needs, Trans. 15th World Congr. Soil Science, Acapulco, Mexico, pp. 151–162.

Kisselle, K. W., Garrett, C. J., Fu, S., Hendrix, P. F., Crossley, D. A., Jr., Coleman, D. C., and Potter, R. L. (2001). Budgets for root-derived C and litter—derived C: comparison between conventional tillage and no-tillage soils. *Soil Biol. Biochem.* 33, 1067–1075.

Kjøller, A., and Struwe, S. (1994). Analysis of fungal communities on decomposing beech litter. *In* "Beyond the Biomass" (K. Ritz, J. Dighton, and K. E. Giller, eds.), pp. 191–200. Wiley-Sayce, Chichester.

Kleiber, M. (1961). "The Fire of Life." Wiley, New York.

Klironomos, J. N., and Kendrick, W. B. (1995). Palatability of microfungi to soil arthropods in relation to the functioning of arbuscular mycorrhizal fungi. *Biol. Fertil. Soils* 21, 43–52.

Klironomos, J. N., and Hart, M. M. (2001). Animal nitrogen swap for plant carbon. *Nature* 410, 651–652.

Knoepp, J. D., Coleman, D. C., Crossley, D. A., Jr., and Clark, J. (2000). Biological indices of soil quality: an ecosystem case study of their use. *For. Ecol. Manag.* 138, 357–368.

Kourtev, P. S., Ehrenfeld, J. G., and Häggblom, M. (2002). Exotic plant species alter the microbial community structure and function in the soil. *Ecology* 83, 3152–3166.

Kozlowska, J., and Wasilewska, L. (1993). Nematoda. In "Methods in Soil Zoology" (M. Górny and L. Grüm, eds.), pp. 163–183. Polish Scientific Publishers, Warsaw.

Krantz, G. W. (1978). "A Manual of Acarology" (2nd edition). Oregon State University Book Stores, Corvallis.

Krantz, G. W., and Ainscough, B. D. (1990). Acarina: Mesostigmata (Gamasida). In "Soil Biology Guide" (D. L. Dindal, ed.), pp. 583–665. Wiley, New York.

Krool, S., and Bauer, T. (1987). Reproduction, development and pheromone secretion in *Heteromurus nitidus* Templeton, 1835 (Collembola, Entomobryidae). *Rev. Écol. Biol. Sol* 24, 187–195.

Kubiëna, W. L. (1938). "Micropedology." Collegiate Press, Ames, Iowa.

Kucey, R. M. N., and Paul, E. A. (1982). Carbon flow, photosynthesis, and N_2 fixation in mycorrhizal and nodulated faba beans (*Vicia faba* L.). *Soil Biol. Biochem.* 14, 407–412.

Kühnelt, W. (1958). Zoogenic crumb-formation in undisturbed soils (in German) *Sonderdruck aus Tagungsberichte.* 13, 193–199.

Kühnelt, W. (1976). "Soil Biology." Michigan State University, East Lansing.

Kuikman, P., and van Veen, J. A. (1989). The impact of protozoa on the availability of bacterial nitrogen to plants. *Biol. Fertil. Soils* 8, 13–18.

Kuikman, P. J., Jansen, A. G., van Veen, J. A., and Zehnder, A. J. B. (1990). Protozoan predation and the turnover of soil organic carbon and nitrogen in the presence of plants. *Biol. Fertil. Soils* 10, 22–28.

Kulman, H. M. (1974). Comparative ecology of North American Carabidae with special reference to biological control. *Entomophaga Memoirs* 7, 61–70.

Kurcheva, G. F. (1960). Role of soil organisms in the breakdown of oak litter. *Pochvovedeniye* 4, 16–23 (in Russian).

Kurcheva, G. F. (1964). Wirbellose Tiere als Faktor der Zersetzung von Waldstreu. *Pedobiologia* 4, 7–30.

Kuske, C. R., Barns, S. M., and Busch, J. D. (1997). Diverse uncultivated bacterial groups from soils of the arid Southwestern United States that are present in many geographic regions. *Appl. Environ. Microbiol.* 63, 3858–3865.

Kuzyakov, Y. (2002). Separating microbial respiration of exudates from root respiration in non-sterile soils: a comparison of four methods. *Soil Biol. Biochem.* 34, 1621–1631.

Kuzyakov, Y., and Domanski, G. (2000). Carbon input by plants into the soil. Review. *J. Plant Nutr. Soil Sci.* 163, 421–431.

Labandeira, C. C., Phillips, L. T., and Norton, R. A. (1997). Oribatid mites and the decomposition of plant tissues in Palaeozoic coal-swamp forests. *Palaios* 12, 319–353.

Lal, R. (2002). The potential of soils of the tropics to sequester carbon and mitigate the green house effect. *Adv. Agron.* 76, 1–30.

Lamoncha, K. L., and Crossley, D. A., Jr. (1998). Oribatid mite diversity along an elevation gradient in a southeastern Appalachian forest. *Pedobiologia* 42, 43–55.

Landeweert, R., Hoffland, E., Finlay, R. D., Kuyper, T. W., and van Breemen, N. (2001). Linking plants to rocks: ectomycorrhizal fungi mobilize nutrients from minerals. *Trends Ecol. Evol.* 16, 248–254.

Landeweert, R., Leeflang, P., Kuyper, T. W., Hoffland, E., Rosling, A., Wernars, K., and Smit, E. (2003). Molecular identification of ectomycorrhizal mycelium in soil horizons. *Appl. Environ. Microbiol.* 69, 327–333.

Lane, D. J. (1991). 16S/23S rRNA sequencing. In "Nucleic Acid Techniques in Bacterial Systematics" (E. Stackebrandt and M. Goodfellow, eds.), pp. 115–175. Wiley, New York.

Larink, O. (1997). Springtails and mites: important knots in the food web of soils. In "Fauna in Soil Ecosystems. Recycling Processes, Nutrient Fluxes, and Agricultural Production" (G. Benckiser, ed.), pp. 225–264. Marcel Dekker, New York.

Larsen, K. J., Work, T. T. and Purrington, F. F. (2003). Habitat use patterns by ground bee- tles (Coleoptera: Carabidae) of northeastern Iowa. *Pedobiologia* 47, 288–299.

Lartey, R. T., Curl, E. A., and Peterson, C. M. (1994). Interactions of mycophagous collem- bola and biological control fungi in the suppression of *Rhizoctonia solani*. *Soil Biol. Biochem.* 26, 81–88.

Lashof, D. A. (1989). The dynamic greenhouse: feedback processes that may influence future of atmospheric trace gases and climate change. *Climatic Change* 14, 213–242.

Lavelle, P. (1978). "Les vers de terre de la savane de Lamto (Ivory Coast): peuplements, populations et fonctions dans l'écosystème." Thèse de Doctorat, Université Paris VI.

Lavelle, P. (1983). The structure of earthworm communities. *In* "Earthworm Ecology: From Darwin to Vermiculture" (J. E. Satchell, ed.), pp. 449–466. Chapman & Hall, London.

Lavelle, P. (2000). Ecological challenges for soil science. *Soil Sci.* 165, 73–86.

Lavelle, P., Blanchart, E., and Martin, A. (1992). Impact of soil fauna on the properties of soils in the humid tropics. *In* "Myths and Science of Soils of the Tropics" (R. Lal and P. Sanchez, eds.), pp. 157–185. Soil Science Society of America, Madison, Wisconsin.

Lavelle, P., and Martin, A. (1992). Small-scale and large-scale effects of endogeic earth- worms on soil organic matter dynamics in soils of the humid tropics. *Soil Biol. Biochem.* 24, 1491–1498.

Lavelle, P., Lattaud, D. T., and Barois, I. (1995). Mutualism and biodiversity in soils. *Plant & Soil* 170, 23–33.

Lavelle, P., Pashanasi, B., Charpentier, F., Gilot, C., Rossi, J., Derouard, L., Andre, J., Ponge, J., and Bernier, N. (1998). Influence of earthworms on soil organic matter dynamics, nutrient dynamics and microbiological ecology. *In* " Earthworm Ecology" (C. A. Edwards, ed.), pp. 103–122. Lewis Publisher, Boca Raton, FL.

Lavelle, P., Brussaard, L., and Hendrix, P. (eds.) (1999). "Earthworm Management in Tropical Agroecosystems." CABI Publishing, Wallingford, U.K.

Lavelle, P., and Spain, A. V. (2002). "Soil Ecology." Kluwer, Dordrecht, The Netherlands.

Lawrence, K. L., and Wise, D. H. (2000). Spider predation on forest-floor Collembola and evidence for indirect effects on decomposition. *Pedobiologia* 44, 33–39.

Leake, J. R., and Read, D. J. (1989). The biology of mycorrhiza in the Ericaceae. 13. Some characteristics of the extracellular proteinase activity of the ericoid endophyte *Hymenoscyphus ericae*. *New Phytol.* 112, 69–76.

Lee, K. E. (1959). The earthworm fauna of New Zealand. *N.Z. Dept. Sci. Industr. Res. Bull.* 130.

Lee, K. E. (1985). "Earthworms: Their Ecology and Relationships with Soils and Land Use." Academic Press, Sydney.

Lee, K. E. (1995). Earthworms and sustainable land use. *In* "Earthworm Ecology and Bio- geography" (P. F. Hendrix, ed.), pp. 215–234. Lewis Publishers, Boca Raton, FL.

Lee, K. E., and Wood, T. G. (1971). "Termites and Soils." Academic Press, London and New York.

Lee, K. E., and Foster, R. C. (1991). Soil fauna and soil structure. *Aust. J. Soil Res.* 29, 745–775.

Lee, K. E., and Pankhurst, C. E. (1992). Soil organisms and sustainable productivity. *Aust. J. Soil Res.* 30, 855–892.

Leetham, J. W., McNary, T. J., Dodd, J. L., and Lauenroth, W. K. (1982). Response of soil nematodes, rotifers, and tardigrades to three levels of season-long sulphur dioxide exposure. *Water, Air Soil Pollut.* 18, 343–356.

Levine, N. D., Corliss, J. O., Cox, F. E. G., Deroux, D., Grain, J., Honigberg, B. M., Leedale, G. F., Loeblich, R., Ill, Lorn, J., Lynn, D., Merinfeld, E. G., Page, F. C., Poljansky, G., Sprague, V., Vavra, J., Wallace, F. G., and Wieser, J. (1980). A new revised classifica- tion of the protozoa. *J. Protozool.* 27, 36–58.

Liiri, M., Setälä, H., Haimi, J., Pennanen, T., and Fritze, H. (2002). Soil processes are not influenced by the functional complexity of soil decomposer food webs under disturbance. *Soil Biol. Biochem.* 34, 1009–1020.

Lindeman, R. L. (1942). The trophic-dynamic aspect of ecology. *Ecology* 23, 399–418.

Linden, D. R., Hendrix, P. F., Coleman, D. C., and van Vliet, P. C. J. (1994). Faunal indicators of soil quality. *In* "Defining Soil Quality for a Sustainable Environment." SSSA Special Publication No. 35 (J. W. Doran, D. C. Coleman, D. F. Bezdicek, and B. A. Stewart, eds.), pp. 91–106. American Society of Agronomy, Madison, Wl.

Lloyd, M., and Dybas, H. S. (1966). The periodical cicada problem. I. Population ecology. *Evolution* 20, 133–149.

Lock, K., and Dekoninck, W. (2001). Centipede communities on the inland dunes of eastern Flanders (Belgium). *Eur. J. Soil Biol.* 37, 113–116.

Lotka, A. J. (1925). "Elements of Physical Biology." Williams & Wilkins, Baltimore.

Lousier, J. D., and Parkinson, D. (1981). Evaluation of a membrane filter technique to count soil and litter Testacea. *Soil Biol. Biochem.* 13, 209–213.

Lousier, J. D., and Parkinson, J. (1984). Annual population dynamics and production ecology of Testacea (Protozoa, Rhizopoda) in an aspen woodland soil. *Soil Biol. Biochem.* 16, 103–114.

Lousier, J. D., and Bamforth, S. S. (1990). Soil Protozoa. *In* "Soil Biology Guide" (D. L. Dindal, ed.), pp. 97–136. Wiley, New York.

Lovelock, J. E. (1979). "Gaia: A New Look at Life on Earth." Oxford University Press, Oxford and London.

Lovelock, J. E. (1988). "The Ages of Gaia." W. W. Norton Co., New York.

Lussenhop, J. (1976). Soil arthropod response to prairie burning. *Ecology* 57, 88–98.

Lussenhop, J. (1992). Mechanisms of microarthropod—microbial interactions in soil. *Adv. Ecol. Res.* 23, 1–33.

Lutz, F. E. (1948). "Field Book of Insects of the United States and Canada" (3rd edition). G. P. Putnam's Sons, New York.

Luxton, M. (1967). The ecology of saltmarsh Acarina. *J. Anim. Ecol.* 36, 257–277.

Luxton, M. (1972). Studies on the oribatid mites of a Danish beech forest. I. Nutritional biology. *Pedobiologia* 12, 434–463.

Luxton, M. (1975). Studies on the oribatid mites of a Danish beech forest. II. Biomass, calorimetry and respirometry. *Pedobiologia* 15, 161–200.

Luxton, M. (1979). Food and energy processing by oribatid mites. *Rev. Écol. Biol. Sol* 16, 103–111.

Luxton, M. (1981a). Studies on the astigmatic mites of a Danish beech wood soil. *Pedobiologia* 22, 29–38.

Luxton, M. (1981b). Studies on the prostigmatic mites of a Danish beech wood soil. *Pedobiologia* 22, 277–303.

Lynch, J. M. (ed). (1990). "The Rhizosphere." Wiley-Interscience, Chichester.

MacArthur, R. H. (1972). "Geographical Ecology: Patterns in the Distribution of Species." Harper & Row, New York.

Macfadyen, A. (1969). The systematic study of soil ecosystems. *In* "The Soil Ecosystem" (J. G. Sheals, ed.), pp. 191–197. The Systematics Association, London.

Mackie-Dawson, L. A., and Atkinson, D. (1991). Methodology for the study of roots in field experiments and the interpretation of results. *In* "Plant Root Growth: An Ecological Perspective" (D. Atkinson, ed.), pp. 25–47. Blackwell, London.

Malloch, D. W., Pirozynski, K. A., and Raven, P. A. (1980). Ecological and evolutionary significance of mycorrhizal symbioses in vascular plants (a review). *Proc. Nat. Acad. Sci.* 77, 2113–2118.

Mann, L. K. (1986). Changes in soil organic carbon storage after cultivation. *Soil Science* 142, 279–288.

354   References

Manton, S. M. (1970). Arthropoda: Introduction. *In* "Chemical Ecology," vol. 5 (M. Florkin and B. T. Scheer, eds.). Academic Press, New York.

Marinissen, J. C. Y., and Dexter, A. R. (1990). Mechanisms of stabilization of earthworm casts and artificial casts. *Biol. Fertil. Soils* 9, 163–167.

Marshall, V. G. (1977). "Effects of manures and fertilizers on soil fauna: a review." Special Publication 3. Commonwealth Bureau of Soils, Slough, UK.

Marshall, V. G., Reeves, R. M., and Norton, R. A. (1987). Catalogue of the Oribatida (Acari) of the continental United States and Canada. *Mem. Ent. Soc. Canada* 139, 418 pp.

Martin, M. M. (1984). The role of ingested enzymes in the digestive processes of insects. *In* "Invertebrate—Microbial Interactions" (J. M. Anderson, A. D. M. Rayner, and D. W. H. Walton, eds.), pp. 155–172. Cambridge University Press, Cambridge.

Martin, J. K., and Kemp, J. R. (1986). The measurement of carbon transfers within the rhizosphere of wheat grown in field plots. *Soil Biol. Biochem.* 18, 103–107.

Mason, C. F. (1974). Mollusca. *In* "Biology of Plant Litter Decomposition," vol. 2 (C. H. Dickinson and G. J. F. Pugh, eds.), pp. 555–591. Academic Press, London.

Matamala, R., Gonzàlez—Meler, M. A., Jastrow, J. D., Norby, R. J., and Schlesinger, W. H. (2003). Impacts of fine root turnover on forest NPP and soil C sequestration potential. *Science* 302, 1385–1387.

Matsuko, K. (1994). Specialized predation on oribatid mites by two species of the ant genus *Myrmecina* (Hymenoptera, Formicidae). *Psyche* 101, 159–173.

Maynard, D. G., Stewart, J. W. B., and Bettany, J. R. (1984). Sulfur cycling in grassland and parkland soils. *Biogeochemistry* 1, 97–111.

Maynard, E. A. (1951). "The Collembola of New York State." Comstock Publishing, Ithaca, New York.

McAlpine, J. F. (1990). Insecta: Diptera Adults. *In* "Soil Biology Guide" (D. L. Dindal, ed.), pp. 1211–1252. Wiley, New York.

McBrayer, J. F. (1973). Exploitation of deciduous leaf litter by *Apheloria montana*. *Pedobiologia* 13, 90–98.

McDowell, W. H. (2003). Dissolved organic matter in soils—future directions and unanswered questions. *Geoderma* 113, 179–186.

McGill, W. B., and Cole, C. V. (1981). Comparative aspects of cycling of organic C, N, S, and P through soil organic matter. *Geoderma* 26, 267–286.

McNaughton, S. J. (1976). Serengeti migratory wildebeest: facilitation of energy flow by grazing. *Science* 191, 92–94.

McNaughton, S. J., Banyikawa, F. F., and McNaughton, M. M. (1998). Root biomass and productivity in a grazing ecosystem: the Serengeti. *Ecology* 79, 587–592.

McSorley, R. (1987). Extraction of nematodes and sampling methods. *In* "Principles and Practice of Nematode Control in Crops" (R. H. Brown and B. R. Kerry, eds.), pp. 13–41. Academic Press, Sydney, Australia.

Meentemeyer, V. (1978). Macroclimate and lignin control of decomposition. *Ecology* 59, 465–472.

Melillo, J. M., Aber, J. D., and Muratore, J. F. (1982). Nitrogen and lignin control of hardwood leaf litter decomposition dynamics. *Ecology* 63, 621–626.

Metcalf, C. L., and Flint, W. P. (1939). "Destructive and Useful Insects" (2nd edition). McGraw-Hill, New York.

Meyer, E. (1996). Mesofauna. *In* "Methods in Soil Biology" (F. Schinner, R. Öhlinger, E. Kandeler, and R. Margesin, eds.), pp. 338–345. Springer-Verlag, Berlin.

Michener, C. D., and Michener, M. H. (1951). "American Social Insects." Van Nostrand, New York.

Mikhail, W. Z. A. (1993). Effect of soil structure on soil fauna in a desert wadi in southern Egypt. *J. Arid Environ.* 24, 321–331.

Milchunas, D. G., Lauenroth, W. K., Singh, J. S., Cole, C. V., and Hunt, H. W. (1985). Root turnover and production by ^{14}C dilution: implications of carbon partitioning in plants. *Plant & Soil* 88, 353–365.

Mills, A. L. (2003). Keeping in touch: microbial life on soil particle surfaces. *Adv. Agron.* 78, 1–43.

Mitchell, M. J., and Parkinson, D. (1976). Fungal feeding of oribatid mites (Acari: Cryptostigmata) in an aspen woodland soil. *Ecology* 57, 302–312.

Molleman, F., and Walter, D. E. (2001). Niche separation and can-openers: Scydmaenid beetles as predators of armoured mites in Australia. *In* "Acarology: Proceedings of the 10th International Congress" (R. B. Halliday, D. E. Walter, H. C. Proctor, R. A. Norton, and M. J. Collof, eds.), pp. 283–288. CSIRO Pub., Melbourne.

Monz, C. A., Reuss, D. E., and Elliott, E. T. (1991). Soil microbial biomass carbon and nitrogen estimates using 2450 MHz microwave irradiation or chloroform fumigation followed by direct extraction. *Agric. Ecosys. Environ.* 34, 55–63.

Moore, J. C., St. John, T. V., and Coleman, D. C. (1985). Ingestion of vesicular arbuscular mycorrhizal hyphae and spores by soil microarthropods. *Ecology* 66, 1979–1981.

Moore, J. C., Ingham, E. R., and Coleman, D. C. (1987). Inter- and intraspecific feeding selectivity of *Folsomia candida* (Willem) (Collembola, Isotomidae) on fungi. *Biol. Fertil. Soils* 5, 6–12.

Moore, J. C., Walter, D. E., and Hunt, H. W. (1988). Arthropod regulation of micro- and mesobiota in belowground food webs. *Ann. Rev. Entomol.* 33, 419–439.

Moore, J. C., and de Ruiter, P. C. (1991). Temporal and spatial heterogeneity of trophic interactions within below-ground food webs. *Agric. Ecosys. Environ.* 34, 371–397.

Moore, J. C., de Ruiter, P. C., and Hunt, H. W. (1993). Influence of productivity on the stability of real and model ecosystems. *Science* 261, 906–908.

Moore, J. C., de Ruiter, P. C., Hunt, H. W., Coleman, D. C., and Freckman, D. W. (1996). Microcosms and soil ecology: critical linkages between field studies and modeling food webs. *Ecology* 77, 694–705.

Moore, J. C., and de Ruiter, P. C. (2000). Invertebrates in detrital food webs along gradients of productivity. *In* "Invertebrates as Webmasters in Ecosystems" (D. C. Coleman and P. F. Hendrix, eds.), pp. 161–184. CABI Publishing, Wallingford, U.K.

Moore, J. C., McCann, K., Setälä, H., and de Ruiter, P. C. (2003). Top-down is bottom-up: does predation in the rhizosphere regulate aboveground dynamics? *Ecology* 84, 846–857.

Morita, R. Y. (1997). "Bacteria in Oligotrophic Environments." Chapman & Hall, New York.

Mosier, A., Schimel, D., Valentine, D., Bronson, K., and Parton, W. (1991). Methane and nitrous oxide fluxes in native, fertilized and cultivated grasslands. *Nature* 350, 330–332.

Muchmore, W. B. (1990). Pseudoscorpionida. *In* "Soil Biology Guide" (D. L. Dindal, ed.), pp. 503–527. Wiley, New York.

Mueller, B. R., Beare, M. H., and Crossley, D. A., Jr. (1990). Soil mites in detrital food webs of conventional and no-tillage agroecosystems. *Pedobiologia* 34, 389–401.

Muma, M. H. (1966). Feeding behavior of North American Solpugida (Arachnida). *Florida Entomol.* 49, 199–216.

Mundel, P. (1990). Chilopoda. *In* "Soil Biology Guide" (D. L. Dindal, ed.), pp. 819–833. Wiley, New York.

Murphy, K. L., Klopatek, J. M., and Klopatek, C. C. (1998). The effects of litter quality and climate on decomposition along an elevation gradient. *Ecol. Appls.* 8, 1061–1071.

Nabholz, J. V., Reynolds, L. J., and Crossley, D. A., Jr. (1977). Range extension of *Pardosa lapidicina* Emerton (Araneia: Lycosidae) to Georgia. *J. Ga. Ent. Soc.* 12, 241–243.

Nadelhoffer, K. J., Aber, J. D., and Melillo, J. M. (1985). Fine roots, net primary production, and soil nitrogen availability: a new hypothesis. *Ecology* 66, 1377–1390.

Nadelhoffer, K. J., and Raich, J. W. (1992). Fine root production estimates and below-ground carbon allocation in forest ecosystems. *Ecology* 73, 1139–1147.

Nadkarni, N., Schaefer, D., Matelson, T. J., and Solano, R. (2002). Comparison of arboreal and terrestrial soil characteristics in a lower montane forest, Monteverde, Costa Rica. *Pedobiologia* 46, 24–33.

Nannipieri, P. (1994). The potential use of soil enzymes as indicators of productivity, sustainability and pollution. *In* "Soil Biota: Management in Sustainable Farming" (C. E. Pankhurst, B. M. Doube, V. V. S. R. Gupta, and P. R. Grace, eds.), pp. 238–244. CSIRO, Melbourne.

Nannipieri, P., Grego, S., and Ceccanti, B. (1990). Ecological significance of the biological activity in soil. *In* "Soil Biochemistry," vol. 6 (J.-M. Bollag and G. Stotzky, eds.), pp. 293–356. Marcel Dekker, New York.

Nannipieri, P., Kandeler, E., and Ruggiero, P. (2002). Enzyme activities and microbiological and biochemical processes in soil. *In* "Enzymes in the Environment" (R. G. Burns and R. P. Dick, eds.), pp. 1–33. Marcel Dekker, New York.

Nekola, J. (2003). Large-scale terrestrial gastropod community composition patterns in the Great Lakes region of North America. *Diversity and Distributions* 9, 55–71.

Nelson, D. R., and Higgins, R. P. (1990). Tardigrada. *In* "Soil Biology Guide" (D. L. Dindal, ed.), pp. 393–419. Wiley, New York.

Nelson, D. R., and Adkins, R. G. (2001). Distribution of tardigrades within a moss cushion: do tardigrades migrate in response to changing moisture conditions? *Zool. Anzeiger* 240, 493–500.

Neuhauser, E. F., and Hartenstein, P. (1978). Phenolic content and palatability of leaves and wood to soil isopods and diplopods. *Pedobiologia* 18, 99–109.

Neutel, A.-M., Heesterbeek, J. A. P., and de Ruiter, P. C. (2002). Stability in real food webs: weak links in long loops. *Science* 296, 1120–1123.

Newell, K. (1984a). Interaction between two decomposer basidiomycetes and a collembolan under Sitka spruce: distribution, abundance and selective grazing. *Soil Biol. Biochem.* 16, 227–233.

Newell, K. (1984b). Interaction between two basidiomycetes and Collembola under Sitka spruce: grazing and its potential effects on fungal distribution and litter decomposition. *Soil Biol. Biochem.* 16, 235–240.

Newell, S. Y. and Fallon, R. D. (1991). Toward a method for measuring instantaneous fungal growth rates in field samples. *Ecology* 72, 1547–1559.

Newton, A. F., Jr. (1990). Insects: Coleoptera Staphylinidae adults and larvae. *In* "Soil Biology Guide" (D. L. Dindal, ed.), pp. 1137–1174. Wiley, New York.

Newman, A. S., and Norman, A. G. (1943). An examination of thermal methods for following microbiological activity in soil. *Soil Sci. Soc. Amer. Proc.* 8, 250–253.

Newman, R. H., and Tate, K. R. (1980). Soil phosphorus characterization by 31P nuclear magnetic resonance. *Commun. Soil Sci. Plant Anal.* 11, 835–842.

Nielsen, C. O., and Christensen, B. (1959). The Enchytraeidae: critical revision and taxonomy of European species. *Natura Jutlandica* 8–9, 1–160.

Nielsen, C. O., and Christensen, B. (1961). The Enchytraeidae: critical revision and taxonomy of European species. *Natura Jutlandica (Suppl. 1)* 10, 1961:1–23.

Nielsen, C. O., and Christensen, B. (1963). The Enchytraeidae: critical revision and taxonomy of European species. *Natura Jutlandica (Suppl. 2)* 10, 1–19.

Nielson, G. A., and Hole, F. D. (1964). Earthworms and the development of coprogenous A1-horizons in forest soils of Wisconsin. *Soil Sci. Soc. Amer. Proc.* 28, 426–430.

Northup, R. R., Yu, Z., Dahlgren, R. A., and Vogt, K. A. (1995). Polyphenol control of nitrogen release from pine litter. *Nature* 377, 227–229.

Norton, R. A. (1994). Book review: primitive oribatids of the Palaearctic region (J. Balogh and S. Mahunka). *Systematic Zool.* 33, 472–474.

Norton, R. A., Bonamo, P. M., Grierson, J. D., and Shear, W. A. (1987). Fossil mites from the Devonian of New York State. *In* "Progress in Acarology," vol. 1 (G. P. Channabasavanna and C. A. Viraktamath, eds.), pp. 271–277. Oxford & IBM Publishing Co., New Delhi.

Norton, R. A., and Behan-Pelletier, V. M. (1991). Calcium carbonate and calcium oxalate as cuticular hardening agents in oribatid mites (Acari: Oribatida). *Can. J. Zool.* 69, 1504–1511.

Norton, R. A., and Alberti, G. (1997). Porose integumental organs of oribatid mites (Acari, Oribatida). Evolutionary and ecological aspects. *Zoologica (Stuttgart)* 146, 115–143.

Norton, R. A., Alberti, G., Weigmann, G., and St. Woas, S. (1997). Porose integumental organs of oribatid mites (Acari, Oribatida). 1. Overview of types and distribution. *Zoologica (Stuttgart)* 146, 1–31.

Nosek, J. (1973). "The European Protura: their taxonomy, ecology and distribution with keys for determination." Muséum d'Histoire Naturelle, Genéve.

Noti, M.-I., André, H. M., Ducarme, X., and Lebrun, P. (2003). Diversity of soil oribatid mites (Acari: Oribatida) from High Katanga (Democratic Republic of Congo): a multiscale and multifactor approach. *Biodiversity and Conservation* 12, 767–785.

Nunan, N., Wu, K., Young, I. M., Crawford, J. W., and Ritz, K. (2002). *In situ* spatial patterns of soil bacterial populations, mapped at multiple scales, in an arable soil. *Microb. Ecol.* 44, 296–305.

Oades, J. M. (1984). Soil organic matter and structural stability: mechanisms and implications for management. *Plant & Soil* 76, 319–337.

Oades, J. M., Gillman, G. P., Uehara, G., Hue, N. V., van Noordwijk, M., Robertson, G. P., and Wada, K. (1989). Interactions of soil organic matter and variable-charge clays. *In* "Dynamics of Soil Organic Matter in Tropical Ecosystems" (D. C. Coleman, J. M. Oades, and G. Uehara, eds.), pp. 69–95. University of Hawaii Press, Honolulu.

Oades, J. M., and Waters, A. G. (1991). Aggregate hierarchy in soils. *Aust. J. Soil Res.* 29, 815–828.

O'Connor, F. B. (1955). Extraction of enchytraeid worms from a coniferous forest soil. *Nature* 175, 815–816.

Odum, E. P. (1971). "Fundamentals of Ecology," (3rd edition). W.B. Saunders Company, Philadelphia.

Odum, E. P. and Biever, L. J. (1984). Resource quality, mutualism and energy partitioning in food chains. *Am. Nat.* 124, 360–376.

Ogram, A., and Sharma, K. (2002). Methods of soil microbial community analysis. *In* "Manual of Environmental Microbiology," (2nd edition). (C. J. Hurst, R. L. Crawford, G. R. Knudsen, M. J. McInerney, and L. D. Stetzenbach, eds.), pp.554–563, American Society for Microbiology, Washington, D.C.

O'Lear, H. A., and Blair, J. M. (1999). Responses of soil microarthropods to changes in soil water availability in tallgrass prairie. *Biol. Fertil. Soils* 29, 207–217.

Olson, J. S. (1963). Energy storage and the balance of producers and decomposers in ecological systems. *Ecology* 44, 322–331.

Olson, J. S., Watts, J. A., and Allison, L. J. (1983). "Carbon in Live Vegetation of the Major World Ecosystems." National Technical Information Service, Springfield, Virginia.

O'Neill, R. V., DeAngelis, D. L., Waide, J. B., and Allen, T. F. H. (1986). "A Hierarchical Concept of Ecosystems." Princeton University Press, Princeton, New Jersey.

Pace, N. R. (1997). A molecular view of microbial diversity and the biosphere. *Science* 276, 735–740.

Pace, N. R. (1999). Microbial ecology and diversity. *ASM News* 65, 328–333.

Padmanabhan, P., Padmanabhan, S., DeRito, C., Gray, A., Gannon, D., Snape, J. R., Tsai, C. S., Park, W., Jeon, C., and Madsen, E. L. (2003). Respiration of ^{13}C-labeled substrates added to soil in the field and subsequent 16S rRNA gene analysis of ^{13}C-labeled soil DNA. *Appl. Environ. Microbiol.* 69, 1614–1622.

Page, A. L., Miller, R. H., and Keeney, D. R. (1982). "Methods of Soil Analysis: Part 2— Chemical and Microbiological Properties." ASA, SSSA, Madison, WI.

Pallant, D. (1974). Assimilation of the grey field slug (*Agriolimax reticulatus* [Mueller]). *Proc. Malacol. Soc. London* 41, 99.

Palm, C. A., Gachengo, C. N., Delve, R. J., Cadisch, G., and Giller, K. E. (2001). Organic inputs for soil fertility management in tropical agroecosystems: application of an organic resource database. *Agric. Ecosys. Environ.* 83, 27–42.

Pantastico-Caldas, M., Duncan, K. E., Istock, C. A., and Bell, J. A. (1992). Population dynamics of bacteriophage and *Bacillus subtilis* in soil. *Ecology* 73, 1888–1902.

Park, O. (1947). Observations on *Batrisodes* (Coleoptera: Pselaphidae), with particular reference to the American species east of the Rocky Mountains. *Bull. Chicago Acad. Sci.* 8, 45–132.

Parker, L. W., Santos, P. F., Phillips, J., and Whitford, W. G. (1984). Carbon and nitrogen dynamics during the decomposition of litter and roots of a Chihuahuan desert annual *Lepidium lasiocarpum*. *Ecol. Monogr.* 54, 339–360.

Parkinson, D., Gray, T. R. G., and Williams, S. T. (1971). "Methods for Studying the Ecology of Soil Microorganisms." IBP Handbook No. 19. Blackwell, Oxford.

Parkinson, D., and Coleman, D. C. (1991). Microbial populations, activity and biomass. *Agric. Ecosys. Environ.* 34, 3–33.

Parmelee, R. W., Beare, M. H., Cheng, W., Hendrix, P. F., Rider, S. J., Crossley, D. A., Jr., and Coleman, D. C. (1990). Earthworms and enchytraeids in conventional and no-tillage agroecosystems: a biocide approach to assess their role in organic matter breakdown. *Biol. Fertil. Soils* 10, 1–10.

Parmelee, R. W., Bohlen, P. J., and Blair, J. M. (1998). Earthworms and nutrient cycling processes: integrating across the ecological hierarchy. *In* "Earthworm Ecology" (C. Edwards, ed.), pp. 123–143. St. Lucie Press, Boca Raton, FL.

Parton, W. J., Schimel, D. S., Cole, C. V., and Ojima, D. S. (1987). Analysis of factors controlling soil organic matter levels in Great Plains grasslands. *Soil Sci. Soc. Am. J.* 51, 1173–1179.

Parton, W. J., Cole, C. V., Stewart, J. W. B., Schimel, D. S., and Ojima, D. (1989a). Simulating the long-term dynamics of C, N and P in soils. *In* "Ecology of Arable Land— Perspectives and Challenges" (M. Clarholm and L. Bergström, eds.), Martinus Nijhoff, Dordrecht.

Parton, W. J., Sanford, R. L., Sanchez, P. A., and Stewart, J. W. B. (1989b). Modeling soil organic matter dynamics in tropical soils. *In* "Dynamics of Soil Organic Matter in Tropical Ecosystems" (D. C. Coleman, J. M. Oades, and G. Uehara, eds.), pp. 153–171. University of Hawaii, Honolulu.

Pastor, J., and Post, W. M. (1988). Responses of northern forests to CO_2-induced climate change. *Nature* 334, 55–58.

Pate, J. S., Layzell, D. B., and Atkins, C. A. (1979). Economy of carbon and nitrogen in a nodulated and non-nodulated (NO_3 grown) legume plant. *Plant Physiol.* 64, 1083–1088.

Paul, E. A., and Clark, F. E. (1989). "Soil Microbiology and Biochemistry." Academic Press, San Diego.

Paul, E. A., and Clark, F. E. (1996). "Soil Microbiology and Biochemistry" (2nd edition). Academic Press, San Diego.

Paustian, K., Andrén, O., Clarholm, M., Hansson, A.-C., Johansson, G., Lagerlöf, J., Lindberg, T., Pettersson, R., and Sohlenius, B. (1990). Carbon and nitrogen budgets of four agro-ecosystems with annual and perennial crops, with and without N fertilization. *J. Appl. Ecol.* 27, 60–84.

Paustian, K., Ågren, G. I., and Bosatta, E. (1997). Modeling litter quality effects on decomposition and soil organic matter dynamics. *In* "Driven by Nature: Plant Litter Quality and Decomposition" (G. Cadisch and K. E. Giller, eds.), pp. 313–335. CABI Publishing, Wallingford, U.K.

Pawluk, S. (1987). Faunal micromorphological features in moder humus of some western Canadian soils. *Geoderma* 40, 3–16.

Payne, J. A. (1965). A summer carrion study of the baby pig, *Sus scrofa* L. *Ecology* 46, 592–602.

Payne, W. J. (1970). Energy yields and growth of heterotrophs. *Ann. Rev. Microbiol.* 24, 17–52.

Pearce, M. J. (1997). "Termites: Biology and Pest Management." CABI Publishing, Wallingford, U.K.

Perdue, J. C. (1987). "Population dynamics of mites (Acari) in conventional and conservation tillage agroecosystems." Ph.D. Dissertation, University of Georgia, Athens, GA.

Perdue, J. C., and Crossley, D. A., Jr. (1989). Seasonal abundance of soil mites (Acari) in experimental agroecosystems: effects of drought in no-tillage and conventional tillage. *Soil Till. Res.* 15, 117–124.

Perez-Moreno, J., and Read, D. J. (2001). Nutrient transfer from soil nematodes to plants: a direct pathway provided by the mycorrhizal mycelial network. *Plant, Cell Environ.* 24, 1219–1226.

Persson, T. (ed.) (1980). "Structure and function of northern coniferous forests—an ecosystem study." Ecological Bulletins (Stockholm) 32.

Pesek, J. C. (ed.) (1989). "Alternative Agriculture." National Research Council Press, Washington, D.C.

Petersen, H. A., and Luxton, M. (1982). A comparative analysis of soil fauna populations and their role in decomposition processes. *Oikos* 39, 287–388.

Petersen, H. (2002). Collembolan ecology at the turn of the millennium. *Pedobiologia* 46, 246–260.

Philips, J. R. (1990). Acarina: Astigmata (Acaridida). *In* "Soil Biology Guide" (D. L. Dindal, ed.), pp. 757–778. Wiley, New York.

Phillips, R. E., and Phillips, S. H. (1984). "No-Tillage Agriculture: Principles and Practices." Van Nostrand Reinhold, New York.

Piearce, T. G., and Phillips, M. J. (1980). The fate of ciliates in the earthworm gut: an in vitro study. *Microb. Ecol.* 5, 313–320.

Pimm, S. L. (1982). "Food Webs." Chapman and Hall, London.

Pimm, S. L., and Lawton, J. H. (1980). Are food webs divided into compartments? *J. Anim. Ecol.* 49, 879–898.

Pirozynski, K. A., and Malloch, D. W. (1975). The origin of land plants: a matter of mycotrophism. *BioSystems* 6, 153–164.

Pirozynski, K. A., and Hawksworth, D. L. (1988). "Coevolution of Fungi with Plants and Animals." Academic Press, London.

Plass, G. N. (1956). Carbon dioxide and the climate. *Am. Sci.* 44, 302–319.

Poinar, G. O., Jr. (1983). "The Natural History of Nematodes." Prentice Hall, Englewood Cliffs, NJ.

Pokarzhevskii, A. D., van Straalen, N. M., Zaboev, D. P., and Zaitsev, A. S. (2003). Microbial links and element flows in nested detrital food-webs. *Pedobiologia* 47, 213–224.

Polis, G. A. (1991). Complex trophic interactions in deserts: an empirical critique of food-web theory. *Am. Nat.* 138, 123–155.

Pomeroy, L. R. (1974). The ocean's food web, a changing paradigm. *Bioscience* 24, 499–504.

Ponge, J.-F. (1991). Food resources and diets of soil animals in a small area of Scots pine litter. *Geoderma* 49, 33–62.

Ponge, J. -F. (2003). Humus forms in terrestrial ecosystems: a framework to diversity. *Soil Biol. Biochem.* 35, 935–945.

Ponge, J.-F., Chevalier, R., and Loussot, P. (2002). Humus Index: An integrated tool for the assessment of forest floor and topsoil properties. *Soil Sci. Soc. Am. J.* 66, 1996–2001.

Porazinska, D. L., Bardgett, R. D., Blaauw, M. B., Hunt, H. W., Parsons, A. N., Seastedt, T. R., and Wall, D. H. (2003). Relationships at the aboveground–belowground interface: plants, soil biota, and soil processes. *Ecol. Monogr.* 73, 377–395.

Porter, K. G. (1975). Enhancement of algal growth and productivity by grazing zooplankton. *Science* 192, 1332–1334.

Post, W. H., Peng, T. H., Emanuel, W. R., King, A. W., Dale, V. H., and De Angelis, D. L. (1990). The global carbon cycle. *Am. Sci.* 78, 310–326.

Post, W. H., King, A. W., and Wullschleger, S. D. (1996). Soil organic matter models and global estimates of soil organic carbon. *In* "Evaluation of Soil Organic Matter Models" (D. S. Powlson, P. Smith, and J. U. Smith, eds.), pp. 201–222. NATO ASI Series, vol. 38. London.

Postgate, J. R. (1987). "Nitrogen Fixation" (2nd edition). Studies in Biology, No. 92. Arnold, London.

Postma, J., and Altemüller, H.-J. (1990). Bacteria in thin soil sections stained with the fluorescence brightener Calcofluor M2R. *Soil Biol. Biochem.* 22, 89–96.

Powlson, D. S. (1975). Effects of biocidal treatments on soil organisms. *In* "Soil Microbiology" (N. Walker, ed.), Butterworths, London.

Powlson, D. S. (1994). The soil microbial biomass: before, beyond and back. *In* "Beyond the Biomass" (K. Ritz, J. Dighton, and K. E. Giller, eds.), pp. 3–20. Wiley-Sayce, Chichester.

Pratt, H. D., and Stojanovich, C. J., (1967). Hymenoptera: key to some common species which sting man. *In* "Pictorial Keys to Arthropods, Reptiles, Birds and Mammals of Public Health Significance" (Anon.), pp. 102–118. National Communicable Disease Center, Atlanta, GA.

Pregitzer, K. S., DeForest, J. L., Burton, A. J., Allen, M. F., Ruess, R. W., and Hendrick, R. L. (2002). Fine root architecture of nine North American trees. *Ecol. Monogr.* 72, 293–309.

Price, D. W. (1975). Vertical distribution of small arthropods in a California pine forest soil. *Ann. Ent. Soc. Amer.* 68, 174–180.

Prosser, J. I. (2002). Molecular and functional diversity in soil micro-organisms. *Plant & Soil* 244, 9–17.

Publicover, D. A., and Vogt, K. A. (1993). A comparison of methods for estimating forest fine root production with respect to sources of error. *Can. J. For. Res.* 23, 1179–1186.

Punzo, F. (1998). "The Biology of Camel Spiders." Kluwer Academic Pub., Boston.

Purvis, G., and Fadel, A. (2002). The influence of cropping rotations and soil cultivation practice on the population ecology of carabids (Coleoptera: Carabidae) in arable land. *Pedobiologia* 46, 452–474.

Pussard, M., Alabouvette, C., and Levrat, P. (1994). Protozoan interactions with the soil microflora and possibilities for biocontrol of plant pathogens. *In* "Soil Protozoa" (J. Darbyshire, ed.), pp. 123–146. CABI Publishing, Wallingford, U.K.

Radajewski, S., Ineson, P., Parekh, N. R., and Murrell, J. C. (2000). Stable-isotope probing as a tool in Microbial Ecology. *Nature* 403, 646–649.

Rao, V. R., Ramakrishnan, B., Adhya, T. K., Kanungo, P. K., and Nayak, D. N. (1998). Current status and future prospects of associative nitrogen fixation in rice. *World J. Microbiol. Biotechnol.* 14, 621–633.

Read, D. J. (1991). Mycorrhizas in ecosystems. *Experientia* 47, 376–3911.

Read, D. J., Francis, R., and Finlay, R. D. (1985). Mycorrhizal mycelia and nutrient cycling in plant communities. In "Ecological Interactions in Soil: Plants, Microbes and Animals" (A. H. Fitter, D. Atkinson, D. J. Read, and M. B. Usher, eds.), pp. 193–217. Blackwell, Oxford.

Read, D. J., Duckett, J. G., Francis, R., Ligrone, R., and Russell, A. (2000). Symbiotic fungal associations in "lower" land plants. *Phil. Trans. Roy. Soc. London B Biol. Sci.* 355, 815–830.

Redeker, D. (2002). Molecular identification and phylogeny of arbuscular mycorrhizal fungi. *Plant & Soil* 244, 67–73.

Redeker, D., Morton, J. B., and Bruns, T. D. (2000). Ancestral lineages of arbuscular mycorrhizal fungi (Glomales). *Molecular Phylogeny and Evolution* 14, 276–284.

Reeves, R. M. (1967). Seasonal distribution of some forest soil Oribatei. In "Proceedings 2nd International Congress of Acarology" (G. O. Evans, ed.), pp. 23–30. Akademiai Kiado, Budapest.

Reeves, R. M. (1973). Oribatid ecology. In "Proceedings of the First Soil Microcommunities Conference" (D. L. Dindal, ed.), pp. 157–175. CONF-711-76, U.S. Atomic Energy Commission, Technical Information Center, Washington, D.C.

Reichle, D. E., and Crossley, D. A., Jr. (1965). Radiocesium dispersal in a cryptozoan food web. *Health Physics* 11, 1375–1384.

Reichle, D. E., Ausmus, B. S., and McBrayer, J. F. (1975). Ecological energetics of decomposer invertebrates in a deciduous forest and total respiration budget. In "Progress in Soil Zoology" (J. Vanek, ed.), pp. 283–292. Proc. 5th Int. Coll. Soil Zool., Prague.

Reid, J. B., and Goss, M. J. (1982). Suppression of decomposition of ^{14}C-labelled plant roots in the presence of living roots of maize and perennial ryegrass. *J. Soil Sci.* 33, 387–395.

Reynolds, J. W. (1973). Earthworm (Annelida: Oligochaeta) ecology and systematics. In D. L. Dindal (ed.), Proceedings of the 1st Soil Microcommunity Conference, pp. 95–120. U.S. Atomic Energy Commission, Washington, D.C.

Reynolds, J. W. (1977). "The Earthworms (Lumbricidae and Sparganophilidae) of Ontario." Life Sci. Misc. Publ., Royal Ontario Museum, Toronto.

Reynolds, J. W., and Cooke, D. G. (1993). "Nomenclatura Oligochaetologica. Supplementum Tertium." New Brunswick Museum Monograph Series, No. 9, 33 pp.

Reynoldson, T. B. (1939). Enchytraeid worms and the bacteria bed method of sewage treatment. *Ann. Appl. Biol.* 26, 139–164.

Richter, D. D., and Markewitz, D. (2001). "Understanding Soil Change: Soil Sustainability Over Millennia, Centuries, and Decades." Cambridge University Press, Cambridge.

Rillig, M. C., Wright, S. F., and Eviner, V. T. (2002). The role of arbuscular mycorrhizal fungi and glomalin in soil aggregation: comparing effects of five plant species. *Plant & Soil* 238, 325–333.

Robertson, G. P., and Gross, K. L. (1994). Assessing the heterogeneity of belowground resources: quantifying pattern and scale. In "Exploitation of Environmental Heterogeneity by Plants" (M. M. Caldwell and R. W. Pearcy, eds.), pp. 237–253. Academic Press, San Diego.

Robertson, G. P., Coleman, D. C., Bledsoe, C. S., and Sollins, P. (eds.) (1999). "Standard Soil Methods For Long-Term Ecological Research." Oxford University Press, New York.

Robinson, D., and Scrimgeour, C. M. (1995). The contribution of plant C to soil CO_2 measured using ^{13}C. *Soil Biol. Biochem.* 27, 1653–1656.

Rogers, J. E., and Whitman, W. B. (1991). "Microbial Production and Consumption of Greenhouse Gasses: Methane, Nitrogen Oxides, and Halomethanes." American Society for Microbiology, Washington, D.C.

Rosomer, W. S., and Stoffolano, J. G., Jr. (1994). "The Science of Entomology. Third Edition." Brown, Dubuque.

Roth, C. H., and Joschko, M. (1991). A note on the reduction of runoff from crusted soils by earthworm burrows and artificial channels. *Ztschr. Pflanzenernähr. Bodenk.* 154, 101–105.

Roth, V. D. (1993). "Spider Genera of North America." The University of Florida, Gainesville.

Rothwell, F. M. (1984). Aggregation of surface mine soil by interaction between VAM fungi and lignin degradation products of lespedeza. *Plant & Soil* 80, 99–104.

Rovira, A. D., Foster, R. C., and Martin, J. K. (1979). Note on terminology: origin, nature and nomenclature of the organic materials in the rhizosphere. *In* "The Soil–Root Interface" (J. L. Harley and R. S. Russell, eds.), pp. 1–4. Academic Press, London.

Ruess, L., Häggblom, M. M., Zapata, E. J. G., and Dighton, J. (2002). Fatty acids of fungi and nematodes—possible biomarkers in the soil food chain? *Soil Biol. Biochem.* 34, 745–756.

Rusek, J. (1975). Die bodenbildende Funktion von Collembolen und Acarina. *Pedobiologia* 15, 299–308.

Rusek, J. (1985). Soil microstructures—contributions on specific soil organisms. *Quaest. Entomol.* 21, 497–514.

Russell, E. J. (1973). "Soil Conditions and Plant Growth." Longmans, London.

Russell, E. J., and Hutchinson, H. B. (1909). The effect of partial sterilization of soil on the production of plant food. *J. Agric. Sci.* 3, 111–144.

Saano, A., and Lindström, K. (1995). Isolation and identification of DNA from soil. *In* "Methods in Applied Soil Microbiology and Biochemistry" (K. Alef and P. Nannipieri, eds.), pp. 440–451. Academic Press, London.

Sanchez, P. A., Palm, C. A., and Buol, S. W. (2003). Fertility capability soil classification: a tool to help assess soil quality in the tropics. *Geoderma* 114, 157–185.

Santos, P. F., and Whitford, W. G. (1981). The effects of microarthropods on litter decomposition in a Chihuahuan Desert ecosystem. *Ecology* 62, 654–663.

Satchell, J. E. (1969). Methods of sampling earthworm populations. *Pedobiologia* 9, 20–25.

Scharpenseel, H. W., Schomaker, M., and Ayoub, A. (1990). "Soils on a Warmer Earth." Elsevier, Amsterdam.

Schauermann, J. (1983). Eine Verbesserung der Extraktionsmethode für terrestrische Enchytraeiden. *In* "New Trends in Soil Biology" (P. Lebrun, H. M. André, A. De Medts, C. Grégoire-Wibo, and G. Wauthy, eds.), pp. 669–670. Dieu-Brichart, Louvain-la-Neuve, Belgium.

Scheller, U. (1988). The Pauropoda of the Savanna River Plant, Aiken, South Carolina. *SRO–NERP* 17, 1–99.

Scheller, U. (1990). Pauropoda. *In* "Soil Biology Guide" (D. L. Dindal, ed.), pp. 861–890. Wiley, New York.

Scheller, U. (2002). Pauropods—the little ones among the Myriapods. *All Taxa Biodiversity Inventory* 3, 3.

Schenker, R. (1984). Spatial and seasonal distribution patterns of oribatid mites (Acari: Oribatei) in a forest soil ecosystem. *Pedobiologia* 27, 133–149.

Scheu, S. (2002). The soil food web: structure and perspectives. *Eur. J. Soil Biol.* 38, 11–20.

Scheu, S., and Setälä, H. (2002). Multitrophic interactions in decomposer food-webs. *In* "Multitrophic Level Interactions" (B. Tscharntke and B. A. Hawkins, eds.), pp. 223–264. Cambridge University Press, Cambridge.

Schimel, D. S. (1993). Population and community processes in the response of terrestrial ecosystems to global change. *In* "Biotic Interactions and Global Change" (P. M. Kareiva, J. G. Kingsolver, and R. B. Huey, eds.), pp. 45–54. Sinauer Associates, Sunderland, MA.

Schimel, D. S., Braswell, B. H., Holland, E. A., McKeown, R., Ojima, D. S., Painter, T. H., Parton, W. J., and Townsend, A. R. (1994). Climatic, edaphic, and biotic controls over storage and turnover of carbon in soils. *Global Biogeochem. Cycles* 8, 279–293.

Schimel, J. P., and Gulledge, J. M. (1998). Microbial community structure and global trace gases. *Global Change Biology* 4, 745–758.

Schimel, J. P., and Weintraub, M. N. (2003). The implications of exoenzyme activity on microbial carbon and nitrogen limitation in soil: a theoretical model. *Soil Biol. Biochem.* 35, 549–563.

Schinner, F., Öhlinger, R., Kandeler, E., and Margesin, R. (eds.) (1996). "Methods in Soil Biology." Springer-Verlag, Berlin.

Schlesinger, W. H. (1991). "Biogeochemistry: An Analysis of Global Change." Academic Press, San Diego.

Schlesinger, W. H. (1996). "Biogeochemistry: An Analysis of Global Change" (2nd edition). Academic Press, San Diego.

Schmidt, O. (2001). Appraisal of the electrical octet method for estimating earthworm populations in arable land. *Ann. Appl. Biol.* 138, 231–241.

Schneider, S. H. (1989). The greenhouse effect: science and policy. *Science* 243, 771–781.

Schoenholzer, F., Hahn, D., and Zeyer, J. (1999). Origins and fate of fungi and bacteria in the gut of *Lumbricus terrestris* L. studied by image analysis. *FEMS Microbiol. Ecol.* 28, 235–248.

Schrader, H. S., Schrader, J. O., Walker, J. J., Bruggeman, N. B., Vanderloop, J. M., Shaffer, J. M., and Kokjohn, T. A. (1997). Effects of host starvation on bacteriophage dynamics. *In* "Bacteria in Oligotrophic Environments" (R. Y. Morita, ed.), pp. 368–385. Chapman & Hall, New York.

Schultz, P. A. (1991). Grazing preferences of two collembolan species, *Folsomia candida* and *Proisotoma minuta*, for ectomycorrhizal fungi. *Pedobiologia* 35, 313–325.

Schuster, R. (1956). Der anteil der Oribatiden and den Zersetzungsvorgängen im Böden. *Ztschr. für Morphol. und Oekol. Tiere* 45, 1–33.

Schwert, D. P. (1990). Oligochaeta: Lumbricidae. *In* "Soil Biology Guide" (D. L. Dindal, ed.), pp. 341–356. Wiley, New York.

Scow, K. M. (1997). Soil microbial communities and carbon flow in agroecosystems. *In* "Ecology in Agriculture" (L. E. Jackson, ed.), pp. 367–413. Academic Press, San Diego.

Seastedt, T. R. (1984). The role of microarthropods in the decomposition and mineralization of litter. *Ann. Rev. Ecol. System.* 29, 25–46.

Seastedt, T. R. (1984a). Microarthropods of burned and unburned tallgrass prairie. *J. Kansas Ent. Soc.* 57, 468–476.

Seastedt, T. R. (1984b). The role of microarthropods in decomposition and mineralization processes. *Ann. Rev. Entomol.* 29, 25–46.

Setälä, H., Rissanen, J., and Markkola, M. (1997). Conditional outcomes in the relationship between pine and ectomycorrhizal fungi in relation to biotic and abiotic environment. *Oikos* 80, 112–122.

Sgardelis, S., Stamou, G., and Margaris, N. S. (1981). Structure and spatial distribution of soil arthropods in a Phryganic (East Mediterranean) ecosystem. *Rev. Écol. Biol. Sol* 18, 221–230.

Shamoot, S., McDonald, L., and Bartholomew, W. V. (1968). Rhizo-deposition of organic debris in soil. *Soil Sci. Soc. Amer. Proc.* 32, 817–820.

Shain, D. H., Carter, M. R., Murray, K. P., Maleski, K. A., Smith, N. R., McBride, T. R., Michalewicz, L. A., and Saidel, W. M. (2000). Morphologic characterization of the ice worm *Mesenchytraeus solifugus*. *J. Morphology* 246, 192–197.

Shanks, R. E., and Olson, J. S. (1961). First-year breakdown of leaf litter in southern Appalachian forests. *Science* 134, 194–195.

Shaw, M. R., Zavaleta, E. S., Chiariello, N. R., Cleland, E. E., Mooney, H. A., and Field, C. B. (2002). Grassland responses to global environmental changes suppressed by elevated CO_2. *Science* 298, 1987–1990.

Shelley, R. M. (2002). "A Synopsis of the North American Centipedes of the Order Scolopendromorpha (Chilopoda)." Memoir 5, Virginia Museum of Natural History, Martinsville.

Shelley, R. M., and Sisson, W. D. (1995). Distributions of the scorpions *Centruroides vittatus* (Say) and *Centruroides hentzi* (Banks) in the United States and Mexico (Scorpiones, Buthidae). *J. Arachnol.* 23, 100–110.

Shimmel, S. M., and Darley, W. M. (1985). Productivity and density of soil algae in an agricultural system. *Ecology* 66, 1439–1447.

Shipton, P. J. (1986). Infection by foot and root rot pathogens and subsequent damage. *In* "Plant Diseases, Infection, Damage and Loss" (R. K. S. Wood and G. K. Jellis, eds.), pp. 139–155. Blackwell Scientific, Oxford.

Siepel, H., and de Ruiter-Dijkman, E. M. (1993). Feeding guilds of oribatid mites based on their carbohydrase activities. *Soil Biol. Biochem.* 25, 1491–1497.

Simonson, R. W. (1959). Outline of a generalized theory of soil genesis. *Soil Sci. Soc. Amer. Proc.* 23, 152–156.

Sims, R. W., and Gerard, B. M. (1985). "Earthworms: Keys and Notes for the Identification and Study of the Species." The Linnaean Society of London, London.

Sinclair, J. L., and Ghiorse, W. C. (1989). Distribution of aerobic bacteria, protozoa, algae, and fungi in deep subsurface sediments. *Geomicrobiol. J.* 7, 15–31.

Singh, B. N. (1946). A method of estimating the numbers of soil Protozoa especially amoebae, based on their differential feeding on bacteria. *Ann. Appl. Biol.* 33, 112–119.

Singh, J. S., Lauenroth, W. K., Hunt, H. W., and Swift, D. M. (1984). Bias and random errors in estimators of net root production: a simulation approach. *Ecology* 65, 1760–1764.

Singleton, D. R., Furlong, M. A., Peacock, A. D., White, D. C., Coleman, D. C., and Whitman, W. B. (2003). *Solirubrobacter pauli* gen. nov., sp. nov., a mesophilic bacterium within the Rubrobacteridae related to common soil clones. *Int. J. System. Evol. Microbiol.* 53, 485–490.

Sinsabaugh, R. L., Moorhead, D. L., and Linkins, A. E. (1994). The enzymic basis of plant litter decomposition: emergence of an ecological process. *Appl. Soil Ecol.* 1, 97–111.

Sinsabaugh, R. L., and Moorhead, D. L. (1997). Synthesis of litter quality and enzymatic approaches to decomposition modeling. *In* "Driven by Nature: Plant Litter Quality and Decomposition." (G. Cadisch and K. E. Giller, eds.), pp. 363–375. CABI Publishing, Wallingford, U.K.

Six, J., Elliott, E. T., Paustian, K., and Doran, J. W. (1998). Aggregation and soil organic matter accumulation in cultivated and native grassland soils. *Soil Sci. Soc. Am. J.* 62, 1367–1377.

Six, J., Elliott, E. T., and Paustian, K. (1999). Aggregate and soil organic matter dynamics under conventional and no-tillage systems. *Soil Sci. Soc. Am. J.* 63, 1350–1358.

Six, J., Conant, R. T., Paul, E. A., and Paustian, K. (2002a). Stabilization mechanisms of soil organic matter: implications for C-saturation of soils. *Plant & Soil* 241, 155–176.

Six, J., Feller, C., Denef, K., Ogle, S. M., Sa, J. C. D. M., and Albrecht, A. (2002b). Soil organic matter, biota and aggregation in temperate and tropical soils—effects of no-tillage. *Agronomie* 22, 755–775.

Skujins, J. J. (1967). Enzymes in soil. *In* "Soil Biochemistry," vol. 1 (A. D. McLaren and G. H. Peterson, eds.), pp. 371–414. Marcel Dekker, New York.

Smith, M. D., Hartnett, D. C., and Wilson, G. W. T. (1999). Interacting influence of mycorrhizal symbiosis and competition on plant diversity in tallgrass prairie. *Oecologia* 121, 574–582.

Smith, M. L., Bruhn, J. N., and Anderson, J. B. (1992). The root-infecting fungus *Armillaria bulbosa* may be among the largest and oldest living organisms. *Nature* 356, 428–443.

Smith, P., Andrén, O., Brussaard, L., Dangerfield, M., Ekschmitt, K., Lavelle, P., and Tate, K. (1998). Soil biota and global change at the ecosystem level: describing soil biota in mathematical models. *Global Change Biology* 4, 773–784.

Smith, S. J., and Read, D. J. (1997). "Mycorrhizal symbiosis" (2nd edition). Academic Press, Cambridge, U.K.

Smith, T. M., Leemans, R., and Shugart, H. H. (1992). Sensitivity of terrestrial carbon storage to CO_2-induced climate change: comparison of four scenarios based on general circulation models. *Climatic Change* 21, 367–384.

Smucker, A. J. M., Ferguson, J. C., De Bruyn, W. P., Belford, R. K., and Ritchie, J. T. (1987). Image analysis of video-recorded plant root systems. *In* "Minirhizotron Observation Tubes: Methods and Applications for Measuring Rhizosphere Dynamics" (H. M. Taylor, ed.), pp. 67–80. American Society of Agronomy Special Publication No. 50, Madison, Wisconsin.

Snider, R. J. (1967). An annotated list of the Collembola (Springtails) of Michigan. *Michigan Entomologist* 1, 178–234.

Snider, R. J. (1969). New species of *Deuterosminthurus* and *Sminthurus* from Michigan (Collembola: Sminthuridae). *Rev. Ecol. Biol. Sol* 3, 357–376.

Snider, R. J. (1987). Class and order collembola. *In* "Immature Insects" (F. W. Stehr, ed.), pp. 55–64. Kendall-Hunt Pub., Dubuque.

Snider, R. J., Snider, R. M., and Smucker, A. J. M. (1990). Collembolan populations and root dynamics in Michigan agroecosystems. *In* "Rhizosphere Dynamics" (J. E. Box, Jr. and L. C. Hammond, eds.), pp. 168–191. Westview Press, Boulder, CO.

Snider, R. M. (1973). Laboratory observations on the biology of *Folsomia candida* (Willem) (Collembola: Isotomidae). *Rev. Ecol. Biol. Sol* 10, 103–124.

Snider, R. M. (1984). Diplopoda as food for Coleoptera: laboratory experiments. *Pedobiologia* 26, 197–204.

Snider, R. M., and Snider, R. J. (1997). Efficiency of arthropod extraction from soil cores. *Ent. News* 108, 203–208.

Söderström, B. E. (1977). Vital staining of fungi in pure cultures and in soil with fluorescein diacetate. *Soil Biol. Biochem.* 9, 59–63.

Sojka, R. E., and Upchurch, D. R. (1999). Reservations regarding the soil quality concept. *Soil Sci. Soc. Am. J.* 63, 1039–1054.

South, A. (1992). "Terrestrial Slugs: Biology, Ecology and Control." Chapman and Hall, London.

Southey, J. (1986). "Laboratory Methods for Work with Plant and Soil Nematodes." Ministry of Agriculture, Fisheries & Food, Ref. Bk. 402. London.

Southwood, T. R. E. (1978). "Ecological Methods with Particular Reference to the Study of Insect Populations" (2nd edition). Chapman and Hall, London.

Sparling, G. P. (1981). Microcalorimetry and other methods to assess biomass and activity in soil. *Soil Biol. Biochem.* 13, 93–98.

Speiser, B. (2001). Food and feeding behaviour. *In* "The Biology of Terrestrial Molluscs" (G. M. Barker, ed.), pp. 259–288. CAB International, Wallingford, U.K.

St. John, T. V., and Coleman, D. C. (1983). The role of mycorrhizae in plant ecology. *Can. J. Bot.* 61, 1005–1014.

St. John, T. V., Coleman, D. C., and Reid, C. P. P. (1983a). Association of vesicular–arbuscular mycorrhizal hyphae with soil organic particles. *Ecology* 64, 957–959.

St. John, T. V., Coleman, D. C., and Reid, C. P. P. (1983b). Growth and spatial distribution of nutrient-absorbing organs: selective exploitation of soil heterogeneity. *Plant & Soil* 71, 487–493.

Stark, J. M. (1994). Causes of soil nutrient heterogeneity at different scales. *In* "Exploitation of Environmental Heterogeneity by Plants" (M. M. Caldwell and R. W. Pearcy, eds.), pp. 255–284. Academic Press, San Diego.

Steen, E. (1984). Variation of root growth in a grass ley studied with a mesh bag technique. *Swed. J. Agric. Res.* 14, 93–97.

Steen, E. (1991). Usefulness of the mesh bag method in quantitative root studies. *In* "Plant Root Growth: An Ecological Perspective" (D. Atkinson, ed.), pp. 75–86. Blackwell, Oxford.

Steinberger, Y., and Wallwork, J. A. (1985). Composition and vertical distribution patterns of the microarthropod fauna in a Negev desert soil. *J. Zool., London (A)* 206, 329–339.

Sterner, R. W., and Elser, J. J. (2002). "Ecological Stoichiometry." Princeton University Press, Princeton and Oxford.

Stewart, J. W. B., and McKercher, R. B. (1982). Phosphorus cycle. *In* "Experimental Microbial Ecology" (R. G. Burns and J. H. Slater, eds.), pp. 221–238. Blackwell, Oxford.

Stewart, J. W. B., Anderson, D. W., Elliott, E. T., and Cole, C. V. (1990). The use of models of soil pedogenic processes in understanding changing land use and climatic change. *In* "Soils on a Warmer Earth" (H. W. Scharpenseel, M. Schomaker, and A. Ayoub, eds.), pp. 121–131. Elsevier, Amsterdam.

Stinner, B. R., and Crossley, D. A., Jr. (1980). Comparison of mineral elements cycling under till and no-till practices: an experimental approach to agroecosystems analysis. *In* "Soil Biology as Related to Land Use Practices" (D. Dindal, ed.), pp. 180–288. U.S. Environmental Protection Agency, Washington, D.C.

Stockdill, S. M. J. (1966). The effect of earthworms on pastures. *Proc. N. Z. Ecol. Soc.* 13, 68–74.

Stork, N. E., and Eggleton, P. (1992). Invertebrates as determinants and indicators of soil quality. *Am. J. Alt. Agric.* 7, 38–47.

Störmer, K. (1908). Ueber die Wirkung des Schwefelkohlenstoffs und aehnlicher Stoffe auf den Boden. *Zentralblatt für Bakteriol.* 20, 282–286.

Stout, J. D. (1963). The terrestrial plankton. *Tuatara* 11, 57–65.

Strandtmann, R. W. (1967). Terrestrial Prostigmata (Trombidiform mites). *In* "Entomology of Antarctica," Antarctic Research Series vol. 10 (J. L. Gressitt, ed.), pp. 51–80. American Geophysical Union, Washington, D.C.

Strandtmann, R. W., and Crossley, D. A., Jr. (1962). A new species of soil-inhabiting mite, *Hypoaspis marksi* (Acarina, Laelaptidae). *J. Kansas Ent. Soc.* 35, 180–185.

Strong, D. R. (2002). Populations of entomopathogenic nematodes in foodwebs. *In* "Entomopathogenic Nematology" (R. Gaugler, ed.), pp. 225–240. CABI Publishing, Wallingford, U.K.

Strong, D. R., Kaya, H. K., Whipple, A. V., Child, A. L., Kraig, S., Bondonno, M., Dyer, K., and Maron, L. L. (1996). Entomopathogenic nematodes: natural enemies of root-feeding caterpillars on bush lupine. *Oecologia* 108, 167–173.

Strong, D. R., Whipple, A. V., Child, A. L., and Dennis, B. (1999). Model selection for a subterranean trophic cascade: root-feeding caterpillars and entomopathogenic nematodes. *Ecology* 80, 2750–2761.

Sturm, H. (1959). Die Nährung der Proturen. Beobachtungen an *Acerentomon dideroi* Sylv. und *Eosentomon transitorium* Berl. *Die Naturwiss.* 46, 90–91.

Summerhayes, V. S., and Elton, C. S. (1923). Contributions to the ecology of Spitsbergen and Bear Island. *J. Ecol.* 11, 214–286.

Sumner, M. E. (ed.) (2000). "Handbook of Soil Science." CRC Press, Boca Raton.

Swank, W. T., and Crossley, D. A., Jr. (1988). "Forest Hydrology and Ecology at Coweeta." Springer-Verlag, New York.

Swift, M. J., Heal, O. W., and Anderson, J. M. (1979). "Decomposition in Terrestrial Ecosystems." University of California Press, Berkeley.

Swift, M. J. (1999). Integrating soils, systems and society. *Nature & Resources* 35, 12–20.

Syers, J. K., Sharpley, A. N., and Keeney, D. R. (1979a). Cycling of nitrogen by surface-casting earthworms in a pasture ecosystem. *Soil Biol. Biochem.* 11, 181–185.

Tandarich, J. P., Darmody, R. G., Follmer, L. R., and Johnson, D. L. (2002). Historical development of soil and weathering profile concepts from Europe to the United States of America. *Soil Sci. Soc. Am. J.* 66, 335–346.

Tansley, A. G. (1935). The use and abuse of vegetational concepts and terms. *Ecology* 16, 284–307.

Tashiro, H. (1990). Insecta: Coleoptera Scarabaeidae larvae. *In* "Soil Biology Guide" (D. L. Dindal, ed.), pp. 1191–1209. Wiley, New York.

Tate, K. R. (1979). Fraction of soil organic phosphorus in two New Zealand soils by use of sodium borate. *N. Z. J. Sci.* 22, 127–142.

Tate, K. R., and Newman, R. H. (1982). Phosphorus fractions of a climosequence of soils in New Zealand tussock grassland. *Soil Biol. Biochem.* 14, 191–196.

Taylor, H. M. (1987). "Minirhizotron Observation Tubes: Methods and Applications for Measuring Rhizosphere Dynamics." American Society of Agronomy Special Publication No. 50, Madison, WI.

Teskey, H. J. (1990). Insecta: Diptera Larvae. *In* "Soil Biology Guide" (D. L. Dindal ed.) pp. 1253–1276. Wiley, New York.

Tevis, L., Jr., and Newell, I. M. (1962). Studies on the biology and seasonal cycle of the giant red velvet mite, *Dinothrombium pandorae* (Acarina: Trombidiidae). *Ecology* 43, 797–505.

Theng, B. K. G. (1979). "Formation and Properties of Clay–Polymer Complexes." Elsevier, Amsterdam.

Theng, B. K. G., Tate, K. R., Sollins, P., Moris, N., Nadkarni, N., and Tate III, R. L. (1989). Constituents of organic matter in temperate and tropical soils. *In* "Dynamics of Soil Organic Matter in Tropical Ecosystems" (D. C. Coleman, J. M. Oades, and G. Uehara, eds.), pp. 5–32. University of Hawaii Press, Honolulu.

Thielemann, U. (1986). Elektrischer Regenwurmfang mit der Oktett–Metode. *Pedobiologia* 29, 296–302.

Thimm, T., Hoffmann, A., Borkott, H., Munch, J. C., and Tebbe, C. C. (1998). The gut of the soil microarthropod *Folsomia candida* (Collembola) is a frequently changeable but selective habitat and a vector for microorganisms. *Appl. Environ. Microbiol.* 64, 2660–2669.

Tian, G., Brussaard, L., and Kang, B. T. (1995). Breakdown of plant residues with chemically contrasting compositions: effect of earthworms and millipedes. *Soil Biol. Biochem.* 27, 277–280.

Tiedje, J. M., Cho, J. C., Murray, A., Treves, D., Xia, B., and Zhou, J. (2001). Soil teeming with life: new frontiers for soil science. *In* "Sustainable Management of Soil Organic Matter" (R. M. Rees, B. C. Ball, C. D. Campbell and C. A. Watson, eds.), pp. 393–412. CABI Publishing, Wallingford, U.K.

Tinker, P. B., with Goudriaan, J., Teng, P., Swift, M., Linder, S., Ingram, J., and van de Geijn, S. (1996). Global change impacts on agriculture, forestry and soils: the programme of the global change and terrestrial ecosystems core project of IGBP. *In* "Global Climate Change and Agricultural Production" (F. Bazzaz and W. Sombroek, eds.), pp. 295–318. FAO and Wiley & Sons, Chichester, U.K.

Tippkötter, R., Ritz, K., and Darbyshire, J. F. (1986). The preparation of soil thin sections for biological studies. *J. Soil Sci.* 37, 681–690.

Tisdall, J. M. (1991). Fungal hyphae and structural stability of soil. *Aust. J. Soil Res.* 29, 729–743.

Tisdall, J. M., and Oades, J. M. (1979). Stabilization of soil aggregates by the root systems of ryegrass. *Aust. J .Soil Res.* 17, 429–441.

Tisdall, J. M., and Oades, J. M. (1982). Organic matter and waterstable aggregates in soils. *J. Soil Sci.* 33, 141–163.

Titlyanova, A. A. (1987). Ecosystem succession and biological turnover. *Vegetatio* 50, 43–51.

Todd, R. L., Crossley, D. A., Jr., and Stormer, J. A., Jr. (1974). "Chemical composition of microarthropods by electron microprobe analysis: a preliminary report." *Proc. 32nd Ann. Proc. Electron Micros. Soc. Am.*, St. Louis, MO.

Tomlin, A. D. (1977). Pipeline construction—impact on soil micro- and mesofauna (Arthropoda and Annelida) in Ontario. *Proc. Entomol. Soc. Ontario* 108, 13–17.

Tomlin, A. D., Shipitalo, M. J., Edwards, W. M., and Protz, R. (1995). Earthworms and their influence on soil structure and infiltration. *In* "Earthworm Ecology and Biogeography in North America" (P. F. Hendrix, ed.), pp. 159–183. Lewis Publisher, Boca Raton, FL.

Torsvik, V., Salte, K., Sørheim, R., and Goksøyr, J. (1990a). Comparison of phenotypic diversity and DNA heterogeneity in a population of soil bacteria. *Appl. Environ. Microbiol.* 56, 776–781.

Torsvik, V., Goksøyr, J., and Daae, F. L. (1990b). High diversity in DNA of soil bacteria. *Appl. Environ. Microbiol.* 56, 782–787.

Torsvik, V., Goksøyr, J., Daae, R. L., Sørheim, R., Michalsen, J., and Salte, K. (1994). Use of DNA analysis to determine the diversity of microbial communities. *In* "Beyond the Biomass" (K. Ritz, J. Dighton, and K. E. Giller, eds.), pp. 39–49. Wiley-Sayce, Chichester.

Torsvik, V., and Øvreås, L. (2002). Microbial diversity and function in soil: from genes to ecosystems. *Current Opinion in Microbiology* 5, 240–245.

Touchot, F., Kilbertus, G., and Vannier, G. (1983). Role d'un collembole (*Folsomia candida*) au cours de la degradation des litiéres de charme et de cheêne, en presence au en absence d'argile. *In* "New Trends in Soil Biology" (P. Lebrun, H. M. André, A. Demedts, C. Grégoire-Wibo, and G. Wauthy, eds.), pp. 269–280. Dieu-Brichart, Ottignies-Louvain-la-Neuve, Belgium.

Travé, J., André, H. M., Taberly, G., and Bernini, F. (1996). "Les Acariens Oribates." Editions AGAR, Wavre, Belgium.

Treonis, A. M., Wall, D. H., and Virginia, R. A. (1999). Invertebrate biodiversity in Antarctic Dry Valley soils and sediments. *Ecosystems* 2, 483–492.

Trofymow, J. A., and Coleman, D. C. (1982). The role of bacterivorous and fungivorous nematodes in cellulose and chitin decomposition in the context of a root/rhizosphere/soil conceptual model. *In* "Nematodes in Soil Systems" (D. W. Freckman, ed.), pp. 117–137. University of Texas Press, Austin.

Trumbore, S. E., Davidson, E. A., Barbosa de Carnago, P., Nepstad, D. C., and Martinelli, L. A. (1995). Belowground cycling of carbon in forests and pastures of Eastern Amazonia. *Global Biogeochem. Cycles* 9, 515–528.

Tunlid, A., and White, D. C. (1992). Biochemical analysis of biomass, community structure, nutritional status, and metabolic activity of microbial communities in soil. *In* "Soil Biochemistry," vol. 7 (G. Stotzky and J.-M. Bollag, eds.), pp. 229–262. Marcel Dekker, New York.

Tynen, M. J. (1972). "Ice-Worms (Oligochaeta: Enchytraeidae) from Western British Columbia." National Museum of Natural Sciences, Ottawa.

Upchurch, D. R., and Ritchie, J. T. (1983). Root observations using a video recording system in mini-rhizotrons. *Agron. J.* 73, 1009–1015.

Upchurch, D. R., and Taylor, H. M. (1990). Tools for studying rhizosphere dynamics. *In* "Rhizosphere Dynamics" (J. E. Box, Jr. and L. C. Hammond, eds.), pp. 83–115. Westview Press, Boulder, Colorado.

Usher, M. B., and Balogun, R. A. (1966). A defense mechanism in *Onychiurus* (Collembola, Onychiuridae). *Ent. Mon. Mag.* 102, 237–238.

van Breemen, N. (1993). Soils as biotic constructs favouring net primary productivity. *Geoderma* 57, 183–211.

van der Heijden, M. G. A., Klironomos, J. N., Ursic, M., Moutoglis, P., Streitwolf-Engel, R., Boller, T., Wiemken, A., and Sanders, I. R. (1998). Mycorrhizal fungal diversity determines plant biodiversity, ecosystem variability and productivity. *Nature* 396, 69–72.

Van Groenigen, J. W., and Stein, A. (1998). Constrained optimisation of spatial sampling using continuous simulated annealing. *J. Environ. Qual.* 27, 1078–1086.

van Noordwijk, M., de Ruiter, P. C., Zwart, K. B., Bloem, J., Moore, J. C., van Faassen, H. G., and Burgers, S. L. G. E. (1993). Synlocation of biological activity, roots, cracks and recent organic inputs in a sugar beet field. *Geoderma* 56, 265–276.

Van Noordwijk, M., Cerri, C., Woomer, P. L., and Murdiyarso, D. (1998). Criteria and indicators of forest soils used for slash-and-burn agriculture and alternative land uses in Indonesia. *In* "The Contributions of Soil Science to the Development and Implementation of Criteria and Indicators of Sustainable Forest Management," pp. 137–153. SSSA Special Publication No. 53. Soil Science Society of America, Madison, WI.

van Vliet, P. C. J. (2000). Enchytraeids. *In* "Handbook of Soil Science" (M. Sumner, ed.), pp. C70–C77. CRC Press, Boca Raton, FL.

van Vliet, P. C. J., West, L. T., Hendrix, P. F., and Coleman, D. C. (1993). The influence of Enchytraeidae (Oligochaeta) on the soil porosity of small microcosms. *Geoderma* 56, 287–299.

van Vliet, P. C. J., and Hendrix, P. F. (2003). Role of fauna in soil physical processes. *In* "Soil Biological Fertility—A Key to Sustainable Land Use in Agriculture." (L. K. Abbott and D. V. Murphy, eds.). Kluwer Publishers, Dordrecht (in press).

Vance, E. D., Brookes, P. C., and Jenkinson, D. S. (1987). An extraction method for measuring microbial biomass C. *Soil Biol. Biochem.* 19, 703–707.

Vanlauwe, B., Diels, J., Sanginga, N., and Merckx, R. eds. (2002). "Integrated Plant Nutrient Management in Sub-Saharan Africa: From Concept to Practice." CABI Publishing, Wallingford, U.K.

Vannier, G. (1973). Originalité des conditions de vie dans le sol due a la presence de l'eau: importance thermodynamique et biologique de la porosphere. *Ann. Soc. R. Zool. Belg.* 103, 157–167.

Vannier, G. (1981). Signification de la persistance de la pedofaune aprés la point de fletrissement permanent dans les sols. *Rev. Écol. Biol. Sol* 8, 343–365.

Vannier, G. (1987). The porosphere as an ecological medium emphasized in Professor Ghilarov's work on soil animal adaptations. *Biol. Fertil. Soils* 3, 39–44.

Vargas, R., and Hattori, T. (1986). Protozoan predation of bacterial cells in soil aggregates. *FEMS Microbiol. Ecol.* 38, 233–242.

Venette, R. C., and Ferris, H. (1998). Influence of bacterial type and density on population growth of bacterial-feeding nematodes. *Soil Biol. Biochem.* 30, 949–960.

Verhoef, H. A., and DeGoede, R. G. M. (1985). Effects of collembolan grazing on nitrogen dynamics in a coniferous forest. *In* "Ecological Interactions in Soil: Plants, Microbes and Animals" (A. H. Fitter, D. Atkinson, D. J. Read, and M. B. Usher, eds.), pp. 367–376. Blackwell, Oxford.

Vernadsky, V. I. (1944). Problems of biogeochemistry. II. The fundamental matter–energy difference between the living and the inert natural bodies of the biosphere. *Trans. Conn. Acad. Arts Sci.* 35, 483–512.

Vernadsky, V. I. (1998). "The Biosphere." Copernicus, Springer-Verlag, New York.

Vitousek, P. M. (1994). Beyond global warming: ecology and global change. *Ecology* 75, 1861–1876.

Vitousek, P. M., Hattenschwiler, S., Olander, L., and Allison, S. (2002). Nitrogen and nature. *Ambio* 31, 97–101.

Vogt, K. A., Grier, C. C., Meier, C. E., and Edmonds, R. L. (1982). Mycorrhizal role in net primary production and nutrient cycling in *Abies amabilis* ecosystems in western Washington. *Ecology* 63, 370–380.

Vogt, K. A., Grier, C. C., Gower, S. T., Sprugel, D. G., and Vogt, D. J. (1986). Overestimation of net root production: a real or imaginary problem? *Ecology* 67, 577–579.

Volobuev, V. R. (1964). "Ecology of Soils." Israel Program for Science Translations, Davey & Co., New York.

Voroney, R. P., and Paul, E. A. (1984). Determination of Kc and Kn *in situ* for calibration of the chloroform fumigation-incubation method. *Soil Biol. Biochem.* 16, 9–14.

Voroney, R. P., Winter, J. P., and Gregorich, E. G. (1991). Microbe, plant, soil interactions. *In* "Carbon Isotope Techniques" (D. C. Coleman and B. Fry, eds.), pp. 77–99. Academic Press, San Diego.

Vos, W., and Stortelder, A. H. F. (1988). "Vanishing Tuscan Landscapes. Landscape Ecology of a Submediterranean-Montane area (Solano Basi, Tuscany, Italy)." Ph.D. Dissertation, University of Amsterdam, the Netherlands.

Vossbrinck, C. R., Coleman, D. C., and Woolley, T. A. (1979). Abiotic and biotic factors in litter decomposition in a semiarid grassland. *Ecology* 60, 265–271.

Waldorf, E. S. (1974). Sex pheromone in the springtail, *Sinella curviseta. Environ. Entomol.* 3, 916–918.

Wall, D. H., and Moore, J. C. (1999). Interactions underground: soil biodiversity, mutualism, and ecosystem processes. *Bioscience* 49, 109–117.

Wall, D. H., and Virginia, R. A. (1999). Controls on soil biodiversity: insights from extreme environments. *Appl. Soil Ecol.* 13, 137–150.

Wall, D. H., Adams, G., and Parsons, A. N. (2001). Soil biodiversity. *In* "Changing Biodiversity in a Changing Environment" (F. S. Chapin, III, O. E. Sala, and E. Huber-Sannwald, eds.), pp. 47–81. Springer Verlag, New York.

Wallace, D. F. (1994). "Cat-scan assessment of earthworm (*Lumbricus terrestris* and *Lumbricus rubellus*) burrows as macropores." M.S. Thesis, University of Georgia, Athens, Georgia.

Wallwork, J. A. (1970). "Ecology of Soil Animals." McGraw-Hill, London.

Wallwork, J. A. (1976). "The Distribution and Diversity of Soil Fauna." Academic Press, London.

Wallwork, J. A. (1982). "Desert Soil Fauna." Praeger Scientific, New York.

Wallwork, J. A. (1983). Oribatids in forest ecosystems. *Ann. Rev. Entomol.* 28, 109–130.

Walsh, M. I., and Bolger, T. (1990). Effects of diet on the growth and reproduction of some Collembola in laboratory cultures. *Pedobiologia* 34, 161–171.

Walter, D. E. (1988). Predation and mycophagy by endeostigmatid mites (Acariformes: Prostigmata). *Exper. Appl. Acarol.* 4, 159–166.

Walter, D. E., Hunt, H. W., and Elliott, E. T. (1987). The influence of prey type on the development and reproduction of some predatory soil mites. *Pedobiologia* 30, 419–424.

Walter, D. E., and Ikonen, E. K. (1989). Species, guilds and functional groups: taxonomy and behavior in nematophagous arthropods. *J. Nematol.* 21, 315–327.

Walter, D. E., Kaplan, D. T., and Permar, T. A. (1991). Missing links: a review of methods used to estimate trophic links in soil food webs. *Agric. Ecosys. Environ.* 34, 399–405.

Walter, D. E., and Proctor, H. C. (1999). "Mites. Ecology, Evolution and Behavior." CABI Publishing, Wallingford, U.K.

Walter, D. E., and Proctor, H. C. (2000). "Life at the Microscale: Mites and the Study of Ecology, Evolution and Behaviour." New South Wales Press, Sydney.

Walter, D. E., and Proctor, H. C. (2001). "Mites in Soil." CD-ROM, CSIRO Pub., Collingswood, Australia.

Wang, G. M., Coleman, D. C., Freckman, D. W., Dyer, M. I., McNaughton, S. J., Acra, M. A., and Goeschl, J. D. (1989). Carbon partitioning patterns of mycorrhizal versus nonmycorrhizal plants: real-time dynamic measurements using $^{11}CO_2$. *New Phytol.* 112, 489–493.

Wardle, D. A. (1998). Controls of temporal variability of the soil microbial biomass: a global-scale synthesis. *Soil Biol. Biochem.* 30, 1627–1637.

Wardle, D. A. (2002). "Communities and Ecosystems: Linking the Aboveground and Belowground Components." Princeton University Press, Princeton, NJ.

Wardle, D. A., and Yeates, G. W. (1993). The dual importance of competition and predation as regulatory forces in terrestrial ecosystems: evidence from decomposer food-webs. *Oecologia* 93, 303–306.

Wardle, D. A., Bonner, K. I., Barker, G. M., Yeates, G. W., Nicholson, K. S., Bardgett, R. D., Watson, R. N., and Ghani, A. (1999). Plant removals in perennial grassland: vegetation dynamics, decomposers, soil biodiversity, and ecosystem properties. *Ecol. Monogr.* 69, 535–568.

Warnock, A. J., Fitter, A. H., and Usher, M. B. (1982). The influence of a springtail *Folsomia candida* on the mycorrhizal association of leek *Allium porrum* and arbuscular mycorrhizal endophyte *Glomus fasciculatus*. *New Phytol.* 90, 285–292.

Wasylik, A. (1995). Indicatory groups of Acarina in the processes of degradative succession. *In* "Advances of Acarology in Poland" (J. Boczek and S. Ignatowicz, eds.), pp. 90–93. Polish Academy of Science, Siedlec, Poland.

Waters, A. G., and Oades, J. M. (1991). Organic matter in water stable aggregates. *In* "Advances in Soil Organic Matter Research: The Impact on Agriculture and the Environment" (W. S. Wilson, ed.). Royal Society of Chemistry, Cambridge.

Weaver, R., Angle, S., Bottomley, P., Bezdicek, D., Smith, S., Tabatabai, A., and Wollum, A. (eds.) (1994). "Methods of Soil Analysis: Part 2 Microbiological and Biochemical Properties." Soil Science Society of America, Inc., Madison, WI.

Webb, D. P. (1977). Regulation of deciduous forest litter decomposition by soil arthropod feces. *In* "The Role of Arthropods in Forest Ecosystems" (W. J. Mattson, ed.), pp. 57–69. Springer, New York.

"Webster's New Universal Unabridged Dictionary. Deluxe Second Edition." (1983). New World Dictionaries/Simon and Schuster, Cleveland.

Wheeler, G. C., and Wheeler, J. (1990). Insecta: Hymenoptera Formicidae. *In* "Soil Biology Guide" (D. L. Dindal, ed.), pp. 1277–1294. Wiley, New York.

White, R. E. (1983). "A Field Guide to the Beetles." Houghton Mifflin, Boston.

Whitford, W. G. (2000). Keystone arthropods as webmasters in desert ecosystems. *In* "Invertebrates as Webmasters in Ecosystems" (D. C. Coleman and P. F. Hendrix, eds.), pp. 25–41. CAB International, Wallingford, U.K.

Whitford, W. G., and Santos, P. F. (1980). Arthropods and detritus decomposition in desert ecosystems. *In* "Soil Biology as Related to Land Use Practices" (D. L. Dindal, ed.), pp. 770–778. U. S. Environmental Protection Agency, Washington, D.C.

Whitford, W. G., Freckman, D. W., Parker, L. W., Schaefer, D., Santos, P. F., and Steinberger, Y. (1983). The contributions of soil fauna to nutrient cycles in desert systems. *In* "New Trends in Soil Biology" (P. Lebrun, H. M. André, A. de Medts, C. Grégoire-Wibo and G. Wauthy, eds.), pp. 49–59. Dieu-Brichart, Ottignies-Louvain-la-Neuve, Belgium.

Whitman, W. B., Coleman, D. C., and Wiebe, W. J. (1998). Perspective: Prokaryotes—the unseen majority. *Proc. Nat. Acad. Sci. USA* 95, 6578–6583.

Whitney, M. (1925). "Soil and Civilization." Van Nostrand, New York.

Wieder, R. K., and Lang, G. E. (1982). A critique of the analytical methods used in examining decomposition data obtained from litterbags. *Ecology* 63, 1636–1642.

Willerslev, E., Hansen, A. J., Binladen, J., Brand, T. B., Gilbert, M. T. P., Shapiro, B., Bunce, M., Wiuf, C., Gilichinsky, D. A., and Cooper, A. (2003). Diverse plant and animal genetic records from Holocene and Pleistocene sediments. *Science* 300, 791–795.

Williams, S. C. (1987). Scorpion bionomics. *Ann. Rev. Entomol.* 32, 275–295.

Williamson, M. (1996). "Biological Invasions." Chapman & Hall, London.

Wilson, A. T. (1978). Pioneer agriculture explosion and CO_2 levels in the atmosphere. *Nature* 273, 40–41.

Wilson, D. S. (1980). "The Natural Selection of Populations and Communities." Benjamin/Cummings, Menlo Park, CA.

Wilson, E. O. (1987). Causes of ecological success: the case of the ants. *J. Anim. Ecol.* 56, 1–9.

Wilson, E. O. (1992). "The Diversity of Life." Norton, New York.

Wilson, K. J., Sessitsch, A., and Akkermans, A. (1994). Molecular markers as tools to study the ecology of microorganisms. *In* "Beyond the Biomass" (K. Ritz, J. Dighton, and K. E. Giller, eds.), pp. 149–156. Wiley-Sayce, Chichester.

Winchester, N. N. (1997). Canopy arthropods of coastal Sitka spruce on Vancouver Island, British Columbia, Canada. *In* "Canopy Arthropods" (N. E. Stork, J. Adis, and R. K. Didham, eds.), pp. 151–168. Chapman & Hall, London.

Winter, J. P., Voroney, R. P., and Ainsworth, D. A. (1990). Soil microarthropods in long-term no-tillage and conventional corn production. *Can. J. Soil Sci.* 70, 641–653.

Wise, D. H. (1993). "Spiders in Ecological Webs." Cambridge University Press, Cambridge.

Withington, J. M., Elkin, A. D., Bulaj, B., Olesinski, J., Tracy, K. N., Bouma, T. J., Oleksyn, J., Anderson, L. J., Modrzynski, J., Reich, P. B., and Eissenstat, D. M. (2003). The impact of material used for minirhizotron tubes for root research. *New Phytol.* 160, 533–544.

Witkamp, M., and van der Drift, J. (1961). Breakdown of forest litter in relation to environmental factors. *Plant & Soil* 15, 295–311.

Wolters, V. (1988). Effects of *Mesenchytraeus glandulosus* (Oligochaeta, Enchytraeidae) on decomposition processes. *Pedobiologia* 32, 387–398.

Wolters, V. (1991). Soil invertebrates—effects on nutrient turnover and soil structure: a review. *Z. Pflanzenernähr. Bodenk.* 154, 389–402.

Wolters, V., and Ekschmitt, K. (1997). Gastropods, isopods, diplopods and chilopods: neglected groups of the decomposer food web. *In* "Fauna in Soil Ecosystems" (G. Benckiser, ed.), pp. 265–306. Dekker, New York.

Wolters, V., W. H. Silver, D. E. Bignell, D. C. Coleman, P. Lavelle, W. H. van der Putten, P. de Ruiter, J. Rusek, D. H. Wall, D. A. Wardle, L. Brussaard, J. M. Dangerfield, V. K. Brown, K. E. Giller, D. U. Hooper, O. Sala, J. Tiedje, and J. A. van Veen (2000). Effects of global changes on above- and below-ground biodiversity in terrestrial ecosystems: implications for ecosystem functioning. *BioScience* 50, 1089–1098.

Wood, T. G., Johnson, R. A., and Anderson, J. M. (1983). Modification of soils in Nigerian Savanna by soil-feeding *Cubitermes* (Isoptera, Termitidae). *Soil Biol. Biochem.* 15, 575–579.

Woomer, P. L., and Swift, M. J. (1994). "Report of the Tropical Soil Biology and Fertility Programme." TSBF, Nairobi, Kenya.

Wright, D. H. (1988). Inverted microscope methods for counting soil mesofauna. *Pedobiologia* 31, 409–411.

Wright, D. H., Huhta, V., and Coleman, D. C. (1989). Characteristics of defaunated soil. II. Effects of reinoculation and the role of the mineral soil. *Pedobiologia* 33, 427–435.

Wright, J. C. (2001). Cryptobiosis 300 years on from van Leeuwenhoek: what have we learned about tardigrades? *Zool. Anzeiger* 240, 563–582.

Wright, S. F., Starr, J. L., and Paltineau, I. C. (1999). Changes in aggregate stability and concentration of glomalin during tillage management transition. *Soil Sci. Soc. Am. J.* 63, 1825–1829.

Yeates, G. W. (1981). Soil nematode populations depressed in the presence of earthworms. *Pedobiologia* 22, 191–195.

Yeates, G. W. (1988). Earthworm and enchytraeid populations in a 13-year-old agroforestry system. *N. Z. J. For. Sci.* 18, 304–310.

Yeates, G. W. (1998). Feeding in free-living soil nematodes: a functional approach. *In* "The Physiology and Biochemistry of Free–Living and Plant–Parasitic Nematodes" (R. N. Perry and D. J. Wright, eds.), pp. 245–269. CABI Publishing, Wallingford, U.K.

Yeates, G. W. (1999). Effects of plants on nematode community structure. *Ann. Rev. Phytopathol.* 37, 127–149.

Yeates, G. W., and Coleman, D. C. (1982). Role of nematodes in decomposition. *In* "Nematodes in Soil Ecosystems" (D. W. Freckman, ed.), pp. 55–80. University of Texas Press, Austin.

Yeates, G. W., Bongers, T., de Goede, R. G. M., Freckman, D. W., and Georgieva, S. S. (1993). Feeding habits in soil nematode families and genera: an outline for soil ecologists. *J. Nematol.* 25, 101–313.

Yeates, G. W., Ross, C. W., and Shepherd, T. G. (1999). Populations of terrestrial planarians affected by crop management: implications for long-term management. *Pedobiologia* 42, 360–363.

Yeates, G. W., Dando, J. L., and Shepherd, T. G. (2002). Pressure plate studies to determine how moisture affects access of bacterial-feeding nematodes to food in soil. *Eur. J. Soil Sci.* 53, 355–365.

Zaborski, E. R. (2003). Allyl isothiocyanate: an alternative chemical expellant for sampling earthworms. *Appl. Soil Ecol.* 22, 87–95.

Zachariae, G. (1963). Was leisten Collembolen für den Waldhumus? *In* "Soil Organisms" (J. Van der Drift and J. Doeksen, eds.), pp. 109–114. North Holland, Amsterdam.

Zachariae, G. (1964). Welche Bedeutung haben Enchyträus in Waldboden? *In* "Soil Micromorphology" (A. Jongerius, ed.), pp. 57–68. Elsevier, Amsterdam.

Zachariae, G. (1965). Spuren tierischer Tätigkeit im Boden des Buchenwaldes. *Forstwiss. Forsch.* 20, 1–68.

Zelles, L., and Alef, K. (1995). Biomarkers. *In* "Methods in Applied Soil Microbiology and Biochemistry" (K. Alef and P. Nannipieri, eds.), pp. 422–439. Academic Press, London.

Zhang, B. G., Rowland, C., Lattaud, C., and Lavelle, P. (1993). Activity and origin of digestive enzymes in gut of the tropical earthworm *Pontoscolex corethrurus*. *Eur. J. Soil Biol.* 29, 7–11.

Zhou, J., Bruns, M. A., and Tiedje, J. M. (1996). DNA recovery from soils of diverse composition. *Appl. Environ. Microbiol.* 62, 316–322.

Zimmer, M. (2002). Nutrition in terrestrial isopods (Isopoda: Oniscidea): an evolutionary–ecological approach. *Biol. Revs.* 77, 455–493.

Zimmer, M., Kautz, G., and Topp, W. (1996). Olfaction in terrestrial isopods (Isopoda: Oniscidea): responses of *Porcellio scaber* to the odour of litter. *Eur. J. Soil Biol.* 32, 141–147.

Zimmerman, P. R., Greenberg, J. P., Wandiga, S. O., and Crutzen, P. J. (1982). Termites: a potentially large source of atmospheric methane, carbon dioxide and molecular hydrogen. *Science* 218, 563–565.

Zinke, P., Stangenberger, A., Post, W., Emanuel, W., and Olson, J. (1984). "Worldwide Organic Carbon and Nitrogen Data." Report ORNL/TM–8857, Oak Ridge National Laboratory, Oak Ridge.

Zonn, S. V., and Eroshkina, A. N. (1996). V. V. Dokuchaev, his disciples and followers. *Eurasian Soil Sci.* 29, 111–120.

Zwart, K. B., and Darbyshire, J. F. (1991). Growth and nitrogenous excretion of a common soil flagellate, *Spumella* sp. *J. Soil Sci.* 43, 145–157.

Zwart, K. B., Kuikman, P. J., and van Veen, J. A. (1994). Rhizosphere protozoa: their significance in nutrient dynamics. *In* "Soil Protozoa" (J. F. Darbyshire, ed.), pp. 93–121. CABI Publishing, Wallingford, U.K.

Index